INTERNATIONAL UNION OF PURE AND APPLIED CHEMISTRY

ANALYTICAL CHEMISTRY DIVISION
COMMISSION ON SOLUBILITY DATA

SOLUBILITY DATA SERIES

Volume 12

SULFUR DIOXIDE, CHLORINE, FLUORINE AND CHLORINE OXIDES

SOLUBILITY DATA SERIES

Volume	1	H. L. Clever, *Helium and Neon*
Volume	2	H. L. Clever, *Krypton, Xenon and Radon*
Volume	3	M. Salomon, *Silver Azide, Cyanide, Cyanamides, Cyanate, Selenocyanate and Thiocyanate*
Volume	4	H. L. Clever, *Argon*
Volume	5/6	C. L. Young, *Hydrogen and Deuterium*
Volume	7	R. Battino, *Oxygen and Ozone*
Volume	8	C. L. Young, *Oxides of Nitrogen*
Volume	9	W. Hayduk, *Ethane*
Volume	10	R. Battino, *Nitrogen and Air*
Volume	11	B. Scrosati and C. A. Vincent, *Alkali Metal, Alkaline-Earth Metal and Ammonium Halides. Amide Solvents*
Volume	12	C. L. Young, *Sulfur Dioxide, Chlorine, Fluorine and Chlorine Oxides*
Volume	18	O. Popovych, *Tetraphenylborates*

Volumes in preparation

H. L. Clever and W. Gerrard, *Hydrogen Halides in Non-Aqueous Solvents*

H. Miyamoto and M. Salomon, *Alkaline-Earth Metal Halates*

S. Siekierski and J. K. Navratil, *Actinide Nitrates*

A. L. Horvath and F. W. Getzen, *Halogenated Benzenes, Toluenes, Xylenes and Phenols with Water*

E. M. Woolley, *Aqueous Silver Chloride Systems*

Z. Galus and C. Guminski, *Metals and Intermetallic Compounds in Mercury*

H. Miyamoto and E. M. Woolley, *Silver, Copper and Gold Halates*

J. W. Lorimer, *Beryllium, Strontium, Barium and Radium Sulfates*

H. L. Clever, D. M. Mason and C. L. Young, *Carbon Dioxide*

H. L. Clever and C. L. Young, *Methane*

SOLUBILITY DATA SERIES

Volume 12

SULFUR DIOXIDE, CHLORINE, FLUORINE AND CHLORINE OXIDES

Volume Editor

COLIN L. YOUNG
University of Melbourne
Parkville, Victoria, Australia

Evaluators

RUBIN BATTINO
Wright State University
Dayton, Ohio, USA

WILLIAM GERRARD
Polytechnic of North London
Holloway, London, UK

P. G. T. FOGG
Polytechnic of North London
Holloway, London, UK

Compilers

H. LAWRENCE CLEVER
Emory University
Atlanta, Georgia, USA

SUSAN A. JOHNSON
Emory University
Atlanta, Georgia, USA

PERGAMON PRESS

OXFORD · NEW YORK · TORONTO · SYDNEY · PARIS · FRANKFURT

U.K.	Pergamon Press Ltd., Headington Hill Hall, Oxford OX3 0BW, England
U.S.A.	Pergamon Press Inc., Maxwell House, Fairview Park, Elmsford, New York 10523, U.S.A.
CANADA	Pergamon Press Canada Ltd., Suite 104, 150 Consumers Road, Willowdale, Ontario M2J 1P9, Canada
AUSTRALIA	Pergamon Press (Aust.) Pty. Ltd., P.O. Box 544, Potts Point, N.S.W. 2011, Australia
FRANCE	Pergamon Press SARL, 24 rue des Ecoles, 75240 Paris, Cedex 05, France
FEDERAL REPUBLIC OF GERMANY	Pergamon Press GmbH, Hammerweg 6, D-6242 Kronberg-Taunus, Federal Republic of Germany

First edition 1983

British Library Cataloguing in Publication Data

Sulfur dioxide, chlorine, fluorine and chlorine oxides.
—(Solubility data series, ISSN 0191-5622; v.12)
I. Young, Colin L. II. Series
546'.7212 QD181.01

ISBN 0-08-026218-X

In order to make this volume available as economically and as rapidly as possible the authors' typescripts have been reproduced in their original forms. This method unfortunately has its typographical limitations but it is hoped that they in no way distract the reader.

Printed in Great Britain by A. Wheaton & Co. Ltd., Exeter

CONTENTS

SOLUBILITY DATA SERIES

Editor-in-Chief

INTERNATIONAL UNION OF PURE AND APPLIED CHEMISTRY

IUPAC Secretariat: Bank Court Chambers, 2-3 Pound Way,
Cowley Centre, Oxford OX4 3YF, UK

FOREWORD

If the knowledge is
undigested or simply wrong,
more is not better.

How to communicate and disseminate numerical data effectively in chemical science and technology has been a problem of serious and growing concern to IUPAC, the International Union of Pure and Applied Chemistry, for the last two decades. The steadily expanding volume of numerical information, the formulation of new interdisciplinary areas in which chemistry is a partner, and the links between these and existing traditional subdisciplines in chemistry, along with an increasing number of users, have been considered as urgent aspects of the information problem in general, and of the numerical data problem in particular.

Among the several numerical data projects initiated and operated by various IUPAC commissions, the *Solubility Data Project* is probably one of the most ambitious ones. It is concerned with preparing a comprehensive critical compilation of data on solubilities in all physical systems, of gases, liquids and solids. Both the basic and applied branches of almost all scientific disciplines require a knowledge of solubilities as a function of solvent, temperature and pressure. Solubility data are basic to the fundamental understanding of processes relevant to agronomy, biology, chemistry, geology and oceanography, medicine and pharmacology, and metallurgy and materials science. Knowledge of solubility is very frequently of great importance to such diverse practical applications as drug dosage and drug solubility in biological fluids, anesthesiology, corrosion by dissolution of metals, properties of glasses, ceramics, concretes and coatings, phase relations in the formation of minerals and alloys, the deposits of minerals and radioactive fission products from ocean waters, the composition of ground waters, and the requirements of oxygen and other gases in life support systems.

The widespread relevance of solubility data to many branches and disciplines of science, medicine, technology and engineering, and the difficulty of recovering solubility data from the literature, lead to the proliferation of published data in an ever increasing number of scientific and technical primary sources. The sheer volume of data has overcome the capacity of the classical secondary and tertiary services to respond effectively.

While the proportion of secondary services of the review article type is generally increasing due to the rapid growth of all forms of primary literature, the review articles become more limited in scope, more specialized. The disturbing phenomenon is that in some disciplines, certainly in chemistry, authors are reluctant to treat even those limited-in-scope reviews exhaustively. There is a trend to preselect the literature, sometimes under the pretext of reducing it to manageable size. The crucial problem with such preselection - as far as numerical data are concerned - is that there is no indication as to whether the material was excluded by design or by a less than thorough literature search. We are equally concerned that most current secondary sources, critical in character as they may be, give scant attention to numerical data.

On the other hand, tertiary sources - handbooks, reference books, and other tabulated and graphical compilations - as they exist today, are comprehensive but, as a rule, uncritical. They usually attempt to cover whole disciplines, thus obviously are superficial in treatment. Since they command a wide market, we believe that their service to advancement of science is at least questionable. Additionally, the change which is taking place in the generation of new and diversified numerical data, and the rate at which this is done, is not reflected in an increased third-level service. The emergence of new tertiary literature sources does not parallel the shift that has occurred in the primary literature.

With the status of current secondary and tertiary services being as briefly stated above, the innovative approach of the *Solubility Data Project* is that its compilation and critical evaluation work involve consolidation and reprocessing services when both activities are based on intellectual and scholarly reworking of information from primary sources.

It comprises compact compilation, rationalization and simplification, and the fitting of isolated numerical data into a critically evaluated general framework.

The *Solubility Data Project* has developed a mechanism which involves a number of innovations in exploiting the literature fully, and which contains new elements of a more imaginative approach for transfer of reliable information from primary to secondary/tertiary sources. *The fundamental trend of the Solubility Data Project is toward integration of secondary and tertiary services with the objective of producing in-depth critical analysis and evaluation which are characteristic to secondary services, in a scope as broad as conventional tertiary services.*

Fundamental to the philosophy of the project is the recognition that the basic element of strength is the active participation of career scientists in it. Consolidating primary data, producing a truly critically-evaluated set of numerical data, and synthesizing data in a meaningful relationship are demands considered worthy of the efforts of top scientists. Career scientists, who themselves contribute to science by their involvement in active scientific research, are the backbone of the project. The scholarly work is commissioned to recognized authorities, involving a process of careful selection in the best tradition of IUPAC. This selection in turn is the key to the quality of the output. These top experts are expected to view their specific topics dispassionately, paying equal attention to their own contributions and to those of their peers. They digest literature data into a coherent story by weeding out what is wrong from what is believed to be right. To fulfill this task, the evaluator must cover *all* relevant open literature. No reference is excluded by design and every effort is made to detect every bit of relevant primary source. Poor quality or wrong data are mentioned and explicitly disqualified as such. In fact, it is only when the reliable data are presented alongside the unreliable data that proper justice can be done. The user is bound to have incomparably more confidence in a succinct evaluative commentary and a comprehensive review with a complete bibliography to both good and poor data.

It is the standard practice that any given solute-solvent system consists of two essential parts: I. Critical Evaluation and Recommended Values, and II. Compiled Data Sheets.

The Critical Evaluation part gives the following information:

(i) a verbal text of evaluation which discusses the numerical solubility information appearing in the primary sources located in the literature. The evaluation text concerns primarily the quality of data after consideration of the purity of the materials and their characterization, the experimental method employed and the uncertainties in control of physical parameters, the reproducibility of the data, the agreement of the worker's results on accepted test systems with standard values, and finally the fitting of data, with suitable statistical tests, to mathematical functions;

(ii) a set of recommended numerical data. Whenever possible, the set of recommended data includes weighted average and standard deviations, and a set of smoothing equations derived from the experimental data endorsed by the evaluator;

(iii) a graphical plot of recommended data.

The compilation part consists of data sheets of the best experimental data in the primary literature. Generally speaking, such independent data sheets are given only to the best and endorsed data covering the known range of experimental parameters. Data sheets based on primary sources where the data are of a lower precision are given only when no better data are available. Experimental data with a precision poorer than considered acceptable are reproduced in the form of data sheets when they are the only known data for a particular system. Such data are considered to be still suitable for some applications, and their presence in the compilation should alert researchers to areas that need more work.

The typical data sheet carries the following information:

(i) components - definition of the system - their names, formulas and Chemical Abstracts registry numbers;

(ii) reference to the primary source where the numerical information is reported. In cases when the primary source is a less common periodical or a report document, published though of limited availability, abstract references are also given;

(iii) experimental variables;

(iv) identification of the compiler;

(v) experimental values as they appear in the primary source. Whenever available, the data may be given both in tabular and graphical form. If auxiliary information is available, the experimental data are converted also to SI units by the compiler.

Under the general heading of Auxiliary Information, the essential experimental details are summarized:

(vi) experimental method used for the generation of data;

(vii) type of apparatus and procedure employed;

(viii) source and purity of materials;

(ix) estimated error;

(x) references relevant to the generation of experimental data as cited in the primary source.

This new approach to numerical data presentation, developed during our four years of existence, has been strongly influenced by the diversity of background of those whom we are supposed to serve. We thus deemed it right to preface the evaluation/compilation sheets in each volume with a detailed discussion of the principles of the accurate determination of relevant solubility data and related thermodynamic information.

Finally, the role of education is more than corollary to the efforts we are seeking. The scientific standards advocated here are necessary to strengthen science and technology, and should be regarded as a major effort in the training and formation of the next generation of scientists and engineers. Specifically, we believe that there is going to be an impact of our project on scientific-communication practices. The quality of consolidation adopted by this program offers down-to-earth guidelines, concrete examples which are bound to make primary publication services more responsive than ever before to the needs of users. The self-regulatory message to scientists of 15 years ago to refrain from unnecessary publication has not achieved much. The literature is still, in 1982, cluttered with poor-quality articles. The Weinberg report (in "Reader in Science Information", Eds. J. Sherrof and A. Hodina, Microcard Editions Books, Indian Head, Ins., 1973, p.292) states that "admonition to authors to restrain themselves from premature, unnecessary publication can have little effect unless the climate of the entire technical and scholarly community encourages restraint..." We think that projects of this kind translate the climate into operational terms by exerting pressure on authors to avoid submitting low-grade material. The type of our output, we hope, will encourage attention to quality as authors will increasingly realize that their work will not be suited for permanent retrievability unless it meets the standards adopted in this project. It should help to dispel confusion in the minds of many authors of what represents a permanently useful bit of information of an archival value, and what does not.

If we succeed in that aim, even partially, we have then done our share in protecting the scientific community from unwanted and irrelevant, wrong numerical information.

A. S. Kertes

PREFACE

This volume is concerned with the solubility of sulfur dioxide, chlorine, fluorine and the oxides of chlorine in pure liquids, liquid mixtures, aqueous solutions and miscellaneous fluids and fluid mixtures. The solubility in molten salts, slags, glasses, metals, alloys and polymeric material have been specifically excluded from this volume. There is always a problem in attempting to distinguish between gas solubilities and vapor-liquid equilibria and this problem is particularly troublesome in the case of sulfur dioxide solubilities. There is no absolute distinction between gas solubility and vapor-liquid equilibria. We have been fairly liberal in making the distinction and have included systems which some might classify as vapor-liquid equilibrium. The literature has been searched up until the middle of 1981. The editor or Professor H. L. Clever, Director, Solubility Data Center, Department of Chemistry, Emory University, Atlanta, Georgia USA would appreciate being notified of any omissions or more recent data.

Unlike most of the previous volumes in this series concerned with the solubility of gases in liquids, this volume deals with gases which react with the solvent in several cases. We have not attempted to evaluate equilibrium constants for reactions which take place in solution because the available data are not of sufficient quality or quantity to make a meaningful evaluation of such constants. Despite the industrial importance of both sulfur dioxide and chlorine, the solubility of these gases in most solvents is not known to better than a few per cent. Much of the data were obtained several decades ago and much of the data in this volume is barely accurate enough for modern applications.

In many cases meaningful evaluation of the solubility data has not been possible due to a lack of comparable data. Furthermore, in many cases it is not clear how the quoted solubilities have been calculated from the experimental results. Earlier in this series, Hayduk, in the preface to the ethane volume, made a plea for inclusion of all pertinent data used in the calculation of gas solubility in all future (original) publications and stated that in some instances compilers and evaluators have had to guess which data were used by the authors in their calculation of solubility. A similar plea and statement are relevant to this volume.

In several tables compilers have calculated solubilities not given in the original paper. In most cases the source of additional data needed for such calculations is given in the compiled table. The tables given by Riddick and Bunger in *Technique of Chemistry*, ed. Weissberger, A. Vol. II, Wiley-Interscience, New York, 1970, 3rd Ed. have been used but not specifically referenced in compiled tables.

The Editor acknowledges the advice and helpful suggestions of the members of the IUPAC Commission V8 on Solubility Data. Acknowledgement is also made to the University of Melbourne for travel support under the travel for research and special studies program. Finally, the Editor acknowledges the help of Lesley Flanagan who typed the major part of this volume.

Colin Young

Melbourne, December 1982.

THE SOLUBILITY OF GASES AND LIQUIDS

Introductory Information

C. L. Young, R. Battino, and H. L. Clever

INTRODUCTION

The Solubility Data Project aims to make a comprehensive search of the literature for data on the solubility of gases, liquids and solids in liquids. Data of suitable accuracy are compiled into data sheets set out in a uniform format. The data for each system are evaluated and where data of sufficient accuracy are available values are recommended and in some cases a smoothing equation is given to represent the variation of solubility with pressure and/or temperature. A text giving an evaluation and recommended values and the compiled data sheets are published on consecutive pages. The following paper by E. Wilhelm gives a rigorous thermodynamic treatment on the solubility of gases in liquids.

DEFINITION OF GAS SOLUBILITY

The distinction between vapor-liquid equilibria and the solubility of gases in liquids is arbitrary. It is generally accepted that the equilibrium set up at 300K between a typical gas such as argon and a liquid such as water is gas-liquid solubility whereas the equilibrium set up between hexane and cyclohexane at 350K is an example of vapor-liquid equilibrium. However, the distinction between gas-liquid solubility and vapor-liquid equilibrium is often not so clear. The equilibria set up between methane and propane above the critical temperature of methane and below the critical temperature of propane may be classed as vapor-liquid equilibrium or as gas-liquid solubility depending on the particular range of pressure considered and the particular worker concerned.

The difficulty partly stems from our inability to rigorously distinguish between a gas, a vapor, and a liquid; a subject which has been discussed in numerous textbooks. We have taken a fairly liberal view in these volumes and have included systems which may be regarded, by some workers, as vapor-liquid equilibria.

UNITS AND QUANTITIES

The solubility of gases in liquids is of interest to a wide range of scientific and technological disciplines and not solely to chemistry. Therefore a variety of ways for reporting gas solubility have been used in the primary literature. Sometimes, because of insufficient available information, it has been necessary to use several quantities in the compiled tables. Where possible, the gas solubility has been quoted as a mole fraction of the gaseous component in the liquid phase. The units of pressure used are bar, pascal, millimeters of mercury, and atmosphere. Temperatures are reported in Kelvins.

EVALUATION AND COMPILATION

The solubility of comparatively few systems is known with sufficient accuracy to enable a set of recommended values to be presented. This is true both of the measurements near atmospheric pressure and at high pressures. Although a considerable number of systems have been studied by at least two workers, the range of pressures and/or temperatures is often sufficiently different to make meaningful comparison impossible.

Occasionally, it is not clear why two groups of workers obtained very different sets of results at the same temperature and pressure, although both sets of results were obtained by reliable methods and are internally consistent. In such cases, sometimes an incorrect assessment has been given. There are several examples where two or more sets of data have been classified as tentative although the sets are mutually inconsistent.

Many high pressure solubility data have been published in a smoothed form. Such data are particularly difficult to evaluate, and unless specifically discussed by the authors, the estimated error on such values can only be regarded as an "informed guess".

Many of the high pressure solubility data have been obtained in a more general study of high pressure vapor-liquid equilibrium. In such cases a note is included to indicate that additional vapor-liquid equilibrium data are given in the source. Since the evaluation is for the compiled data, it is possible that the solubility data are given a classification which is better than that which would be given for the complete vapor-liquid data (or vice versa). For example, it is difficult to determine coexisting liquid and vapor compositions near the critical point of a mixture using some widely used experimental techniques which yield accurate high pressure solubility data. As another example, conventional methods of analysis may give results with an expected error which would be regarded as sufficiently small for vapor-liquid equilibrium data but an order of magnitude too large for acceptable high pressure gas-liquid solubility.

It is occasionally possible to evaluate data on mixtures of a given substance with a member of a homologous series by considering all the available data for the given substance with other members of the homologous series. In this study the use of such a technique has been limited.

The estimated error is often omitted in the original article and sometimes the errors quoted do not cover all the variables. In order to increase the usefulness of the compiled tables *estimated* errors have been included even when absent from the original article. If the error on *any* variable has been inserted by the compiler, this has been noted.

PURITY OF MATERIALS

The purity of materials has been quoted in the compiled tables where given in the original publication. The solubility is usually more sensitive to impurities in the gaseous component than to liquid impurities in the liquid component. However, the most important impurities are traces of a gas dissolved in the liquid. Inadequate degassing of the absorbing liquid is probably the most often overlooked serious source of error in gas solubility measurements.

APPARATUS AND PROCEDURES

In the compiled tables brief mention is made of the apparatus and procedure. There are several reviews on experimental methods of determining gas solubilities and these are given in References 1-7.

METHODS OF EXPRESSING GAS SOLUBILITIES

Because gas solubilities are important for many different scientific and engineering problems, they have been expressed in a great many ways:

The Mole Fraction, x(g)

The mole fraction solubility for a binary system is given by:

$$x(g) = \frac{n(g)}{n(g) + n(l)}$$

$$= \frac{W(g)/M(g)}{\{W(g)/M(g)\} + \{W(l)/M(l)\}}$$

here n is the number of moles of a substance (an *amount* of substance), W is the mass of a substance, and M is the molecular mass. To be unambiguous, the partial pressure of the gas (or the total pressure) and the temperature of measurement must be specified.

The Weight Per Cent Solubility, wt%

For a binary system this is given by

$$wt\% = 100\, W(g)/\{W(g) + W(l)\}$$

where W is the weight of substance. As in the case of mole fraction, the pressure (partial or total) and the temperature must be specified. The weight per cent solubility is related to the mole fraction solubility by

$$x(g) = \frac{\{wt\%/M(g)\}}{\{wt\%/M(g)\} + \{(100 - wt\%)/M(1)\}}$$

The Weight Solubility, C_W

The weight solubility is the number of moles of dissolved gas per gram of solvent when the partial pressure of gas is 1 atmosphere. The weight solubility is related to the mole fraction solubility at one atmosphere partial pressure by

$$x(g)\ \text{(partial pressure 1 atm)} = \frac{C_W M(1)}{1 + C_W M(1)}$$

where M(1) is the molecular weight of the solvent.

The Moles Per Unit Volume Solubility, n

Often for multicomponent systems the density of the liquid mixture is not known and the solubility is quoted as moles of gas per unit volume of liquid mixture. This is related to the mole fraction solubility by

$$x(g) = \frac{n\,v^{o}(1)}{1 + n\,v^{o}(1)}$$

where $v^{o}(1)$ is the molar volume of the liquid component.

The Bunsen Coefficient, α

The Bunsen coefficient is defined as the volume of gas reduced to 273.15K and 1 atmosphere pressure which is absorbed by unit volume of solvent (at the temperature of measurement) under a partial pressure of 1 atmosphere. If ideal gas behavior and Henry's law are assumed to be obeyed, then

$$\alpha = \frac{V(g)}{V(1)}\ \frac{273.15}{T}$$

where V(g) is the volume of gas absorbed and V(1) is the original (starting) volume of absorbing solvent. The mole fraction solubility x is related to the Bunsen coefficient by

$$x(g, 1\ \text{atm}) = \frac{\alpha}{\alpha + \dfrac{273.15}{T}\ \dfrac{v^{o}(g)}{v^{o}(1)}}$$

where $v^{o}(g)$ and $v^{o}(1)$ are the molar volumes of gas and solvent at a pressure of one atmosphere. If the gas is ideal,

$$x(g) = \frac{\alpha}{\alpha + \dfrac{273.15R}{v^{o}(1)}}$$

Real gases do not follow the ideal gas law and it is important to establish the real gas law used for calculating α in the original publication and to make the necessary adjustments when calculating the mole fraction solubility.

The Kuenen Coefficient, S

This is the volume of gas, reduced to 273.15K and 1 atmosphere pressure, dissolved at a partial pressure of gas of 1 atmosphere by 1 gram of solvent.

The Ostwald Coefficient, L

The Ostwald coefficient, L, is defined as the ratio of the volume of gas absorbed to the volume of the absorbing liquid, all measured at the same temperature:

$$L = \frac{V(g)}{V(l)}$$

If the gas is ideal and Henry's Law is applicable, the Ostwald coefficient is independent of the partial pressure of the gas. It is necessary, in practice, to state the temperature and total pressure for which the Ostwald coefficient is measured. The mole fraction solubility, x(g), is related to the Ostwald coefficient by

$$x(g) = \left[\frac{RT}{P(g)\ L\ v^{o}(l)} + 1\right]^{-1}$$

where P is the partial pressure of gas. The mole fraction solubility will be at a partial pressure of P(g). (See the following paper by E. Wilhelm for a more igorous definition of the Ostwald coefficient.)

The Absorption Coefficient, β

There are several "absorption coefficients", the most commonly used one being defined as the volume of gas, reduced to 273.15K and 1 atmosphere, absorbed per unit volume of liquid when the total pressure is 1 atmosphere. β is related to the Bunsen coefficient by

$$\beta = \alpha(1 - P(l))$$

where P(l) is the partial pressure of the liquid in atmospheres.

The Henry's Law Constant

A generally used formulation of Henry's Law may be expressed as

$$P(g) = K_H x(g)$$

where K_H is the Henry's Law constant and x(g) the mole fraction solubility. Other formulations are

$$P(g) = K_2 C(l) \qquad \text{or} \qquad C(g) = K_c C(l)$$

where K_2 and K_c are constants, C the concentration, and (l) and (g) refer to the liquid and gas phases. Unfortunately, K_H, K_2 and K_c are all sometimes referred to as Henry's Law constants. Henry's Law is a *limiting law* but can sometimes be used for converting solubility data from the experimental pressure to a partial gas pressure of 1 atmosphere, provided the mole fraction of the gas in the liquid is small, and that the difference in pressures is small. Great caution must be exercised in using Henry's Law.

The Mole Ratio, N

The mole ratio, N, is defined by

$$N = n(g)/n(l)$$

Table 1 contains a presentation of the most commonly used inter-conversions not already discussed.

For gas solubilities greater than about 0.01 mole fraciton at a partial pressure of 1 atmosphere there are several additional factors which must be taken into account to unambiguously report gas solubilities. Solution densities or the partial molar volume of gases must be known. Corrections should be made for the possible non-ideality of the gas or the non-applicability of Henry's Law.

TABLE 1 Interconversion of parameters used for reporting solubility

$$L = \alpha(T/273.15)$$

$$C_w = \alpha/v_o\rho$$

$$K_H = \frac{17.033 \times 10^6 \rho(\text{soln})}{\alpha\, M(1)} + 760$$

$$L = C_w\, v_{t,gas}\, \rho$$

where v_o is the molal volume of the gas in $cm^3 mol^{-1}$ at 0°C, ρ the density of the solvent at the temperature of the measurement, ρ_{soln} the density of the solution at the temperature of the measurement, and $v_{t,gas}$ the molal volume of the gas ($cm^3 mol^{-1}$) at the temperature of the measurement.

SALT EFFECTS

Salt effect studies have been carried out for many years. The results are often reported as Sechenov (Setchenow) salt effect parameters. There appears to be no common agreement on the units of either the gas solubility, or the electrolyte concentration.

Many of the older papers report the salt effect parameter in a form equivalent to

$$k_{scc}/\text{mol dm}^{-3} = (1/(c_2/\text{mol dm}^{-3}))\ \log\ ((c_1^{\circ}/\text{mol dm}^{-3})/(c_1/\text{mol dm}^{-3}))$$

where the molar gas solubility ratio, c_1°/c_1, is identical to the Bunsen coefficient ratio, α°/α, or the Ostwald coefficient ratio, L°/L. One can designate the salt effect parameters calculated from the three gas solubility ratios as k_{scc}, $k_{sc\alpha}$, k_{scL}, respectively, but they are identical, and $k_{scc}/dm^3 mol^{-1}$ describes all of them. The superzero refers to the solubility in the pure solvent.

Recent statistical mechanical theories favor a molal measure of the electrolyte and gas solubility. Some of the more recent salt effects are reported in the form

$$k_{smm}/\text{kg mol}^{-1} = (1/(m_2/\text{mol kg}^{-1})\ \log\ ((m_1^{\circ}/\text{mol kg}^{-1})/(m_1/\text{mol kg}^{-1})$$

In this equation the m_1°/m_1 ratio is identical to the Kuenen coefficient ratio, s_1°/s_1, or the solvomolality ratio referenced to water, A_{sm}°/A_{sm}. Thus the salt effect parameters k_{smm}, k_{sms}, and $k_{smA_{sm}}$ are well represented by the $k_{smm}/\text{kg mol}^{-1}$.

Some experimentalists and theoreticians prefer the gas solubility ratio as a mole fraction ratio, x_1°/x_1. It appears that most calculate the mole fraction on the basis of the total number of ions. The salt effect parameters

$$k_{scx}/\text{dm}^3\ \text{mol}^{-1} = (1/(c_2/\text{mol dm}^{-3}))\ \log\ (x_1^{\circ}/x_1)$$

and

$$k_{smx}/\text{kg mol}^{-1} = (1/(m_2/\text{mol kg}^{-1}))\ \log\ (x_1^{\circ}/x_1)$$

are both in the literature, but k_{scx} appears to be the more common.

The following conversions were worked out among the various forms of the salt effect parameter from standard definitions of molarity, molality, and mole fraction assuming the gas solubilities are small.

$$k_{smc} = (c_2/m_2)\ k_{scc} = (c_2/m_2)\ k_{scm} + F_{1m}$$

$$k_{scm} = k_{scc} - F_{1c} = (m_2/c_2)\ k_{smc} - F_{1c} = (m_2/c_2)\ k_{smm}$$

$$k_{scx} = (m_2/c_2)\ k_{smx} = (m_2/c_2)\ k_{smm} + F_{2c}$$

$$k_{smm} = k_{smx} - F_{2m} = (c_2/m_2)\ k_{scx} - F_{2m}$$

$$k_{smx} = (c_2/m_2)\ k_{scx} = (c_2/m_2)\ k_{scc} + F_{3m}$$

$$k_{scc} = k_{scx} - F_{3c} = (m_2/c_2)\ k_{smx} - F_{3c}$$

where

$$F_{1m} = (1/m_2)\ \log\ [(\rho°/\rho)\ (1000 + m_2M_2)/1000]$$

$$F_{1c} = (m_2/c_2)\ F_{1m}$$

$$F_{2m} = (1/m_2)\ \log\ [(1000 + \nu m_3\ M_3)/1000]$$

$$F_{2c} = (m_2/c_2)\ F_{2m}$$

$$F_{3m} = (1/m_2)\ \log\ [(1000\rho + (\nu M_3 - M_2)\ c_2)/1000\rho°)$$

$$F_{3c} = (m_2/c_2)\ F_{3m}$$

The factors F_{1m}, F_{1c}, F_{2m}, F_{2c}, F_{3m}, and F_{3c} can easily be calculated from aqueous electrolyte data such as weight per cent and density as found in Volume III of the International Critical Tables. The values are small and change nearly linearly with both temperature and molality. The factors normally amount to no more than 10 to 20 per cent of the value of the salt effect parameter.

The symbols in the equations above are defined below:

Component	Molar Concentration c/mol dm^{-3}	Molal Concentration m/mol kg^{-1}	Mole Fraction x	Molecular Weight M/g mol^{-1}
Nonelectrolyte	c_1^o, c_1	m_1^o, m_1	x_1^o, x_1	M_1
Electrolyte	c_2	m_2	x_2	M_2
Solvent	c_3	m_3	x_3	M_3

The superscript "°" refers to the nonelectrolyte solubility in the pure solvent. The pure solvent and solution densities are $\rho°$/g cm^{-3} and ρ/g cm^{-3}, respectively. They should be the densities of gas saturated solvent (water) and salt solution, but the gas free densities will differ negligibly in the $\rho°/\rho$ ratio. The number of ions per formula of electrolyte is symbolized by ν.

The following table gives estimated errors in k_{scc} for various salt concentrations and a range of random errors in the gas solubility measurement

Error in k_{scc}/dm^3 mol^{-1} [a]

c_2/ mol dm^{-3}	Random Error in gas solubility Measurement ±2%	±1%	±0.5%	±0.1%	±0.05%
1	±18%	±9%	±5%	±1.5%	±1%
0.1	±175%	±87%	±43%	±9%	±4%
0.05	±350%	±174%	±87%	±17%	±9%
0.01	±1750%	±870%	±435%	±87%	±43%

[a] Based on a k_{scc} value of 0.100.

AQUAMOLAL OR SOLVOMOLAL, A_{sm} or $m_i^{(s)}$

The term aquamolal was suggested by R. E. Kerwin (9). The unit was first used in connection with D_2O and $H_2O + D_2O$ mixtures. It has since been extended in use to other solvents. The unit represents the numbers of moles of solute per 55.51 moles of solvent. It is represented by

$m_i^{(s)}/\text{mol kg}^{-1} = (n_1 M_2/w_2)(w_2/M_0) = m_i(M_2/M_0)$ where an amount of n_i of solute i is dissolved in a mass w_2 of solvent of molar mass M_2; M_0 is the molar mass of a reference solvent and $m_i/\text{mol kg}^{-1}$ is the conventional molality in the reference solvent. The reference solvent is normally water.

TEMPERATURE DEPENDENCE OF GAS SOLUBILITY

In a few cases it has been found possible to fit the mole fraction solubility at various temperatures using an equation of the form

$$\ln x = A + B / (T/100K) + C \ln (T/100K) + DT/100K$$

It is then possible to write the thermodynamic functions $\Delta\bar{G}_1^\circ$, $\Delta\bar{H}_1^\circ$, $\Delta\bar{S}_1^\circ$ and $\Delta\bar{C}_{p_1}$ for the transfer of the gas from the vapor phase at 101,325 Pa partial pressure to the (hypothetical) solution phase of unit mole fraction as:

$$\Delta\bar{G}_1^\circ = -RAT - 100\ RB - RCT \ln (T/100) - RDT^2/100$$

$$\Delta\bar{S}_1^\circ = RA + RC \ln (T/100) + RC + 2\ RDT/100$$

$$\Delta\bar{H}_1^\circ = -100\ RB + RCT + RDT^2/100$$

$$\Delta\bar{C}_{p_1}^\circ = RC + 2\ RDT/100$$

In cases where there are solubilities at only a few temperatures it is convenient to use the simpler equations

$$\Delta\bar{G}_1^\circ = -RT \ln x = A + BT$$

in which case $A = \Delta\bar{H}_1^\circ$ and $-B = \Delta\bar{S}_1^\circ$

REFERENCES

1. Battino, R.; Clever, H. L. *Chem. Rev.* 1966, *66*, 395.
2. Clever, H. L.; Battino, R. in *Solutions and Solubilities*, Ed. M. R. J. Dack, J. Wiley & Sons, New York, 1975, Chapter 7.
3. Hildebrand, J. H.; Prausnitz, J. M.; Scott, R. L. *Regular and related Solutions*, Van Nostrand Reinhold, New York, 1970, Chapter 8.
4. Markham, A. E.; Kobe, K. A. *Chem. Rev.* 1941, *63*, 449.
5. Wilhelm, E.; Battino, R. *Chem. Rev.* 1973, *73*, 1.
6. Wilhelm, E.; Battino, R.; Wilcock, R. J. *Chem. Rev.* 1977, *77*, 219.
7. Kertes, A. S.; Levy, O.; Markovits, G. Y. in *Experimental Thermochemistry* Vol. II, Ed. B. Vodar and B. LeNaindre, Butterworth, London, 1974, Chapter 15.
8. Long, F. A.; McDevit, W. F. *Chem. Rev.* 1952, *51*, 119.
9. Kerwin, K. E., Ph.D. Thesis, University of Pittsburgh 1964.

Revised: April 1982 (R.B., H.L.C.)

COMPONENTS:	EVALUATOR:
1. Sulfur dioxide; SO_2; [7446-09-5] 2. Miscellaneous compounds of similar volatility to sulfur dioxide.	Colin L. Young, School of Chemistry, University of Melbourne, Parkville, Victoria 3052, Australia. March 1982

EVALUATION:

In general measurements generally referred to as vapor-liquid equilibrium determinations are not considered in the Solubility Data Series since it is necessary to limit the scope of the mixtures considered. Many studies of mixtures containing sulfur dioxide, which has a boiling point of 263 K are usually classified as vapor-liquid equilibrium determinations rather than gas solubility measurements.

Below is a list of systems which have been excluded from the compilation on the grounds that

(a) the experimental method was such that it was only suitable if both components were of similar volatility;

(b) the data were presented only in graphical form;

(c) the vapor pressure of the two pure components did not vary by a factor greater than three at the temperature of the measurements.

The list, which includes vapor-liquid equilibrium determinations, is given so that the user of this volume can refer to the original article if further information is required.

Solvent			Reference
1,2,3-Trichloropropane	$C_3H_5Cl_3$	[96-18-4]	1
1,2-Dichloropropane	$C_3H_6Cl_2$	[78-87-5]	1
Tetrachloromethane	CCl_4	[56-23-5]	1
Trichloromethane	$CHCl_3$	[67-66-3]	1
Dichloromethane	CH_2Cl_2	[75-09-2]	1,2
Tetrachloroethene	C_2Cl_4	[127-18-4]	1,6
1,1,2-Trichloroethene	C_2HCl_3	[79-01-6]	1
1,1-Dichloroethene	$C_2H_2Cl_2$	[75-35-4]	1
cis-1,2-Dichloroethene	$C_2H_2Cl_2$	[156-59-2]	1
trans-1,2-Dichloroethene	$C_2H_2Cl_2$	[156-60-5]	1
1,1,2,2-Tetrachloroethane	$C_2H_2Cl_4$	[79-34-5]	1
1,1,2-Trichloroethane	$C_2H_3Cl_3$	[79-00-5]	1
1,1,1-Trichloroethane	$C_3H_3Cl_3$	[71-55-6]	1
1,1-Dichloroethane	$C_2H_4Cl_2$	[75-34-3]	1
1,2-Dichloroethane	$C_2H_4Cl_2$	[107-06-2]	1,2
2-Chloropropane	C_3H_7Cl	[75-29-6]	1
1,4-Dichlorobutane	$C_4H_8Cl_2$	[110-56-5]	1
2-Chlorobutane	C_4H_9Cl	[78-86-4]	1
1-Chloro-2-methylpropane	C_4H_9Cl	[513-36-0]	1
2-Chloro-2-methylpropane	C_4H_9Cl	[507-20-0]	1
Chloromethane	CH_3Cl	[74-87-3]	3
Chloroethane	C_2H_5Cl	[75-00-3]	3
1-Chloropropane	C_3H_7Cl	[540-54-5]	3
1-Chlorobutane	C_4H_9Cl	[109-69-3]	3
1-Chloropentane	$C_5H_{11}Cl$	[543-59-9]	3
1-Chlorohexane	$C_6H_{13}Cl$	[544-10-5]	3
1-Chloroheptane	$C_7H_{15}Cl$	[629-06-1]	3
Thionyl chloride	Cl_2OS	[7719-09-7]	3
Sulfuryl chloride	Cl_2O_2S	[7791-25-5]	3
2-Methyl-1-propene	C_4H_8	[115-11-7]	4
3-Methyl-1-butene	C_5H_{10}	[563-45-1]	5
Hydrogen fluoride	HF	[7664-39-3]	7
Benzenamine (Aniline)	C_6H_7N	[62-53-3]	8

(cont.)

COMPONENTS:	EVALUATOR:
1. Sulfur dioxide; SO_2; [7446-09-5] 2. Miscellaneous compounds of similar volatility to sulfur dioxide.	Colin L. Young, School of Chemistry, University of Melbourne, Parkville, Victoria 3052, Australia. March 1982

In addition to the above systems the work of Komissarov *et al.* (9) have not been compiled. These workers studied the system sulfur dioxide + oxygen + tetradecane. It is, however, not clear as to what conditions the reported numerical data refer.

References

1. Gilot, B.; Aubry, M.; Jayles, C. *Chim. Ing. Genie. Chim.*, 1972, *105*, 1797.

2. Aubry, M.; Gilot, B.; Jayles, C.; *Ann. Genie. Chim.*, 1967, *3*, 73.

3. Aubry, M.; Jayles, C.; Guiglion, C. *Chim. Ind. Genie. Chim.*, 1972, *105*, 1429.

4. Ayscough, P. B.; Ivin, K. J.; O'Donnell, J. H. *Trans. Faraday Soc.*, 1965, *61*, 1601.

5. Brady, B. H. G.; O'Donnell, J. H. *Trans. Faraday Soc.*, 1968, *64*, 23.

6. Gilot, B.; Aubry, M.; Jayles, C. *Chim. Ind. Genie. Chim.*, 1969, *101*, 821.

7. Kaiser, V. *Chem. Ing. Techn.*, 1966, *38*, 151.

8. Westra, J. G.; Peters, C. J.; Diepen, G. A. M. *J. Chem. Thermodynamics*, 1978, *10*, 343.

9. Komissarov, V. D.; Saitova, M. A.; Kuramshin, E. M.; Timirova, R.G. *Khim. Prom.*, 1973, *49*, 738.

COMPONENTS:	EVALUATOR:
1. Sulfur dioxide; SO_2; [7446-09-5] 2. Water; H_2O; [7732-18-5]	Rubin Battino, Department of Chemistry, Wright State University, Dayton, Ohio 45435, U.S.A. July 1981

CRITICAL EVALUATION:

For this critical evaluation of the solubility of sulfur dioxide in water at around atmospheric pressure, we examined the literature from 1855 until the present. Surprisingly, the earliest work (1) (that of Schönfeld as reported by Bunsen in several papers) is very close to the best modern values — in the middle of the temperature range studied (273-313 K). These early results are within 2-4 per cent of the values to be recommended in this section.

Reactive gases like sulfur dioxide pose a special problem with respect to gas solubilities. SO_2(g) is in equilibrium (at saturation) with SO_2(aq), but the SO_2(aq) is also in equilibrium with other species in solution. The solubilities reported here as mole fractions at 101.325 kPa partial pressure of gas are bulk solubilities, i.e., they are taken as if all of the SO_2-containing species in solution were simply SO_2(aq). There are many papers dealing with the remaining ionization and equilibrium constants.

Many of the measurements made on sulfur dioxide solubilities were carried out at different partial pressures. Due to the high solubilities (mole fractions at 101.325 kPa partial pressure of gas of 0.003 to 0.06) exhibited by this system and the chemical interactions involved, the system does not obey Henry's law even over relatively small extrapolations. We limited our interpolations to within about 13 kPa either side of 101.325 kPa.

Of the more modern determinations we judged the work of four workers to be sufficiently reliable to use for smoothing via least squares analysis. Individual data points which were more than about two standard deviations from the smoothed values were rejected. Thus, 24 points were used for the final smoothing and were obtained as follows (reference — number of data points used from that reference): 2-10, 3-8, 4-1, 5-5. The best fit for the 24 points is

$$\ln x_1 = -25.26286 + 45.75521/(T/100K) + 5.685450 \ln (T/100K)$$

where x_1 is the mole fraction solubility at 101.325 kPa partial pressure of gas. The standard deviation in x_1 was 1.7×10^{-4} or about ±1.8 per cent at the mid-range of 283 to 380 K. The following table gives the recommended smoothed values of the mole fraction and the Ostwald coefficient at 5 K intervals calculated from the above equation. (Thermodynamic properties are not presented - due to the complexity of the aqueous solution. They would be meaningless.)

T/K	x_1 [a]	L [b]	T/K	$10^2 x_1$ [a]	L [b]
278.15	0.0499	66.6	333.15	9.22	13.9
283.15	0.0413	55.6	338.15	8.19	12.5
288.15	0.0345	46.9	343.15	7.31	11.3
293.15	0.0290	39.8	348.15	6.55	10.2
298.15	0.0246	34.1	353.15	5.90	9.27
303.15	0.0210	29.5	358.15	5.33	8.47
308.15	0.0180	25.6	363.15	4.84	7.76
313.15	0.0156	22.4	368.15	4.41	7.14
318.15	0.0136	19.7	373.15	4.03	6.59
323.15	0.0119	19.7	378.15	3.70	6.10
328.15	0.0104	15.5	383.15	3.40	5.66

[a] Mole fraction solubility at 101.325 kPa partial pressure of gas.

[b] Ostwald coefficient.

(cont.)

COMPONENTS:	EVALUATOR:
1. Sulfur dioxide; SO_2; [7446-09-5] 2. Water; H_2O; [7732-18-5]	Rubin Battino, Department of Chemistry, Wright State University, Dayton, Ohio 45435, U.S.A. July 1981

CRITICAL EVALUATION:

The results of several workers were not used in the smoothing process. Bunsen/Schönfeld's early measurements (1) were surprisingly good at the middle of their temperature range of 273 to 313 K but differed strongly at the extremes. Sims' early work (6) ranged 4-5 per cent high and low. Fox's two values (7) were high and very high. Lindner's values (8) were all at lower pressures. Baume and Tykociner's values at three temperatures (9) were very high when interpolated to 101.325 kPa. Smith and Parkhurst's results (10) were erratically low and high. Öman's results (11) were very low at all temperatures he investigated (313-368 K). The interpolated Maass and Maass results (12) were over 10 per cent low. The interpolated Morgan and Maass (13) results were also quite low (4-7 per cent). The extrapolared value for Conrad and Beuschlein (14) from data over a pressure range of 41 to 97 kPa at 298 K was low. Johnstone and Leppla's results (15) were at too low pressures to be reasonably extrapolated. Otuka's results (16) were mostly at pressures above atmospheric and not readily extrapolated. Parkinson's determinations (17) were at rather low partial pressures of sulfur dioxide (0.08-3.5 kPa). The Bodor *et al.* (18) results were quite low (7-14 per cent) after extrapolation and interpolation to 101.325 kPa. Tokunaga's measurements (19) were of reasonable precision but systematically higher than the recommended values. Douabul and Riley's results (20) were 6 to 14 per cent higher than the recommended values. Byerley *et al.*'s two values (21) were 8 and 12 per cent low. Vosolsobe *et al.*'s results (22) were at too low pressure to be extrapolated. Sano and Nakamoto's value agrees with the value in the table. This system awaits a definitive study.

References:

1. Bunsen, R. *Ann.* 1855, *93*, 1-50; Schonfeld, M. *Ann.* 1855, *93*, 26.
2. Hudson, J. C. *J. Chem. Soc.* 1925, 1332-47.
3. Beuschlein, W. L.; Simenson, L. O. *J. Am. Chem. Soc.* 1940, *62*, 610-2.
4. Rabe, A. E.; Harris, J. F. *J. Chem. Eng. Data* 1963, *8*, 333-6.
5. Lavrova, E. M.; Tudorovskaya, L. L. *Zhur. Prikl. Khim.* 1977, *50*, 1146-51.
6. Sims, T. H. *J. Chem. Soc.* 1862, *14*, 1-22.
7. Fox, C. J. J. *Z. Phys. Chem.* 1902, *41*, 458-82.
8. Lindner, J. *Monatsh.* 1912, *33*, 613-72.
9. Baume, G.; Tykociner, A. *J. Chim. Phys.* 1914, *12*, 270-5.
10. Smith, W. T.; Parkhurst, R. B. *J. Am. Chem. Soc.* 1922, *44*, 1918-27.
11. Öman, *Teknisk. Tidskrift Uppl. D.* 1924, *54*, 81. Cited in International Critical Tables, Vol. III, p.258.
12. Maass, C. E.; Maass, O. *J. Am. Chem. Soc.* 1928, *50*, 1352-68.
13. Morgan, O. M.; Maass, O. *Can. J. Res.* 1931, *5*, 162-99.
14. Conrad, F. H.; Beuschlein, W. L. *J. Am. Chem. Soc.* 1934, *56*, 2554-62.
15. Johnstone, H. F.; Leppla, P. W. *J. Am. Chem. Soc.* 1934, *56*, 2233-8.
16. Otuka, Y. *J. Soc. Chem. Ind. Japan Suppl.* 1939, *42*, 205B-9B.
17. Parkinson, R. V. *Tappi* 1956, *39*, 517-9.
18. Bodor, E.; Mohai, B.; Papp, S.; Makranczy, J. *Veszpremi Vegyip. Egyet. Kozlemen.* 1961, *5*, 85-97.
19. Tokunaga, J. *J. Chem. Eng. Data* 1974, *19*, 162-5.
20. Douabul, A.; Riley, J. *J. Chem. Eng. Data* 1979, *24*, 274-6.
21. Byérley, J. J.; Rempel, G. L.; Le, V. T. *J. Chem. Eng. Data* 1980, *25*, 55-6.

COMPONENTS:	EVALUATOR:
1. Sulfur dioxide; SO_2; [7446-09-5] 2. Water; H_2O; [7732-18-5]	Rubin Battino, Department of Chemistry, Wright State University, Dayton, Ohio 45435, U.S.A. July 1981

CRITICAL EVALUATION:

22. Vosolsobe, J.; Simecek, S.; Michalek, J.; Kadler, B. *Chem. Prumysl.* 1965, *15*, 401.

23. Sano, H.; Nakamoto, Y. *Nippon Kagaku Zasshi* 1968, *89*, 369.

COMPONENTS:	ORIGINAL MEASUREMENTS:
1. Sulfur dioxide; SO_2; [7446-09-5] 2. Water; H_2O; [7732-18-5]	Bunsen, R. *Ann.* 1855, *93*, 1-50. (Schönfeld, M. *Ann.* 1855, *93*, 26).
VARIABLES:	PREPARED BY:
T/K = 273-313	R. Battino

EXPERIMENTAL VALUES:

t/°C	T^a/K	$x_1^{a,b}$	α^c/cm^3 (STP) cm^{-3} atm^{-1}	t/°C	T^a/K	$x_1^{a,b}$	α^c/cm^3 (STP) cm^{-3} atm^{-1}
0	273	0.0525	68.861	21	294	0.0274	34.989
1	274	0.0511	67.003	22	295	0.0266	33.910
2	275	0.0498	65.169	23	296	0.0258	32.847
3	276	0.0485	63.360	24	297	0.0250	31.800
4	277	0.0472	61.576	25	298	0.0242	30.766
5	278	0.0459	59.816	26	299	0.0234	29.748
6	279	0.0446	58.080	27	300	0.0227	28.744
7	280	0.0434	56.369	28	301	0.0219	27.754
8	281	0.0421	54.683	29	302	0.0212	26.780
9	282	0.0409	53.021	30	303	0.0204	25.819
10	283	0.0397	51.383	31	304	0.0197	24.873
11	284	0.0385	49.770	32	305	0.0190	23.942
12	285	0.0373	48.182	33	306	0.0183	23.025
13	286	0.0361	46.618	34	307	0.0176	22.122
14	287	0.0350	45.079	35	308	0.0169	21.234
15	288	0.0339	43.564	36	309	0.0162	20.361
16	289	0.0327	42.073	37	310	0.0155	19.502
17	290	0.0316	40.608	38	311	0.0149	18.658
18	291	0.0306	39.165	39	312	0.0142	17.827
19	292	0.0295	37.749	40	313	0.0136	17.013
20	293	0.0283	36.216				

[a] Calculated by compiler.
[b] Mole fraction solubility at 101.325 kPa partial pressure of gas.
[c] Bunsen coefficient.

AUXILIARY INFORMATION

METHOD/APPARATUS/PROCEDURE:

Analysed chemically. This is really Schönfeld's data (1) as cited by Bunsen. Data were smoothed by Schönfeld.

SOURCE AND PURITY OF MATERIALS:

No details given.

ESTIMATED ERROR:

$\delta\alpha/\alpha$ = ±0.03 (compiler's estimate).

REFERENCES:

1. Schönfeld, M. *Ann.* 1855, *93*, 26.

COMPONENTS:

1. Sulfur dioxide; SO_2; [7746-09-5]

2. Water; H_2O; [7732-18-5]

ORIGINAL MEASUREMENTS:

Sims, T. H.
J. Chem. Soc.
1862, *14*, 1-22.

VARIABLES:

T/K = 281-323

PREPARED BY:

R. Battino

EXPERIMENTAL VALUES:

t^a/°C	T^b/K	x_1[b,c]	Solubility[d] (volume)	Solubility[e] (weight)
8	281.2	0.0451	58.7	0.168
10	283.2	0.0415	53.9	0.154
12	285.2	0.0384	49.6	0.142
14	287.2	0.0353	45.6	0.130
16	289.2	0.0329	42.2	0.121
18	291.2	0.0305	39.3	0.112
20	293.2	0.0284	36.4	0.104
22	295.2	0.0268	34.2	0.098
24	297.2	0.0252	32.3	0.092
26	299.2	0.0239	30.5	0.087
28	301.2	0.0228	28.9	0.083
30	303.2	0.0215	27.3	0.078
32	305.2	0.0201	25.7	0.073
34	307.2	0.0190	24.3	0.069
36	309.2	0.0180	22.8	0.065
38	311.2	0.0171	21.6	0.062
40	313.2	0.0160	20.4	0.058
42	315.2	0.0152	19.3	0.055
44	317.2	0.0147	18.4	0.053

(cont.)

AUXILIARY INFORMATION

METHOD/APPARATUS/PROCEDURE:

The experimental method is similar to that of Roscoe (1). Accurate analysis was used to determine the quantity of gas absorbed by a known weight of water under known conditions of temperature and pressure. The analysis was iodimetric. Details are given in the paper.

SOURCE AND PURITY OF MATERIALS:

1. Evolved from sulfuric acid and copper and washed with water before freezing into a container.

2. No details given.

ESTIMATED ERROR:

Solubility/Solubility = 0.03 (compiler's estimate).

REFERENCES:

1. Roscoe, H. E.
J. Chem. Soc.
1855, *8*, 14.

COMPONENTS:

1. Sulfur dioxide; SO_2; [7746-09-5]

2. Water; H_2O; [7732-18-5]

ORIGINAL MEASUREMENTS:

Sims, J. H.

J. Chem. Soc.

1862, *14*, 1-22.

EXPERIMENTAL VALUES:

t^a/°C	T^b/K	x_1[b,c]	Solubility[d] (volume)	Solubility[e] (weight)
46	319.2	0.0139	17.4	0.050
48	321.2	0.0130	16.4	0.047
50	323.2	0.0125	15.6	0.045

[a] These are smoothed values presented by Sims at 2 K intervals. The smoothed values are based on extensive pressure-dependent measurements at 7, 20, 39.8, and 50 °C. They were calculated for an SO_2 partial pressure of 760 mmHg. The pressure range at each temperature ranged from about 27 to 2000 mmHg, so the smoothed values at 760 mmHg are interpolations.

[b] Calculated by compiler.

[c] Mole fraction solubility at 101.325 kPa partial pressure of gas.

[d] Solubility in cm^3 SO_2/g H_2O. Effectively an Ostwald coefficient.

[e] Solubility in g SO_2/g H_2O. Mole fraction solubilities were calculated from the values in this column.

COMPONENTS:

1. Sulfur dioxide; SO_2; [7446-09-5]
2. Water; H_2O; [7732-18-5]

ORIGINAL MEASUREMENTS:

Fox, C.J.J.

Z. Phys. Chem. 1902, *41*, 458-482.

VARIABLES:

Temperature

PREPARED BY:

C.L. Young

EXPERIMENTAL VALUES:

Partial pressure = 0.101325 MPa

T/K	Ostwald coefficient, L	Mole fraction of sulfur dioxide, [c] x_{SO_2}
298.2	32.76[a]	0.0257
308.2	22.43[b]	0.0178

[a] average of 6 values.

[b] average of 5 values.

[c] calculated by compiler assuming ideal gas molar volume for sulfur dioxide.

AUXILIARY INFORMATION

METHOD/APPARATUS/PROCEDURE:

Sulfur dioxide was passed through the water until saturated at the barometric pressure. After saturation a sample of solution was analysed, the amount of sulfur dioxide being determined by iodine-thiosulfate titration.

SOURCE AND PURITY OF MATERIALS:

1. Purity 99.8 mole per cent major impurity water.
2. Conductivity grade.

ESTIMATED ERROR:

$\delta x_{SO_2} = \pm 3\%$

REFERENCES:

COMPONENTS:	ORIGINAL MEASUREMENTS:
1. Sulfur dioxide; SO_2; [7446-09-5] 2. Water; H_2O; [7732-18-5]	Lindner, J. *Monatsh.* 1912, *33*, 613-672.
VARIABLES:	**PREPARED BY:**
Temperature, pressure	C. L. Young

EXPERIMENTAL VALUES:

T/K	p/mmHg	p/kPa	Solubility $/(g_{(SO_2)}/10^2\ cm^3$ (sat.soln.)
273.2	0.4	0.05	0.0537
	3.5	0.47	0.237
	29.4	3.92	1.227
	109.4	14.59	3.804
293.2	1.4	0.19	0.0534
	11.75	1.57	0.234
	87.9	11.72	1.212
	313	41.7	3.750
323.2	4.9	0.65	0.0525
	30.5	4.07	0.2276
	204.5	27.26	1.181
	696	92.8	3.628

AUXILIARY INFORMATION

METHOD/APPARATUS/PROCEDURE:

Manometric method. Fairly complicated method, of historical interest only.

SOURCE AND PURITY OF MATERIALS:

No purity data given.

ESTIMATED ERROR:

REFERENCES:

COMPONENTS:

1. Sulfur dioxide; SO_2; [7446-09-5]

2. Water; H_2O; [7732-18-5]

ORIGINAL MEASUREMENTS:

Smith, W.T.; Parkhurst, R.B.

J. Am. Chem. Soc. 1922, *44*, 1918-1927

VARIABLES:

Temperature, pressure

PREPARED BY:

C.L. Young

EXPERIMENTAL VALUES:

T/K	Partial pressure of sulfur dioxide p/mmHg	p/kPa	Wt. of SO_2 per 100 cm^3 water /g	Mole fraction of sulfur dioxide, x_{SO_2}
278.2	474.8	63.30	11.55	0.0315
293.2	198.6	26.48	2.97	0.00828
	224.4	29.92	3.34	0.00931
	366.6	48.88	5.22	0.0145
	1107.8	147.69	15.01	0.0405
313.2	306.4	40.85	2.21	0.00618
	373.8	49.84	2.68	0.00748
333.2	175.7	23.42	0.84	0.00236

AUXILIARY INFORMATION

METHOD/APPARATUS/PROCEDURE:

Static system in which water was saturated with sulfur dioxide at stated pressure. Samples of saturated solutions analysed by iodine-thiosulfate titration. Details in source.

SOURCE AND PURITY OF MATERIALS:

1. Virginia Smelting Co. sample.

2. Degassed.

ESTIMATED ERROR:

$\delta T/K = \pm 0.1$; $\delta x_{SO_2} = \pm 5\%$
(estimated by compiler)

REFERENCES:

COMPONENTS:

1. Sulfur dioxide; SO_2; [7746-09-5]
2. Water; H_2O; [7732-18-5]

ORIGINAL MEASUREMENTS:

Hudson, J. C.

J. Chem. Soc.

1925, 1332-1347.

VARIABLES:

Temperature

PREPARED BY:

C. L. Young

EXPERIMENTAL VALUES:

T/K	Total pressure P/mmHg	Partial pressure p/mmHg	p/kPa	Solubility S_1^*	$S_o^\#$	Mole fraction of sulfur dioxide, x_{SO_2} at a partial pressure of 101.3 kPa
283.15	777.3	764.2	101.9	15.53	15.44	0.04162
	776.3	767.1	102.3	15.50	15.36	0.04141
	776.3	767.0	102.3	15.58	15.44	0.04162
	775.4	766.2	102.2	15.48	15.36	0.04141
	766.2	757.0	100.9	15.28	15.32	0.04131
288.15	769.5	757.6	101.0	12.72	12.76	0.03464
	772.5	760.6	101.4	12.75	12.74	0.03459
	769.5	757.6	101.0	12.64	12.68	0.03443
293.15	787.7	771.3	102.8	10.79	10.63	0.02903
	773.3	765.9	102.1	10.61	10.66	0.02911
	778.6	762.2	101.6	10.68	10.64	0.02905
	779.1	762.7	101.7	10.65	10.61	0.02898
	773.9	757.5	101.0	10.64	10.68	0.02916
303.05	798.5	768.5	102.5	7.640	7.554	0.02080
	789.4	759.4	101.2	7.600	7.607	0.02095
	791.0	761.0	101.5	7.610	7.600	0.02093
	800.9	770.9	102.8	7.674	7.565	0.02083
313.15	800.9	747.3	99.63	5.428	5.520	0.01529
	801.4	747.8	99.70	5.441	5.529	0.01531
	798.1	744.5	99.26	5.406	5.516	0.01528

(cont.)

AUXILIARY INFORMATION

METHOD/APPARATUS/PROCEDURE:

Sulfur dioxide was bubbled through water in a thermostatted tube. Mercury manometer used to measure total pressure at sautration. After saturation, tube weighed and sulfur dioxide content determined by oxidation with hydrogen peroxide in sodium hydroxide solution and the gravimetric estimation of sulfate as barium sulfate. The partial pressure of sulfur dioxide was calculated assuming partial pressure of water could be calculated from Raoult's Law.

SOURCE AND PURITY OF MATERIALS:

1. Obtained from a cylinder; the purity was stated to be high, the major impurity being water, less than 0.4%.

2. No details given.

ESTIMATED ERROR:

$\delta T/K = \pm 0.1$; $\delta p/\text{mmHg} = \pm 0.2$;

$\delta x_{SO_2} = \pm 0.001$ (estimated by compiler).

REFERENCES:

COMPONENTS:

1. Sulfur dioxide; SO_2; [7746-09-5]

2. Water; H_2O; [7732-18-5]

ORIGINAL MEASUREMENTS:

Hudson, J. C.

J. Chem. Soc.

1925, 1332-1347.

EXPERIMENTAL VALUES:

T/K	Total pressure P/mmHg	Partial pressure p/mmHg	p/kPa	Solubility S_1^*	$S_o^\#$	Mole fraction of sulfur dioxide, x_{SO} at a partial pressure of 101.3 kPa
313.15	816.7	763.1	101.7	5.593	5.569	0.01542
	815.7	762.1	101.6	5.585	5.555	0.01538
	815.7	762.1	101.6	5.581	5.550	0.01537
321.30	799.0	717.0	95.59	4.147	4.395	0.01221
	801.5	719.5	95.93	4.132	4.365	0.01213
	794.3	712.3	94.97	4.108	4.382	0.01217
	840.9	758.9	101.2	4.409	4.415	0.01226
333.15	881.5	735	98.0	3.151	3.257	0.009077
	894.5	748	99.7	3.192	3.244	0.009041
	894.5	748	99.7	3.215	3.268	0.009107
	886.5	740	98.7	3.142	3.220	0.008975
343.15	897	666.5	88.86	2.278	2.598	0.007254
	892.5	662	88.3	2.277	2.614	0.007298
	893	662.5	88.33	2.300	2.637	0.007362
	892	661.5	88.19	2.251	2.585	0.007218
353.15	883	531.5	70.86	1.496	2.139	0.005980
	883	531.5	70.86	1.490	2.131	0.005958
	958	606.5	80.86	1.704	2.130	0.005955
	967	616	82.1	1.714	2.114	0.005910
363.15	893	378	50.4	0.901	1.801	0.005040
	954.3	439.5	58.60	1.037	1.785	0.004995
	960	439	58.5	1.056	1.829	0.005118

* g of SO_2 per 100 g of water at stated pressure.

\# g of SO_2 per 100 g of water at pressure of 101.3 kPa.

COMPONENTS:

1. Sulfur dioxide; SO_2; [7446-09-5]
2. Water; H_2O; [7732-18-5]

ORIGINAL MEASUREMENTS:

Maass, C.E.; Maass, O.

J. Am. Chem. Soc. 1928, *50*, 1352-1368.

VARIABLES:

Temperature, pressure

PREPARED BY:

C.L. Young

EXPERIMENTAL VALUES:

T/K	Total Pressure p/cmHg	Total Pressure $p/10^4$Pa	Mass % of SO_2	Mole fraction of sulfur dioxide, x_{SO_2}
283.2	24.3	3.24	4.57	0.0133
	45.2	6.03	8.19	0.0245
	67.4	8.99	11.64	0.0357
	87.4	11.65	14.75	0.0464
	124.5	16.60	18.91	0.0615
	128.8	17.17	19.86	0.0652
	154.3	20.57	23.10	0.0779
289.7	31.0	4.13	4.48	0.0130
	57.2	7.63	8.03	0.0240
	84.4	11.25	11.42	0.0350
	111.4	14.85	14.31	0.0449
	151.9	20.25	18.57	0.0603
	156.0	20.80	19.52	0.0637
	188.4	25.12	22.71	0.0763
295.2	37.8	5.04	4.40	0.0128
	69.3	9.24	7.88	0.0235
	102.2	13.63	11.17	0.0342
	132.6	17.68	14.04	0.0439
	178.3	23.77	18.22	0.0590
	186.1	24.81	19.14	0.0624
	222.2	29.62	22.32	0.0748
298.2	66.2	8.83	6.9	0.0204
	246.9	32.92	22.4	0.0751
300.2	44.8	5.97	4.32	0.0125
	81.0	10.80	7.71	0.0230

AUXILIARY INFORMATION

METHOD/APPARATUS/PROCEDURE:

The vapor pressure of solutions containing a known amount of sulfur dioxide were measured.

SOURCE AND PURITY OF MATERIALS:

1. Commercial sample, dried and distilled several times in vacuo.
2. Degassed.

ESTIMATED ERROR:

$\delta T/K = \pm 0.1$; $\delta x_{SO_2} = \pm 2\%$
(estimated by compiler)

REFERENCES:

COMPONENTS:	ORIGINAL MEASUREMENTS:
1. Sulfur dioxide; SO_2; [7746-09-5]	Maass, C.E.; Maass, O.
2. Water; H_2O; [7732-18-5]	*J. Am. Chem. Soc.* 1928, *50*, 1352-1368.

EXPERIMENTAL VALUES:

T/K	Total Pressures p/cmHg	$p/10^4$Pa	Mass % of SO_2	Mole fraction of sulfur dioxide, x_{SO_2}
300.2	118.3	15.77	10.95	0.0334
	152.8	20.37	13.74	0.0429
	206.9	27.58	17.85	0.0576
	255.6	34.08	21.86	0.0729

T/K	x_{SO_2} [a]
283.2	0.03838
290.0	0.03053
295.2	0.02518
298.2	0.02157
300.2	0.02141

[a] Calculated by compiler for a partial pressure of sulfur dioxide of 101.3 kPa.

COMPONENTS:

1. Sulfur dioxide; SO_2; [7746-09-5]
2. Water; H_2O; [7732-18-5]

ORIGINAL MEASUREMENTS:

Morgan, O. M.; Maass, O.

Can. J. Res.

1931, *5*, 162-99.

VARIABLES:

T/K = 273-298

PREPARED BY:

R. Battino

EXPERIMENTAL VALUES:

P/cm Hg	P^a/kPa	% SO_2[b]	P/cm Hg	P^a/kPa	% SO_2[b]
t/°C = 0	T/K = 273		t/°C = 18	T/K = 291	
0.17	0.23	0.1369	0.32	0.43	0.0816
0.63	0.84	0.3374	1.10	1.47	0.2605
1.97	2.63	0.8888	2.68	3.57	0.5563
6.05	8.07	2.284	5.92	7.89	1.103
11.55	15.40	3.978	14.83	19.77	2.458
17.30	22.06	5.597	26.66	35.54	4.150
29.21	38.94	8.630	39.01	52.01	5.773
41.63	55.50	11.52	51.44	68.58	7.340
55.02	73.35	14.23	63.69	84.91	8.831
			76.04	101.38	10.30
t/°C = 10	T/K = 283		t/°C = 25	T/K = 298	
0.35	0.47	0.1267	0.79	1.05	0.1735
1.00	1.33	0.3414	3.42	4.56	0.5448
2.12	2.83	0.6360	7.37	9.83	1.056
4.71	6.28	1.261	13.61	18.15	1.809
10.98	14.64	2.593	25.28	33.70	3.116
19.52	26.02	4.256	40.00	53.33	4.672
28.37	37.82	5.872	55.00	73.33	6.184
46.62	62.15	8.937	69.88	93.17	7.640
64.18	85.57	11.68	81.17	108.22	8.756
73.58	98.10	13.03			

(cont.)

AUXILIARY INFORMATION

METHOD/APPARATUS/PROCEDURE:

The amount of sulfur dioxide dissolved was determined via conductivity measurements.

SOURCE AND PURITY OF MATERIALS:

1. From Ansul Chemical Co., Marinette, Wisconsin. Distilled three times.
2. "Conductivity water".

ESTIMATED ERROR:

δ% SO_2/% SO_2 = ±0.05 (compiler's estimate).

REFERENCES:

COMPONENTS:	ORIGINAL MEASUREMENTS:
1. Sulfur dioxide; SO_2; [7746-09-5]	Morgan, O. M.; Maass, O.
2. Water; H_2O; [7732-18-5]	*Can. J. Res.* 1931, *5*, 162-99.

EXPERIMENTAL VALUES:

[a] Calculated by compiler.

[b] Presumably a weight per cent of the total amount of sulfur dioxide as SO_2 in the water.

T/K	H^a	$x_1{}^b$
283	0.01080	0.0386
291	0.00870	0.0295
298	0.00758	0.0235

[a] Henry's law constant calculated by authors from $[SO_2] = Hp$ where $[SO_2]$ is in mol dm^{-3} and p is in cm Hg. These are smoothed and averaged values.

[b] Mole fraction solubility at 101.325 kPa partial pressure of gas calculated by compiler from data around 101.325 kPa given in the paper. Mole fractions calculated from the Henry's law constants given by the authors are quite different since for this system H (as calculated) is concentration-dependent.

COMPONENTS:

1. Sulfur dioxide; SO_2; [7446-09-5]
2. Water; H_2O; [7732-18-5]

ORIGINAL MEASUREMENTS:

Conrad, F. H.; Beuschlein, W. L.

J. Am. Chem. Soc.

1934, *56*, 2554-2562.

VARIABLES:

Pressure

PREPARED BY:

C. L. Young

EXPERIMENTAL VALUES:

T/K	Partial Pressure of sulfur dioxide p/mmHg	p/kPa	Solubility, S /($gSO_2/100gH_2O$)	Mole fraction of sulfur dioxide in solution, x_{SO_2}
298.2	308	41.1	3.72	0.0104
	299	39.9	3.63	0.0101
	593	79.1	6.89	0.0190
	640	85.3	7.28	0.0201
	611	81.5	6.99	0.0193
	730	97.3	8.37	0.0230
	760 [a]	101.3	-	0.0237[a]

[a] Extrapolated by compiler.

AUXILIARY INFORMATION

METHOD/APPARATUS/PROCEDURE:

Sulfur dioxide was passed into a measuring pipet and then into a flask containing water. The flask was shaken until a constant total pressure was achieved. The pressure was measured using a mercury manometer. The solution was analysed by titration using the iodine-thiosulfate procedure.

SOURCE AND PURITY OF MATERIALS:

1. Refrigeration grade from Virginia Smelting Co. No trace of sulfur trioxide detected.
2. Distilled and freshly boiled.

ESTIMATED ERROR:

$\delta T/K = \pm 0.05$; $\delta x_{SO_2} = \pm 4\%$ (estimated by compiler).

REFERENCES:

COMPONENTS:

1. Sulfur dioxide; SO_2; [7446-09-5]

2. Water; H_2O; [7732-18-5]

ORIGINAL MEASUREMENTS:

Johnstone, H.F.; Leppla, P.W.

J. Am. Chem. Soc. 1934, *56*, 2233-2238.

VARIABLES:

Pressure

PREPARED BY:

C.L. Young

EXPERIMENTAL VALUES:

T/K	Partial pressure of sulfur dioxide 10^3p/atm	p/kPa	Moles of SO_2 per 1000g water	Mole fraction of sulfur dioxide $10^3x_{SO_2}$
298.2	0.27	0.027	0.002484	0.0448
	1.20	0.122	0.006203	0.1117
	2.29	0.232	0.009546	0.1720
	2.67	0.271	0.01084	0.1953
	6.71	0.680	0.02059	0.3708
	9.11	0.923	0.02561	0.4612
	13.50	1.368	0.03328	0.5992
308.2	1.03	0.104	0.00419	0.0755
	2.49	0.252	0.00745	0.1342
	4.12	0.417	0.00999	0.1799
	5.80	0.588	0.01356	0.2442
	9.55	0.968	0.01987	0.3579
323.2	2.30	0.233	0.00467	0.0841
	3.83	0.388	0.00637	0.1147
	5.38	0.545	0.00838	0.1510
	8.79	0.891	0.01057	0.1904

AUXILIARY INFORMATION

METHOD/APPARATUS/PROCEDURE:

The low partial pressure of sulfur dioxide was obtained by using a mixture of nitrogen and sulfur dioxide. This mixture was bubbled through water in a "Pyrex saturator" (Bichowsky and Starch, ref (1)) until equilibrium was reached. Samples of solution were analysed by adding excess iodate and titrating with thiosulfate solution.

SOURCE AND PURITY OF MATERIALS:

1. Stated to be of high purity.

2. Conductivity water, degassed.

ESTIMATED ERROR:

REFERENCES:

1. Bichowsky, F.R.; Starch, H.

J. Am. Chem. Soc. 1915, *37*, 2695.

COMPONENTS:

1. Sulfur dioxide; SO_2; [7746-09-5]
2. Water; H_2O; [7732-18-5]

ORIGINAL MEASUREMENTS:

Otuka, Y.

J. Soc. Chem. Ind. Japan. Suppl. 1939, *42*, 205B-209B

VARIABLES:

Temperature, pressure

PREPARED BY:

W. Gerrard

EXPERIMENTAL VALUES:

760 mmHg = 1 atm = 101.325 kPa

T/K	p_{Total}/atm	p_{SO_2}/atm	Mole SO_2 /dm^3 soln	H
373	1.72	0.73	0.1763	0.242
	1.94	0.95	0.2443	0.257
378	2.01	0.83	0.1728	0.208
	2.26	1.08	0.2401	0.222
383	2.29	0.89	0.1708	0.192
	2.63	1.24	0.2352	0.190
388	2.70	1.05	0.1656	0.158
	3.03	1.38	0.2311	0.168
393	3.12	1.18	0.1616	0.137
	3.50	1.56	0.2259	0.145
398	3.58	1.31	0.1578	0.120
	4.00	1.74	0.2210	0.127
403	4.10	1.45	0.1535	0.106
	4.55	1.91	0.2164	0.113
408	4.63	1.57	0.1502	0.096
	5.12	2.07	0.2123	0.103
413	5.24	1.71	0.1464	0.086
	5.73	2.21	0.2087	0.094
418	5.91	1.85	0.1428	0.077
	6.41	2.36	0.2050	0.087
423	6.60	1.96	0.1400	0.071
	7.08	2.46	0.2028	0.082

The author defined the "solubility" as $H = (H_2SO_3)/p_{SO_2}$, the pressure being in atm. (H_2SO_3) evidently means mole SO_2/dm^3 of solution. $p_{Total} = p_{SO_2} + p_{H_2O}$. NOTE: The author refers to "pure sulfurous acid, H_2SO_3".

AUXILIARY INFORMATION

METHOD/APPARATUS/PROCEDURE:

The total pressure, p_{Total}, was determined by means of a glass spring manometer. It was stated that the partial pressures were approximated from the dimensions of the apparatus, but details were not given.

SOURCE AND PURITY OF MATERIALS:

Not stated.

ESTIMATED ERROR:

REFERENCES:

COMPONENTS:

1. Sulfur dioxide; SO_2; [7746-09-5]
2. Water; H_2O; [7732-18-5]

ORIGINAL MEASUREMENTS:

Beuschlein, W. L.; Simenson, L. O.
J. Am. Chem. Soc.
1940, *62*, 610-2.

VARIABLES:

T/K = 296-386
P/kPa = 7-192

PREPARED BY:

R. Battino

EXPERIMENTAL VALUES:

t/°C	T^a/K	$x_1^{a,b}$	P^c/mmHg	P^a/kPa
		0.51 g SO_2/100 g H_2O		
35.6	308.8	0.02227	49	6.53
41.0	314.2	0.01914	57	7.60
47.0	320.2	0.01559	70	9.33
52.0	325.2	0.01364	80	10.67
62.6	335.8	0.00925	118	15.73
71.0	344.2	0.00727	150	20.00
78.6	351.8	0.00590	185	24.66
85.0	358.2	0.00506	215.5	28.73
92.2	365.4	0.004364	250	33.33
99.0	372.2	0.003822	285.5	38.06
105.2	378.4	0.003404	320.5	42.73
111.8	385.0	0.003048	358	47.73
113.0	386.2	0.002969	367.5	49.00
		1.09 g SO_2/100 g H_2O		
26.8	300.0	0.03308	70.5	9.40
33.6	306.8	0.02429	96	12.80
39.4	312.6	0.01873	124.5	16.60
44.2	317.4	0.01586	147	19.60
50.6	323.8	0.01310	178	23.73

(cont.)

AUXILIARY INFORMATION

METHOD/APPARATUS/PROCEDURE:

Solutions were prepared by passing sulfur dioxide through freshly boiled, distilled water until saturated and then diluting to the desired concentration. Analysis was by running the sulfur dioxide solution into an excess of iodine and back titrating with sodium thiosulfate. Details and a drawing are in the paper.

SOURCE AND PURITY OF MATERIALS:

1. "Refrigeration dry", 99.99 per cent.
2. Distilled.

ESTIMATED ERROR:

$\delta T/K = \pm 0.2$; $\delta P/kPa = \pm 0.07$;
$\delta x_1/x_1 = \pm 0.03$ (compiler's estimate).

REFERENCES:

COMPONENTS:	ORIGINAL MEASUREMENTS:
1. Sulfur dioxide; SO_2; [7746-09-5] 2. Water; H_2O; [7732-18-5]	Beuschlein, W. L.; Simenson, L. O. *J. Am. Chem. Soc.* 1940, *62*, 610-2.

EXPERIMENTAL VALUES:

t/°C	T^a/K	$x_1^{a,b}$	p^c/mmHg	p^a/kPa
		1.09 g SO_2/100 g H_2O		
61.6	334.8	0.00952	245	32.66
67.4	340.6	0.00831	280.5	37.40
73.4	346.6	0.00712	327.5	43.66
79.2	352.4	0.00627	372	49.60
86.4	359.6	0.005234	445.5	59.39
98.4	371.6	0.004572	510	67.99
100.6	373.8	0.003886	600	79.99
106.6	379.8	0.003491	668	89.06
		4.36 g SO_2/100 g H_2O		
23.2	296.4	0.02809	332	44.26
27.2	300.4	0.02501	373	49.73
30.0	303.2	0.02221	420	56.00
33	306.2	0.02004	465.5	62.06
36	309.2	0.01818	513	68.39
38.2	311.4	0.01691	551.5	73.53
42.2	315.4	0.01507	619	82.53
45.2	318.4	0.01371	680.5	90.73
48.2	321.4	0.01259	741	98.79
52	325.2	0.01141	818	109.06
55.3	328.5	0.01041	896	119.46
58.4	331.6	0.00961	970.5	129.39
61.2	334.4	0.00900	1036.5	138.19
63	336.2	0.00859	1085.5	144.72
67.2	340.4	0.00777	1200.5	160.05
70.8	344.0	0.00722	1292	172.25
72.4	345.6	0.00705	1323.5	176.45
		7.45 g SO_2/100 g H_2O		
25.2	298.4	0.02463	647	86.26
31.4	304.6	0.02042	780.5	104.06
34.2	307.4	0.01902	838	111.72
37.4	310.6	0.01741	915.5	122.06
41	314.2	0.01612	989	131.86
44	317.2	0.01491	1069	142.52
47.2	320.4	0.01383	1152.5	153.65
50.4	323.6	0.01269	1255.5	167.39
54.4	327.6	0.01146	1391.0	185.45
55.6	328.8	0.01104	1443.5	192.45

[a] Calculated by compiler.

[b] Mole fraction solubility at 101.325 kPa partial pressure of gas. These values are only reasonable at pressures close to 101.325 kPa.

[c] Partial pressure of SO_2.

COMPONENTS:

1. Sulfur dioxide; SO_2; [7446-09-5]

2. Water; H_2O; [7732-18-5]

ORIGINAL MEASUREMENTS:

Parkison, R. V.

Tappi

1956, *39*, 517-9.

VARIABLES:

T/K = 283-305

P/kPa = 0.08-3.6

PREPARED BY:

R. Battino

EXPERIMENTAL VALUES:

t/°F	T^a/K	P^b/atm	$P^{a,b}$/kPa	Solubility/ (lb moles/ ft^3)	Solubility^{a,c}/ (mol dm^{-3})
50	283	0.000809	0.0820	0.000511	0.00819
		0.00193	0.196	0.000912	0.0146
		0.00292	0.296	0.00119	0.0191
		0.00440	0.446	0.00144	0.0231
		0.00480	0.486	0.00164	0.0263
		0.00537	0.544	0.00167	0.0268
		0.00668	0.677	0.00212	0.0340
		0.00916	0.928	0.00265	0.0425
		0.00987	1.00	0.00279	0.0447
		0.0174	1.76	0.00441	0.0707
		0.0185	1.88	0.00446	0.0715
		0.0261	2.64	0.00599	0.0960
		0.0281	2.85	0.00622	0.0997
		0.0332	3.36	0.00712	0.114
		0.0350	3.55	0.00745	0.119
60	289	0.00659	0.668	0.00168	0.0269
		0.00910	0.922	0.00214	0.0343
		0.00979	0.992	0.00225	0.0361
70	294	0.00102	0.103	0.000326	0.00523
		0.00201	0.204	0.000568	0.00911
		0.00209	0.212	0.000642	0.0103
		0.00465	0.472	0.00109	0.0175
		0.00507	0.514	0.00116	0.0186

(cont.)

AUXILIARY INFORMATION

METHOD/APPARATUS/PROCEDURE:

A nitrogen/sulfur dioxide mixture of varying composition was bubbled through water. The analysis of the gas and liquid phase was done by a modified Palmrose technique (1,2) via a chemical titration. Details are given in the paper.

SOURCE AND PURITY OF MATERIALS:

1. No details given.

2. Triple distilled.

ESTIMATED ERROR:

δSolubility/Solubility = ±0.03 (compiler's estimate);

δT/K = ±0.02.

REFERENCES:

1. Palmrose, G. V. *Paper Trade J.* 1935, *100*, 38-9.

2. White, C. K.; Vivian, J. E.; Whitney, R. P. *Paper Trade J.* 1948, *126*, 26-8.

COMPONENTS:	ORIGINAL MEASUREMENTS:
1. Sulfur dioxide; SO_2; [7446-09-5] 2. Water; H_2O; [7732-18-5]	Parkison, R. V. *Tappi* 1956, *39*, 517-9.

EXPERIMENTAL VALUES:

t/°F	T^a/K	P^b/atm	$P^{a,b}$/kPa	Solubility/ (lb moles/ ft^3)	Solubility[a,c]/ (mol dm^{-3})
70	294	0.00660	0.669	0.00141	0.0226
		0.00734	0.744	0.00133	0.0213
		0.00913	0.925	0.00176	0.0282
		0.00983	0.996	0.00188	0.0301
		0.0172	1.74	0.00288	0.0462
		0.0280	2.84	0.00417	0.0668
		0.0349	3.54	0.00498	0.0798
80	263	0.00150	0.153	0.000320	0.00513
		0.00362	0.367	0.000580	0.00930
		0.00469	0.475	0.000722	0.0116
		0.00633	0.641	0.000921	0.0148
		0.00641	0.649	0.000918	0.0147
		0.00812	0.823	0.00111	0.0178
		0.00885	0.897	0.00120	0.0192
		0.00952	0.965	0.00126	0.0202
		0.0110	1.11	0.00138	0.0221
		0.0178	1.80	0.00204	0.0327
		0.0251	2.54	0.00265	0.0425
		0.0336	3.40	0.00334	0.0535

[a] Calculated by compiler.

[b] Partial pressure of SO_2.

[c] Molar solubility calculated by multiplying previous column by 16.03.

COMPONENTS:	ORIGINAL MEASUREMENTS:
1. Sulfur dioxide; SO_2; [7446-09-5] 2. Water; H_2O; [7732-18-5]	Rabe, A. E.; Harris, J. F. *J. Chem. Engng. Data* 1963, *8*, 333-336.
VARIABLES:	**PREPARED BY:**
Temperature, pressure	C. L. Young

EXPERIMENTAL VALUES:

T/K	p_{SO_2}/mmHg	p_{SO_2}/kPa	Conc. g_{SO_2}/100 g H_2O	Mole fraction* of sulfur dioxide
303.15	443.5	59.13	4.46	0.01238
	443.9	59.18	4.46	0.01238
	442.4	58.98	4.46	0.01238
	444.5	59.26	4.48	0.01245
	451.6	60.21	4.48	0.01245
	308.6	41.14	3.21	0.00894
	304.9	40.65	3.21	0.00894
	308.4	41.12	3.21	0.00894
	309.0	41.20	3.21	0.00894
	312.0	41.60	3.21	0.00894
	172.2	22.96	1.92	0.00537
	181.0	24.13	1.92	0.00537
	177.0	23.60	1.92	0.00537
	39.2	5.23	0.59	0.00166
	42.7	5.69	0.59	0.00166
313.15	624.4	83.25	4.48	0.01245
	638.8	85.17	4.48	0.01245
	418.0	55.73	3.21	0.00894
	420.0	56.00	3.21	0.00894
	242.0	32.26	1.92	0.00537

(cont.)

* calculated by compiler

AUXILIARY INFORMATION

METHOD/APPARATUS/PROCEDURE:	SOURCE AND PURITY OF MATERIALS:
Partial vapor pressure of sulfur dioxide above solution containing known masses of sulfur dioxide determined by estimating amount of sulfur dioxide vapor in known volume. Sulfur dioxide estimated by dissolving gas sample up in dilute aqueous acetaldehyde soln. and titrating α-hydroxy sulfonic acid against sodium hydroxide potentiometrically. Details in source.	No details given. **ESTIMATED ERROR:** $\delta T/K = \pm 0.02$ or less; $\delta p/p$ = 0.0044 (std. dev.). **REFERENCES:**

COMPONENTS:	ORIGINAL MEASUREMENTS:
1. Sulfur dioxide; SO_2; [7446-09-5] 2. Water; H_2O; [7732-18-5]	Rabe, A. E.; Harris, J. F. *J. Chem. Engng. Data* 1963, *8*, 333-336.

EXPERIMENTAL VALUES:

T/K	p_{SO_2}/mmHg	p_{SO_2}/kPa	Conc. g_{SO_2}/100 g H_2O	Mole fraction* of sulfur dioxide
313.15	249.0	33.20	1.92	0.00537
	57.8	7.71	0.59	0.00166
	60.6	8.08	0.59	0.00166
323.15	583.3	77.77	3.20	0.00892
	592.9	79.05	3.20	0.00892
	608.4	81.11	3.20	0.00892
	583.3	77.77	3.21	0.00894
	585.7	78.09	3.21	0.00894
	324.6	43.28	1.92	0.00537
	328.5	43.80	1.92	0.00537
	330.9	44.12	1.92	0.00537
	335.1	44.68	1.92	0.00537
	85.8	11.44	0.59	0.00166
	86.7	11.56	0.59	0.00166
333.15	742.0	98.93	3.21	0.00894
	756.0	100.8	3.21	0.00894
	447.5	59.66	1.92	0.00537
	437.1	58.28	1.92	0.00537
	114.0	15.20	0.59	0.00166
	116.0	15.47	0.59	0.00166
343.15	151.0	20.13	0.59	0.00166
	153.0	20.40	0.59	0.00166
353.15	189.0	25.20	0.59	0.00166

* calculated by compiler

T/K	x_{SO_2} [a]
313.15	0.01498
323.15	0.01149
333.15	0.00907

[a] Interpolated mole fraction solubility at a partial pressure of sulfur dioxide of 101.325 kPa; calculated by compiler.

COMPONENTS:

1. Sulfur dioxide; SO_2; [7446-09-5]

2. Water; H_2O; [7732-18-5]

ORIGINAL MEASUREMENTS:

Vosolsobe, J.; Simecek, S.; Michalek, J.; Kadler, B.

Chem. Prumysl.

1965, *15*, 401-404.

VARIABLES:

PREPARED BY:

C. L. Young

EXPERIMENTAL VALUES:

T/K (T/°C)	p_{SO_2}/mmHg	p/kPa	g SO_2 / 100 g H_2O	$10^3 x_{SO_2}$
293.2	14.8	1.97	0.33	0.927
(20)	29.7	3.96	0.57	1.60
	44.5	5.93	0.78	2.19
	59.4	7.92	1.02	2.86
	74.2	9.89	1.26	3.53
	89.1	11.88	1.48	4.15
	126	16.80	2.04	5.70
303.2	14.6	1.95	0.25	0.703
(30)	29.1	3.88	0.45	1.26
	43.7	5.83	0.61	1.71
	58.2	7.76	0.77	2.16
	72.8	9.71	0.92	2.58
	87.4	11.65	1.06	2.97
	123.8	16.51	1.45	4.06
313.2	14.1	1.88	0.19	0.534
(40)	28.2	3.76	0.32	0.899
	43.3	5.77	0.46	1.29
	57.4	7.65	0.54	1.52
	70.5	9.40	0.67	1.88
	84.6	11.28	0.78	2.19
	119.8	15.97	1.08	3.03

(cont.)

AUXILIARY INFORMATION

METHOD/APPARATUS/PROCEDURE:

Flow system apparatus in which a sulfur dioxide plus nitrogen gas mixture was bubbled through water. The partial pressure of sulfur dioxide could be varied by changing the composition of the gas mixture. Samples of water saturated with sulfur dioxide were analysed by iosometry.

SOURCE AND PURITY OF MATERIALS:

No details given.

ESTIMATED ERROR:

$\delta T/K = \pm 0.1$; $\delta x_{SO_2} = \pm 3\%$ (estimated by compiler).

REFERENCES:

COMPONENTS:

1. Sulfur dioxide; SO_2; [7446-09-5]

2. Water; H_2O; [7732-18-5]

ORIGINAL MEASUREMENTS:

Vosolsobe, J.; Simecek, S.; Michalek, J.; Kadler, B.

Chem. Prumysl.

1965, *15*, 401-404.

EXPERIMENTAL VALUES:

T/K (T/°C)	p_{SO_2}/mmHg	p/kPa	g SO_2 / 100 g H_2O	$10^3 x_{SO_2}$
323.2	13.3	1.77	0.16	0.450
(50)	26.7	3.56	0.26	0.731
	40.0	5.33	0.34	0.955
	53.4	7.12	0.42	1.18
	66.7	8.89	0.50	1.40
	80.1	10.68	0.60	1.68
	113.5	15.13	0.80	2.24
333.2	12.2	1.63	0.11	0.309
(60)	24.4	3.25	0.20	0.562
	36.6	4.88	0.24	0.674
	48.8	6.51	0.33	0.927
	61.1	8.15	0.36	1.01
	73.3	9.77	0.42	1.18
	103.8	13.84	0.58	1.63

COMPONENTS:

1. Sulfur dioxide; SO_2; [7446-09-5]
2. Water; H_2O; [7732-18-5]

ORIGINAL MEASUREMENTS:

Sano, H.; Nakamoto, Y.
Nippon Kagaku Zasshi
1968, *89*, 369-373.

VARIABLES:

PREPARED BY:

C. L. Young

EXPERIMENTAL VALUES:

$T/°C = 30$ $T/K = 303$

Vapor pressure of water = 0.040 atm

Absorption coefficient, $S^{a,b}$	Bunsen coefficient,[b] α	Bunsen coefficient,[c] α	$\ell^{c,d}$	Mole ratio	Mole fraction[e]
29.8	28.3	25.6	24.5	0.023	0.022

[a] Volume of sulfur dioxide absorbed by unit volume of solvent at experimental temperature and a total pressure of 1 atmosphere.

[b] Determined by static absorption method.

[c] Determined by flow method.

[d] $\alpha = \ell(1 - p_s^\circ)$ where p_s° is the vapor pressure of solvent.

[e] Calculated by compiler.

AUXILIARY INFORMATION

METHOD/APPARATUS/PROCEDURE:

Two techniques were used: (a) static absorption method and (b) flow method. Static method consisted of equilibrating solvent and sulfur dioxide in a 50 ml flask and measuring the absorption volumetrically. The flow method consists of passing sulfur dioxide through a micro gas absorption tube and estimating sulfur dioxide by iodometric titration. The vapor pressure of the solvents were measured roughly and were assumed to be equal to the partial pressure of the solvent by the authors when calculating the Bunsen coefficient.

SOURCE AND PURITY OF MATERIALS:

1. Purity 99.0-99.8 mole per cent.

ESTIMATED ERROR:

$\delta T/K = \pm 0.5$; $\delta x_{SO_2} = \pm 5\%$ (estimated by compiler).

REFERENCES:

COMPONENTS:

1. Sulfur dioxide; SO_2; [7446-09-5]
2. Water; H_2O; [7732-18-5]

ORIGINAL MEASUREMENTS:

Tokunaga, J.

J. Chem. Engng. Data

1974, *19*, 162-165.

VARIABLES:

PREPARED BY:

C. L. Young

EXPERIMENTAL VALUES:

T/K	Solubility#*	Ostwald coefficient	Mole fraction* of sulfur dioxide, x_{SO_2}
283.15	0.1566	56.8	0.0422
293.15	0.1090	40.9	0.0297
303.15	0.0777	30.0	0.0214
313.15	0.0576	22.9	0.0159

(g of SO_2/g H_2O).

* at partial pressure of 101.3 kPa.

AUXILIARY INFORMATION

METHOD/APPARATUS/PROCEDURE:

Sulfur dioxide bubbled through water. Saturated solution analysed by iodometric titration. Details in source.

The apparatus employed was similar to that of Hudson, ref. (1).

SOURCE AND PURITY OF MATERIALS:

1. High purity sample with purity of 99.96 mole per cent.

2. No details given.

ESTIMATED ERROR:

$\delta T/K = \pm 0.1$; $\delta x_{SO_2} = \pm 1\%$ (estimated by compiler).

REFERENCES:

1. Hudson, J. C. *J. Chem. Soc.* 1925, 1332.

COMPONENTS:

1. Sulfur dioxide; SO_2; [7746-09-5]

2. Water; H_2O; [7732-18-5]

ORIGINAL MEASUREMENTS:

Lavrova, E. M.; Tudorovskaya, L. L.

Zhur. Prikl. Khim.

1977, *50*, 1146-51.

VARIABLES:

T/K = 299-363

PREPARED BY:

R. Battino

EXPERIMENTAL VALUES:

t/°C	T^a/K	x_1 [b]	wt-% [c]
26	299.2	0.02387	8.00
40	313.2	0.01534	5.25
60	333.2	0.009063	3.15
80	353.2	0.0059	2.08
90	363.2	0.00504	1.77

[a] Calculated by compiler.

[b] Mole fraction solubility at 101.325 kPa partial pressure of gas.

[c] Presumably at 101.325 kPa partial pressure of gas.

AUXILIARY INFORMATION

METHOD/APPARATUS/PROCEDURE:

Water was saturated with SO_2. The liquid phase was analyzed acidimetrically after oxidation of the dissolved SO_2 to SO_3 by hydrogen peroxide. The paper also reports on SO_2 solubilities in aqueous HCl solutions.

SOURCE AND PURITY OF MATERIALS:

1. Obtained by the action of cp H_2SO_4 on analytical grade sodium sulfite.

2. No details given.

ESTIMATED ERROR:

T/K = ±0.2; $\delta x_1/x_1$ = ±0.05

REFERENCES:

COMPONENTS:

1. Sulfur dioxide; SO_2; [7446-09-5]
2. Water; H_2O; [7732-18-5]

ORIGINAL MEASUREMENTS:

Douabul, A.; Riley, J.

J. Chem. Engng. Data, 1979, *24*, 274-276.

VARIABLES:

Temperature

PREPARED BY:

C.L. Young.

EXPERIMENTAL VALUES:

T/K	Solubility$^+$/mol dm^{-3}	Mole fraction of sulfur dioxide, x_{SO_2}
278.97	2.959	0.05685
283.12	2.482	0.04693
288.10	2.065	0.03876
292.98	1.697	0.03168
298.15	1.419	0.02644
303.25	1.201	0.02224

\+ Partial pressure of SO_2 = 1 atm = 101.3 kPa.

AUXILIARY INFORMATION

METHOD/APPARATUS/PROCEDURE:

Sulfur dioxide bubbled through water for 21 hours with gentle stirring. Pressure measured with mercury manometer. Samples of about 60 cm^3 were withdrawn and mixed with 150 cm^3 of hydrogen peroxide solution. Solution then titrated with carbonate free sodium hydroxide solution.

SOURCE AND PURITY OF MATERIALS:

1. Refrigeration grade obtained from BDH. Mass spectrometric analysis showed its purity was better than 99.9 mole per cent.

2. Distilled.

ESTIMATED ERROR:

$\delta T/K = \pm 0.01$; $\delta x_{SO_2} = \pm 0.1\%$

REFERENCES:

COMPONENTS:

1. Sulfur dioxide; SO_2; [7746-09-5]
2. Water; H_2O; [7732-18-5]

ORIGINAL MEASUREMENTS:

Byerley, J. J.; Rempel, G. L.; Le, V. T.

J. Chem. Eng. Data 1980, *25*, 55-6.

VARIABLES:

T/K = 298-323

PREPARED BY:

R. Battino

EXPERIMENTAL VALUES:

t/°C	T^a/K	$x_1^{a,b}$	g SO_2 ^{c}/100 g H_2O
25	298.15	0.02265	8.24
50	323.15	0.01049	3.77

[a] Calculated by compiler.

[b] Mole fraction solubility at 101.325 kPa partial pressure of gas.

[c] At 101.325 kPa partial pressure of gas.

AUXILIARY INFORMATION

METHOD/APPARATUS/PROCEDURE:

Sulfur dioxide was bubbled through water and the amount dissolved was determined by titrating excess iodine with standard thiosulfate solution. (The authors also studied the solubility of SO_2 in water-acetonitrile solutions.)

SOURCE AND PURITY OF MATERIALS:

1. Anhydrous from Union Carbide Canada Ltd.
2. Distilled.

ESTIMATED ERROR:

$\delta T/K = \pm 0.1$; $\delta x_1/x_1 = \pm 0.02$ (compiler's estimate).

REFERENCES:

COMPONENTS:	EVALUATOR:
1. Sulfur dioxide; SO_2; [7446-09-5] 2. Seawater	Colin L. Young, School of Chemistry, University of Melbourne, Parkville, Victoria 3052, Australia. December 1981

CRITICAL EVALUATION:

This system has been investigated by Douabul and Riley (1) in some detail. Although their data are self-consistent there is some doubt as to absolute accuracy of the measurements since data on the solubility of sulfur dioxide in pure water given in the same appear appear to be at least several per cent in error. (See sulfur dioxide + water evaluation.) The only other measurements on this system are those of Abdulsattar (2) whose data are very limited in scope and are of questionable accuracy, ±15% at the best. Abdulsattar's data are therefore rejected. The solubility of sulfur dioxide in seawater has been calculated by Abdulsattar *et al.* (3) based on a simplified chemical model. However, such calculated data are for a lower partial pressure of sulfur dioxide than used in the experimental work of Douabul and Riley (1). Under these conditions a relatively large proportion of the sulfur dioxide is accounted for in neutralizing the natural alkalinity of the sea water. It should also be pointed out that the data of Douabul and Riley (1) should not be extrapolated to low partial pressures assuming Henry's Law to be obeyed. Such an assumption is not valid in the present context.

References

1. Douabul, A.; Riley, J.
J. Chem. Eng. Data 1979, *24*, 274.

2. Abdulsattar, A. H. *M.Sc. Thesis*, University of California, Berkeley, 1971.

3. Abdulsattar, A. H .; Sridhar, S.; Bromley, L. A.
Am. Inst. Chem. Engnrs. J. 1977, *23*, 62.

COMPONENTS:

1. Sulfur dioxide; SO_2; [7446-09-5]

2. Seawater

ORIGINAL MEASUREMENTS:

Douabul, A.; Riley, J.

J. Chem. Engng. Data. 1979, *24*, 274-276.

VARIABLES:

Temperature, salinity.

PREPARED BY:

C.L. Young

EXPERIMENTAL VALUES:

T/K	Salinity $/g kg^{-1}$	Specific gravity of soln.	Solubility[+] $/mol\ dm^{-3}$ exptl.	smoothed
278.97	0.0	1.074	2.959	2.959
	10.065	1.080	2.956	2.956
	20.034	1.087	2.954	2.954
	30.074	1.095	2.954	2.953
	40.005	1.103	2.953	2.953
283.12	0.0	1.067	2.482	2.491
	10.065	1.075	2.479	2.488
	20.034	1.082	2.477	2.487
	30.074	1.088	2.477	2.486
	40.005	1.094	2.476	2.485
288.10	0.0	1.055	2.065	2.047
	10.065	1.063	2.062	2.044
	20.034	1.071	2.061	2.042
	30.074	1.078	2.060	2.042
	40.005	1.085	2.059	2.041
292.98	0.0	1.045	1.697	1.706
	10.065	1.052	1.694	1.704
	20.034	1.059	1.693	1.703
	30.074	1.066	1.692	1.701
	40.005	1.074	1.691	1.700
298.15	0.0	1.032	1.419	1.421
	10.065	1.042	1.417	1.420
	20.034	1.050	1.416	1.418
	30.074	1.058	1.414	1.417

AUXILIARY INFORMATION

METHOD/APPARATUS/PROCEDURE:

Sulfur dioxide bubbled through sea-water for 21 hours with gentle stirring. Pressure measured with mercury manometer. Samples of about 60 cm^3 were withdrawn and mixed with 150 cm^3 of hydrogen peroxide solution. Solution then titrated with carbonate free sodium hydroxide solution. Allowance made for acid originally in sample.

SOURCE AND PURITY OF MATERIALS:

1. Refrigeration grade obtained from BDH. Mass spectrometric analysis showed its purity to be better than 99.9 mole per cent.

2. Surface water from Irish sea filtered and samples of salinities ranging from 10-40% prepared by evaporation or dilution with distilled water, acidified to pH 2.8 with sulfuric acid.

ESTIMATED ERROR:

$\delta T/K = \pm 0.01$; $\delta x_{SO_2} = \pm 0.1\%$

REFERENCES:

COMPONENTS:	ORIGINAL MEASUREMENTS:
1. Sulfur dioxide; SO ; [7446-09-5] 2. Seawater	Douabul, A.; Riley, J. *J. Chem. Engng. Data.* 1979, *24*, 274-276.

EXPERIMENTAL VALUES:

T/K	Salinity, $/g kg^{-1}$	Specific gravity of soln.	Solubility^{+} $/mol\ dm^{-3}$ exptl.	smoothed
298.15	40.005	1.064	1.414	1.416
303.25	0.0	1.028	1.201	1.199
	10.065	1.034	1.200	1.198
	20.034	1.040	1.198	1.196
	30.074	1.048	1.197	1.195
	40.005	1.056	1.196	1.194

\+ Partial pressure of SO_2 = 1 atm = 101.3 kPa.

COMPONENTS:	EVALUATOR:
1. Sulfur dioxide; SO_2; [7446-09-5] 2. Water; H_2O; [7732-18-5] 3. Metallic salts	C. L. Young,* School of Chemistry, University of Melbourne, Parkville, Victoria 3052, Australia. August 1980

CRITICAL EVALUATION:

The evaluation of the data for aqueous solutions of 19 salts reported by Fox (1) is uncertain, because of unsatisfactory presentation. Fox expressed the concentration of the salt as "normality," using one half of the molecular weight in the examples of the sulfates of sodium, potassium and ammonium, and of the halides and sulfate of cadmium. The "solubility" was expressed as the Ostwald coefficient, *L*; but it is not absolutely clear if the volume used was that of the original liquid (initial solution of salt) or that of the solution saturated with sulfur dioxide. Densities were not recorded, and therefore mole fractions cannot be accurately estimated. The weights of sulfur dioxide absorbed by a given volume or weight of solution were not reported, although these data were determined. The particular weight was converted into a volume of gas by the use of 22.4 dm^3 as the molar volume of sulfur dioxide at 273.15 K and 1 atm; whereas the real volume is 21.87 dm^3. This means that the *L* values as given are about 2% too large.

In Seidell and Linke (2) the data were given (p.1413) as the "Ostwald Solubility Expression ml SO_2 at t°, 1 atm per ml solvent". Bancroft and Gould (3) gave a table of Bunsen absorption coefficients for sulfur dioxide and aqueous solutions of sodium and potassium salts restricted to "0.5 molar." These data are based entirely on the Fox data, although this source was not acknowledged. The coefficients for the sulfates, given as for "0.5 molar" are based on Fox's *L* values for 0.5 *normal*, *i.e.*, 0.25 molar. In Seidell and Linke (p.1414) the Bancroft and Gould list is given as if these were original data.

Hudson's (4) data were presented in the rudimentary form of g SO_2/ 100 g water content for each concentration of salt (only potassium chloride and sodium sulfate were studied) as g salt/100 g of water content. The main uncertainty in the Hudson values lies in the allowance for the vapor pressure of water over the final solution. Data on the lowering of the vapor pressure of water over the aqueous solution of potassium chloride were taken from the work of Tammann (5), and could be as much as 75 mmHg at 363 K. The lowering due to dissolved sulfur dioxide was calculated on the basis of Raoult's law, and was estimated as about 0.1% at 283 K and 1.25% at 363 K. The author believed the corrections to be only approximate, "as there are grounds for believing that sulphur dioxide combines both with the water and with the added salts." Hudson assumed the application of Henry's law (mole ratio form) in the adjustment of the g SO_2/100 g H_2O value from the estimated p_{SO_2} to 1 atm. As presented, the data for sodium sulfate solutions did not show the breakdown into p_{Total}

---

* based on a more detailed evaluation in terms of the R-line approach (11) by W. Gerrard.

COMPONENTS:	EVALUATOR:
1. Sulfur dioxide; SO_2; [7446-09-5] 2. Water; H_2O; [7732-18-5] 3. Metallic salts	C. L. Young,* School of Chemistry, University of Melbourne, Parkville, Victoria 3052, Australia. August 1980

CRITICAL EVALUATION:

and p_{SO_2} because it was concluded that the estimated p_{SO_2} values were within 10 mmHg of 1 atm. Again the adjustment to 1 atm was based on Henry's law (mole ratio form). Attempts to correlate the Hudson data and the Fox data are frustrated by the lack of density data in the Fox work. Hudson stated that the densities of the salt solutions themselves are insufficient, for "considerable expansion of the solutions occurs when sulphur dioxide dissolves."

The evaluation of the data of Conrad and Beuschlein (6) on the CaO-water-SO_2 system, and of Conrad and Brice (7) on the MgO-water-SO_2 system requires cognition of the purpose of the measurements, and the definition of "total", "combined", and "free" sulfur dioxide, and that a solid phase was present. The amount of sulfur dioxide in the liquid phase at equilibrium was expressed as g SO_2/100 g water. The work was done in connection with the use of the "Sulfite Process" in the manufacture of wood pulp. These authors used the following definition decided upon by the Technical Association of Pulp and Paper Institute: "the free SO_2 is the actual free SO_2 plus half of the SO_2 in the bisulfite of calcium (or magnesium), and is more properly called the *available* SO_2 as it indicates the SO_2 in excess of the amount necessary to form monosulfite." The authors defined the term "combined sulfur dioxide" as that required to form the monosulfite.

Data by White, Vivian and Whitney (8) are for the calcium sulfite system in which there was no solid phase, and the solution may be deemed as unsaturated with respect to the calcium sulfite. These workers stated the partial pressure, p_{SO_2}, and they reported data for water itself to show the reliability of their data for the salt solutions.

The data of Domansky and Rendos (9) for dilute solutions of "ammonium bisulfite" are deemed acceptable (see compilation sheets).

The earlier data of Smith and Parkhurst (10), determined in relation to the paper-pulp industry, were based on a combined sulfur dioxide (C) fixed by the "normality" of $Ca(SO_3H)_2$ or $Mg(SO_3H)_2$, and not on $CaSO_3$ as in the later TAPPI definition. In the former definition, the free SO_2 is the total (T) SO_2 minus the SO_2 combined as $Ca(SO_3H)_2$ or $Mg(SO_3H)_2$, and is therefore less than that based on the second definition for the same value of SO_2 (T).

References:

1. Fox, C. J. J. *Z. Phys. Chem.* 1902, *41*, 458.

2. Seidell, A.; Linke, W. F. *Solubilities of Inorganic and Organic Compounds*, 1958, 1865.

(cont.)

COMPONENTS:	EVALUATOR:
1. Sulfur dioxide; SO_2; [7446-09-5] 2. Water; H_2O; [7732-18-5] 3. Metallic salts	C. L. Young,* School of Chemistry, University of Melbourne, Parkville, Victoria 3052, Australia. August 1980

CRITICAL EVALUATION:

References (cont.)

3. Bancroft, W. D.; Gould, L. P. *J. Phys. Chem.* 1934, *38*, 197.

4. Hudson, J. C. *J. Chem. Soc.* 1925, *127*, 1332.

5. Tammann, G. *Ann. Phys. Chem.* 1885, [2], *24*, 523.

6. Conrad, F. H.; Beuschlein, W. L. *J. Am. Chem. Soc.* 1934, *56*, 2554.

7. Conrad, F. H.; Brice, D. B. *J. Am. Chem. Soc.* 1948, *70*, 2179.

8. White, C. K.; Vivian, J. E.; Whitney, R. P. *Paper Trade J.* 1948, *126*, 46.

9. Domansky, R.; Rendos, F. *Chem. Zvesti* 1957, *11*, 453-460.

10. Smith, W. T.; Parkhurst, R. B. *J. Am. Chem. Soc.* 1922, *44*, 1918.

11. Gerrard, W. *Solubility of Gases and Liquids*, 1976, Plenum Press, New York.

COMPONENTS:	ORIGINAL MEASUREMENTS:
1. Sulfur dioxide; SO_2; [7446-09-5] 2. Water; H_2O; [7732-18-5] 3. Hydrogen chloride; HCl; [7647-01-0]	Lavrova, E. M.; Tudorovskaya, G. L. *Zh. Prikl. Khim.* 1977, *50*, 1146-1151; *J. Appl. Chem. USSR* 1977, *50*, 1102-1106.
VARIABLES:	PREPARED BY:
Temperature, liquid composition	C. L. Young

EXPERIMENTAL VALUES:

T/K	Conc. of HCl in absorbing soln. /wt-%	Solubility of sulfur dioxide /wt-%	Mole fraction of sulfur dioxide x_{SO_2}
299.2	0	8.00	0.02387
313.2		5.25	0.01534
333.2		3.15	0.009063
353.2		2.08	0.0059
363.2		1.77	0.00504
299.2	2.2	8.00	0.02407
313.2		5.42	0.01603
333.2		3.40	0.0099
353.2		2.75	0.00798
363.2		2.24	0.00647
299.2	4.30	7.76	0.02354
313.2		5.25	0.01569
333.2		2.89	0.0084
353.2		2.47	0.00723
363.2		2.24	0.00654
299.2	11.0	8.40	0.02671
313.2		5.35	0.01660
333.2		3.25	0.009353
353.2		2.50	0.00769
363.2		2.00	0.00607
299.2	15.0	8.40	0.02891
313.2		5.50	0.01750
333.2		3.48	0.01089

(cont.)

AUXILIARY INFORMATION

METHOD/APPARATUS/PROCEDURE:	SOURCE AND PURITY OF MATERIALS:
Sulfur dioxide bubbled into hydrochloric acid solution. Solubility determined by acidimetric analysis of liquid phase after oxidation of dissolved sulfur dioxide to sulfuric acid using hydrogen peroxide.	1. Prepared by action of "cp" sulfuric acid on analytical grade sodium sulfite. Dried. 2. No details given. 3. "cp" grade.
	ESTIMATED ERROR:
	$\delta T/K = \pm 0.2$; $\delta x_{SO_2} = \pm 5\%$.
	REFERENCES:

COMPONENTS:	ORIGINAL MEASUREMENTS:
1. Sulfur dioxide; SO_2; [7446-09-5] 2. Water; H_2O; [7732-18-5] 3. Hydrogen chloride; HCl; [7647-01-0]	Lavrova, E. M.; Tudorovskaya, G. L. *Zh. Prikl. Khim.* 1977, *50*, 1146-1151; *J. Appl. Chem. USSR* 1977, *50*, 1102-1106.

EXPERIMENTAL VALUES:

T/K	Conc. of HCl in absorbing soln. /wt-%	Solubility of sulfur dioxide /wt-%	Mole fraction of sulfur dioxide x_{SO_2}
353.2	15.0	2.67	0.00830
363.2		2.06	0.00637
299.2	20.0	13.65	0.0470
313.2		6.28	0.02113
333.2		4.00	0.0130
353.2		3.60	0.01160
363.2		2.67	0.00854
299.2	21.5	13.65	0.00484
313.2		6.47	0.02199
333.2		3.72	0.01267
353.2		3.01	0.00975
363.2		2.73	0.00880
299.2	24.8	18.8	0.07154
313.2		6.14	0.02081
333.2		3.15	0.01040
353.2		2.90	0.00956
363.2		0.909	0.00295
299.2	28.0	13.7	0.05034

Smoothing equations of the form

$$\log x_{SO_2} = -A + B/(T/K)$$

were given.

Conc. of HCl/wt-%	A	B
0	5.47	1.41
2.2	4.77	0.937
4.3	4.85	0.953
11.0	5.1	1.050
15.0	5.1	1.051
22.0	5.19	1.128
21.5	5.43	1.20
24.8	7.8	1.96

COMPONENTS:

1. Sulfur dioxide; SO_2; [7446-09-5]
2. Water; H_2O; [7732-18-5]
3. Sulfuric acid; H_2SO_4; [7664-93-9]

ORIGINAL MEASUREMENTS:

Miles, F.D.; Fenton, J.

J. Chem. Soc. 1920, *117*, 59-61.

VARIABLES:

Concentration of acid.

PREPARED BY:

W. Gerrard.

EXPERIMENTAL VALUES:

T/K	Sulfuric acid per cent	SO_2 dissolved by 100 g of acid / g	*Moles SO_2 /100 g acid	*Moles SO_2/ total number of moles, H_2SO_4 + H_2O in 100 g of acid.
293	55.1	5.13	0.0802	0.0262
	59.6	4.90	0.0766	0.0269
	61.6	4.82	0.0753	0.0273
	68.9	4.16	0.0650	0.0267
	74.1	3.63	0.0567	0.0258
	78.3	3.23	0.0505	0.0252
	80.2	3.12	0.0488	0.0254
	82.5	2.99	0.0467	0.0258
	84.2	2.88	0.0450	0.0259
	85.3	2.83	0.0442	0.0261
	85.8	2.80	0.0438	0.0263
	86.5	2.82	0.0441	0.0270
	88.1	2.9	0.0453	0.0290
	90.8	3.10	0.0484	0.0337
	92.8	3.21	0.0502	0.0372
	93.7	3.27	0.0511	0.0391
	94.0	3.31	0.0517	0.0400
	94.6	3.50	0.0547	0.0432
	95.5	3.69	0.0577	0.0471
	95.6	3.77	0.0589	0.0483
	96.5	3.83	0.0598	0.0508
	98.0	3.98	0.0622	0.0560
	98.5	4.03	0.0630	0.0579

NOTE: "Sulfuric acid per cent" probably means by weight. The term "acid" refers to the aqueous solution of sulfuric acid.

* Calculated by compiler.

AUXILLIARY INFORMATION

METHOD/APPARATUS/PROCEDURE:

Sulfur dioxide was passed into the acid (15-20 g), mechanically stirred in a test-tube held in a thermostat. Samples of the saturated solution were withdrawn into small weighed glass bulbs that were sealed, weighed, and opened under (1) *N*-sodium hydroxide, (2) *N*/10-iodine for chemical titrations.

"A correction was applied when the barometer diverged appreciably from 760 mm."

SOURCE AND PURITY OF MATERIALS:

No information given.

ESTIMATED ERROR:

$\delta T/K = \pm 0.1$

REFERENCES:

COMPONENTS:

1. Sulfur dioxide; SO_2; [7446-09-5]
2. Water; H_2O; [7732-18-5]
3. Sulfuric acid; H_2SO_4; [7664-93-9]

ORIGINAL MEASUREMENTS:

Cupr, V.

Rec. Trav. Chim. 1928, *47*, 55-72.

VARIABLES:

Temperature, concentration of acid.

PREPARED BY:

W. Gerrard.

EXPERIMENTAL VALUES: The values were stated to be for a total pressure of 760 mmHg, i.e., for 1 atm = 101.325 kPa.

T/K 314.15			T/K 335.15		
Weight of H_2SO_4 in 100 g solution / g	Wt. of SO_2 /100 g solution of H_2SO_4/g	*L*	Weight of H_2SO_4 in 100 g solution / g	Wt. of SO^2 /100 g solution of H_2SO_4/g	*L*
0.00	4.88	16.9	0.00	2.16	7.42
0.00	4.75	16.5	0.00	2.13	7.32
7.32	3.65	13.3	10.02	1.71	6.28
18.07	3.35	12.8	30.26	1.26	5.25
18.28	3.29	12.8	30.52	1.20	5.00
44.34	2.57	11.9	54.06	1.20	5.90
71.86	1.68	9.47	79.07	1.19	6.95
79.51	1.54	9.18	84.34	1.15	6.95
86.20	1.36	8.40	84.82	1.17	7.08
86.54	1.45	8.96	90.73	1.19	7.36
86.66	1.47	9.09	91.20	1.17	7.28
93.78	1.81	11.5	93.68	1.16	7.26
96.65	2.06	13.1	96.10	1.16	7.29
97.69	2.15	13.6			

The symbol *L* indicates the "absorption coefficient" which appears to be the volume in cm^3 (adjusted to 273 K) of SO_2 absorbed to give one cm^3 of solution saturated with SO_2 at 760 mmHg and at the temperature recorded in the table. These were calculated from the weight values and the published densities of aqueous solutions of sulfuric acid. See evaluation.

AUXILIARY INFORMATION

METHOD/APPARATUS/PROCEDURE:

Sulfur dioxide was passed into the aqueous solution in an "absorption apparatus." A sample was withdrawn into a weighed flask for the determination of sulfur dioxide by a chemical titration, using the iodine-thiosulfate technique.

The "absorption coefficient, *L* " was determined by calculation from the weight data. The molar volume of 22400 cm^3 appears to have been used to convert weight of SO_2 into volume

SOURCE AND PURITY OF MATERIALS:

(1) Sulfur dioxide was obtained from Merck quality sodium sulfite and concentrated sulfuric acid, and passed through the acid.

(2) Water may be taken as of satisfactory purity.

(3) Sulfuric acid was of satisfactory purity.
The solutions were attested volumetrically.

ESTIMATED ERROR:

REFERENCES:

COMPONENTS:

1. Sulfur dioxide; SO_2; [7446-09-5]
2. Water; H_2O; [7732-18-5]
3. Sulfuric acid; H_2SO_4; [7664-93-9]

ORIGINAL MEASUREMENTS:

Lopatto, E.K.; Savinaev, A.M.
J. Applied Chem. (USSR) 1934, *7*, 881-886.

VARIABLES:

Concentration of acid, partial pressure of gas.

PREPARED BY:

W. Gerrard.

EXPERIMENTAL VALUES:

760 mmHg = 1 atm = 101.325 kPa

Weight % of H_2SO_4 in aqueous solution (initial).	*% SO_2 in mixture of air + SO_2	% SO_2 in aqueous solution of acid at T/K 293	303	313	323	333
61.59	9	-	0.345	0.262	0.194	0.153
	12	-	0.587	0.449	0.314	0.250
	24	-	1.990	1.495	1.00	1.730
66.71	7.2	0.304	0.221	0.167	-	-
	11.2	-	0.342	0.253	-	-
	17.6	0.809	0.560	0.401	-	-
	19.2	-	0.569	0.486	-	-
	23.8	1.172	0.734	-	-	-
	27.9	-	0.891	0.726	-	-
	34.2	1.920	-	-	-	-
	37.6	2.058	1.317	1.043	-	-
70.74	9	-	0.255	0.175	0.137	0.104
	12	-	0.362	0.270	0.199	0.162
	18	-	0.522	0.420	0.344	0.271
	24	-	0.744	0.542	0.418	0.355
73.64	9	-	0.415	0.327	0.249	0.169
	12	-	0.523	0.395	0.270	0.210
	24	-	0.854	0.609	0.474	0.371
78.04	9	-	0.380	0.262	0.209	0.146
	12	-	0.455	0.353	0.252	0.194
	18	-	0.586	0.439	0.340	0.263
	24	-	0.779	0.542	0.452	0.364

* By volume

\+ Typographical error in original. Should be 0.730.
This error was copied without comment in "Solubilities of Inorganic and Organic Compounds", Seidell, A.; Linke, W.F. 1958-1965.

METHOD/APPARATUS/PROCEDURE:

A mixture of air and sulfur dioxide was passed into the bottom of an absorption tube via a porous plate above which the aqueous solution was held. A thermometer reaching below the liquid was fitted into the tube. The tube was immersed in a thermostat.

SOURCE AND PURITY OF MATERIALS:

May be taken as satisfactory.

ESTIMATED ERROR:

REFERENCES:

COMPONENTS:	ORIGINAL MEASUREMENTS:
1. Sulfur dioxide; SO_2; [7446-09-5] 2. Water; H_2O; [7732-18-5] 3. Sulfuric acid; H_2SO_4; [7664-93-9]	Johnstone, H.F.; Leppla, P.W. *J. Am. Chem. Soc.* 1934, *56*, 2233-2238
VARIABLES:	PREPARED BY:
Pressure, concentration of acid	W. Gerrard.

EXPERIMENTAL VALUES:

T/K = 298.15

Conc. of H_2SO_4, /mol kg^{-1} (H_2O)	Pressure, p_{SO_2} /atm	Conc. of SO_2, /mol kg^{-1} (H_2O)	Ratio of conc. of SO_2/p_{SO_2} /mol kg^{-1} atm^{-1}	*Mole fraction x_{SO_2}, referred to H_2O (**)
0.0879	0.00130	0.00193	1.485	0.0260
0.0879	0.00571	0.01129	1.977	0.0344
0.0879	0.01022	0.01878	1.838	0.0320
0.5174	0.00130	0.00161	1.238	0.0218
0.5174	0.00571	0.00796	1.394	0.0245
0.5174	0.01022	0.01345	1.316	0.0232
1.103	0.00131	0.00134	1.023	0.0181
1.103	0.00573	0.00719	1.255	0.0221
1.103	0.01022	0.01253	1.226	0.0216

* Calculated by the compiler

** Assuming all the liquid is pure water.

AUXILIARY INFORMATION

METHOD/APPARATUS/PROCEDURE:

The low partial pressures of sulfur dioxide were attained by using a mixture of nitrogen and sulfur dioxide, the latter being measured from a gas buret. The sulfur dioxide was passed into a "Pyrex saturator (Bichowsky, and Storch) (1)" until equilibrium was reached, as indicated by concordance of successive analyses Samples were withdrawn for titration with iodine and thiosulfate solutions

SOURCE AND PURITY OF MATERIALS:

1. Sulfur dioxide stated to be of high purity from a tank of the liquid.
2. Conductivity water.
3. Not stated.

ESTIMATED ERROR:

REFERENCES:

1. Bichowsky, F.R.; Storch, H. *J. Am. Chem. Soc.* 1915, *37*, 2695.

COMPONENTS:

1. Sulfur dioxide; SO_2; [7446-09-5]
2. Water; H_2O; [7732-18-5]
3. Sulfuric acid; H_2SO_4; [7664-93-9]

ORIGINAL MEASUREMENTS:

Milbauer, J.

Bull. Soc. Chim. 1935, (5), *3*, 221-224

VARIABLES:

Temperature

PREPARED BY:

W. Gerrard

EXPERIMENTAL VALUES:

760 mmHg = 1 atm = 101.325 kPa

T/K	Density of the acid solution, /g cm^{-3}	Weight of SO_2 absorbed by 100 g of the original acid solution/g
296.15	1.824	3.075
303.15	(90.8% H_2SO_4)	2.270
323.15		1.525
373.15		0.436
423.15		0.171
473.15		0.053
510.15		0.040
298.15	1.53	3.1582
318.15	(62% H_2SO_4)	1.4597
333.15		1.0200
353.15		0.5813
373.15		1.4908 *(Probably meant to be 0.4908)

* By compiler.

There was no mention of pressure.

AUXILIARY INFORMATION

METHOD/APPARATUS/PROCEDURE:

Sulfur dioxide was passed into a weighed amount of the aqueous acid solution in a bubbler tube fitted with an inlet tap and an outlet tap. The temperature was stated to be controlled to within ±0.1°C. Pressure was not mentioned. The weight of sulfur dioxide absorbed was determined by a chemical titration using iodine-thiosulfate solutions.

SOURCE AND PURITY OF MATERIALS:

1. Sulfur dioxide was stated to be "pure," and was dried by sulfuric acid.

2. Distilled water was mentioned in the description of the chemical titration.

3. Presumably good quality sulfuric acid was used.

ESTIMATED ERROR:

REFERENCES:

COMPONENTS:

1. Sulfur dioxide; SO_2; [7446-09-5]
2. Sulfuric acid; H_2SO_4; [7664-93-9]

ORIGINAL MEASUREMENTS:

Sano, H.; Nakamoto, Y.
Nippon Kagaku Zasshi 1968, *89*, 369-373.

VARIABLES:

PREPARED BY:

C. L. Young

EXPERIMENTAL VALUES:

T/°C = 30 T/K = 303

Absorption coefficient, $S^{a,b}$	Bunsen coefficient,[b] α	Bunsen coefficient,[c,f] α	$\ell^{c,d}$	Mole ratio	Mole fraction[e]
21.2	19.2	19.2	19.2	0.046	0.044

[a] Volume of sulfur dioxide absorbed by unit volume of solvent at experimental temperature and a total pressure of 1 atmosphere.

[b] Determined by static absorption method.

[c] Determined by flow method.

[d] $\alpha = \ell(1 - p^{\circ}_{s})$ where p°_{s} is the vapor pressure of solvent.

[e] Calculated by compiler.

[f] Assuming vapor pressure of sulfuric acid is negligible.

AUXILIARY INFORMATION

METHOD/APPARATUS/PROCEDURE:

Two techniques were used: (a) static absorption method and (b) flow method. Static method consisted of equilibrating solvent and sulfur dioxide in a 50 ml flask and measuring the absorption volumetrically. The flow method consists of passing sulfur dioxide through a micro gas absorption tube and estimating sulfur dioxide by iodometric titration. The vapor pressure of the solvents were measured roughly and were assumed to be equal to the partial pressure of the solvent by the authors when calculating the Bunsen coefficient.

SOURCE AND PURITY OF MATERIALS:

1. Purity 99.0-99.8 mole per cent.

ESTIMATED ERROR:

$\delta T/K = \pm 0.5$; $\delta x_{SO_2} = \pm 5\%$ (estimated by compiler).

REFERENCES:

COMPONENTS:

1. Sulfur dioxide; SO_2; [7446-09-5]
2. Sulfuric acid (fuming); H_2SO_4; [7664-93-9]

ORIGINAL MEASUREMENTS:

Miles, F.D.; Carson, T.

J. Chem. Soc. 1946, 786-790.

VARIABLES:

PREPARED BY:

W. Gerrard

EXPERIMENTAL VALUES:

760 mmHg = 1 atm = 101.325 kPa.

T/K	SO_3, % Total	SO_2 %	*Mole SO_2/ mole SO_3
293.15	80.21	3.85	0.0600
	80.37	3.81	0.0593
	80.64	3.66	0.0567
	80.81	3.65	0.0565
	81.63	3.88	0.0594
	82.03	3.95	0.0602
	82.07	3.85	0.0586
	82.12	3.91	0.0595
	82.13	3.96	0.0603
	82.17	3.98	0.0605
	82.53	4.66	0.0706
	82.75	4.18	0.0631
	82.8	4.58	0.0691
	82.85	4.33	0.0653
	82.90	4.60	0.0694
	82.90	4.48	0.0675
	82.91	4.28	0.0645
	83.65	4.96	0.0741
	84.57	5.62	0.0831
	85.77	6.18	0.0901
	85.94	6.27	0.0912
	86.47	6.77	0.0979
	86.49	5.99	0.0866
	86.65	5.99	0.0864
	86.71	7.06	0.1018

AUXILIARY INFORMATION

METHOD/APPARATUS/PROCEDURE:

The "absorber" was a glass vessel "5 inches by 1.25 inches", closed by a large tap. A gas inlet was attached to the side, and had a perforated end. "Acid or oleum" was weighed in the tube. Sulfur dioxide was passed into the liquid, which was shaken; and saturation was deemed to be complete after 1-2 hours. Samples of the final liquid were obtained by means of capillary-bulb tubes and the sulfur dioxide was determined by iodine-thiosulfate titrations.

The weight of sulfur dioxide was stated to be corrected to correspond with "an atmosphere of dioxide at 760 mm pressure by multiplying each of the experimental values by the factor $760/(P-p)$," P being the atmospheric pressure recorded during the determination, and p being the vapor pressure of sulfuric trioxide over the oleum at the temperature of saturation .

SOURCE AND PURITY OF MATERIALS:

1. Sulfur dioxide was used from a cylinder.

2. No information.

ESTIMATED ERROR:

REFERENCES:

COMPONENTS:

1. Sulfur dioxide; SO_2; [7446-09-5]
2. Sulfuric acid (fuming); H_2SO_4; [7664-93-9]

ORIGINAL MEASUREMENTS:

Miles, F.D.; Carson, T.

J. Chem. Soc. 1946, 786-790.

EXPERIMENTAL VALUES: 760mmHg = 1 atm = 101.325 kPa

T/K	SO_3 % Total	SO_2 %	*Mole SO_2/ mole SO_3
293.15	86.93	7.52	0.1081
	87.11	6.47	0.0928
	87.90	9.06	0.1288
	88.30	9.17	0.1298
	88.60	9.58	0.1352
	88.90	9.41	0.1323
	90.65	11.98	0.1652
	90.70	12.05	0.1661
	91.10	12.17	0.1670
	91.30	12.14	0.1662
	93.16	18.73	0.2513
	93.30	18.93	0.2536
	93.30	20.42	0.2736
	93.34	20.28	0.2716
	94.20	23.31	0.3093
	94.60	22.13	0.2924
	95.22	25.78	0.3384
	95.02	27.58	0.3628
	95.10	28.34	0.3725
	95.23	28.16	0.3696
313.15	56.75	1.798	0.0396
	56.90	1.747	0.0384
	56.90	1.764	0.0387
	57.0	1.712	0.0375
	57.05	1.702	0.0373
	64.99	1.602	0.0308
	65.04	1.406	0.0270
	65.13	1.431	0.0275
	65.87	1.406	0.0267
	73.64	1.628	0.0276
	73.70	1.634	0.0277
	78.15	1.920	0.0307
	78.20	1.792	0.0286
	78.20	1.928	0.0308
	79.38	1.981	0.0312
	79.39	2.035	0.0320
	79.52	2.050	0.0322
	80.80	2.020	0.0313
	80.95	2.050	0.0317
	82.70	2.468	0.0373
	85.0	3.176	0.0467
	85.0	3.132	0.0461
333.15	57.75	0.983	0.0213
	57.9	0.986	0.0213
	57.91	0.988	0.0213
	65.88	0.855	0.0162
	66.03	0.834	0.0158
	66.14	0.848	0.0160
	66.23	0.858	0.0162
	73.75	0.949	0.0161
	73.82	0.953	0.0161
	73.87	0.990	0.0168
	78.22	1.196	0.0191
	78.31	1.172	0.0187
	79.50	1.346	0.0212
	79.57	1.337	0.0210
	79.60	1.359	0.0213
	79.90	1.371	0.0214
	84.30	1.625	0.0241
	84.40	1.610	0.0238

* Calculated by compiler

(cont'd)

COMPONENTS:

1. Sulfur dioxide; SO_2; [7446-09-5]
2. Sulfuric acid (fuming); H_2SO_4; [7664-93-9]

ORIGINAL MEASUREMENTS:

Miles, F.D.; Carson, T.
J. Chem. Soc. 1946, 786-790.

EXPERIMENTAL VALUES:

% Total SO_3 = 100 $SO_3/(SO_3 + H_2O)$;

% SO_2 = g of SO_2 dissolved by 100 g of acid or oleum, corrected for a pressure of 760 mm of SO_2.

The authors stated "to obtain a rough check on the applicability of Henry's law, some measurements of the equilibrium absorption from a gas mixture containing 6% of sulfur dioxide were also made." The gas mixture was stated to contain 94% of air, p_{SO_2} then being taken as 0.06 atm.

T/K	Liquid composition	Solubility at 1 atm S_1	Solubility at 0.06 atm S_2	S_1 x 0.06
293	80.2% H_2SO_4 65.4% total SO_3	3.10	0.187	0.186
313		1.43	0.0974	0.086
333		0.85	0.0637	0.051
293	96.5% H_2SO_4	3.80	0.222	0.228
313	78.6% total SO_3	1.95	0.126	0.117
333		1.28	0.0863	0.0768
293	104% H_2SO_4	5.70	0.342	0.342
313	17.8% free SO_3	2.90	0.193	0.174
333		1.74	0.120	0.105

S_1 and S_2 are g SO_2/100 g acid.

COMPONENTS:

1. Sulfur dioxide; SO_2; [7446-09-5]
2. Water; H_2O; [7732-18-5]
3. Sulfuric acid; H_2SO_4; [7664-93-9]

ORIGINAL MEASUREMENTS:

Parkison, R.V.

TAPPI, 1956, *39*, (7), 517-519.

VARIABLES:

Temperature, pressure

PREPARED BY:

W. Gerrard

EXPERIMENTAL VALUES: 760 mmHg = 1 atm = 101.325 kPa

T/K	p_{SO_2}/atm	SO_2 /lb mol ft^{-3}	$*x_{SO_2}$	$*x_{SO_2}/p_{SO_2}$
283.15	0.00150	0.000255	0.0000735	0.0467
(50°F)	0.00505	0.000850	0.000245	0.0463
	0.0174	0.00307	0.000885	0.0484
	0.0185	0.00329	0.000949	0.0488
	0.0261	0.00470	0.001355	0.0493
	0.0281	0.00498	0.00144	0.0486
	0.0332	0.00588	0.00170	0.0486
	0.0350	0.00615	0.00177	0.0483
294.26	0.00153	0.000171	0.0000493	0.0322
(70°F)	0.00339	0.000413	0.000119	0.0351
	0.00480	0.000503	0.000145	0.0302
	0.00662	0.000738	0.000213	0.0321
	0.00913	0.00102	0.000294	0.0322
	0.00983	0.00115	0.000332	0.0337
	0.0182	0.00203	0.000585	0.0322
	0.0257	0.00289	0.000833	0.0324
	0.0345	0.00380	0.00110	0.0318
	0.0349	0.00381	0.00110	0.0315
305.37	0.00633	0.000466	0.000134	0.0212
(90°F)	0.00641	0.000494	0.000142	0.0222
	0.00812	0.000601	0.000173	0.0213
	0.00885	0.000600	0.000173	0.0195
	0.00952	0.000759	0.000219	0.0230

AUXILIARY INFORMATION

METHOD/APPARATUS/PROCEDURE:

A mixture of sulfur dioxide and nitrogen (of desired "solute" concentration) was passed into the "equilibrium"solubility cell by an inlet tube shaped as a funnel at the end under the liquid. A tapped exit for sampling was attached to a bottom corner of the cell. The gas outlet was at the top of the cell. The cell was placed in a thermostat. The gas and liquid were analyzed by the Palmrose technique(1), an iodine-thiosulfate type of titration. The gas mixtures were analyzed by drawing samples into gas weighing balloons of 1 dm^3 capacity.

SOURCE AND PURITY OF MATERIALS:

(1) Sulfur dioxide was of refrigerant grade.

(2) Water was "triple-distilled from a Pyrex still."

(3) Probably of good quality.

ESTIMATED ERROR:

REFERENCES:

1. Palmrose, G.V.

   Paper Trade J. 1935, *100*, No. 3, 38.

COMPONENTS:	ORIGINAL MEASUREMENTS:
1. Sulfur dioxide; SO_2; [7446-09-5] 2. Water; H_2O; [7732-18-5] 3. Sulfuric acid; H_2SO_4; [7664-93-9]	Parkison, R.V. *TAPPI*, 1956, *39*, (7), 517-519.

EXPERIMENTAL VALUES: 760mmHg = 1 atm = 101.325 kPa.

T/K	p_{SO_2}/atm	SO_2 /lb mol ft^{-3}	*x_{SO_2}	*x_{SO_2}/p_{SO_2}
305.37	0.0110	0.000801	0.000231	0.0210
(90°F)	0.0178	0.00126	0.000363	0.0204
	0.0251	0.00205	0.000591	0.0235
	0.0336	0.00244	0.0007035	0.0209

* Calculated by compiler

Molality of sulfuric acid was 0.058, i.e., 5.684 g H_2SO_4 in 1 kg of water. *The approximate mole fraction, x_{SO_2} was referred to 1 cu.ft. of water, taking 3.468 lb. moles of water per cu.ft.

NOTE:

At these low partial pressures the average value of 64.08 was used by Parkison for computing the partial pressure of sulfur dioxide from the analysis of the dry gas. The vapor pressure of water over the solutions of sulfur dioxide was taken to be the same as for pure water.

COMPONENTS:

1. Sulfur dioxide; SO_2; [7446-09-5]
2. Water; H_2O; [7732-18-5]
3. Thiocyanic acid; HCNS; [463-56-9]

ORIGINAL MEASUREMENTS:

Hansen, C.J.

Ber. 1933, *66* 447-450

VARIABLES:

PREPARED BY:

W. Gerrard

EXPERIMENTAL VALUES:

At 277.65 K, 76.9 g (76 cm^3, density = 1.012 g cm^{-3}) of solution containing 5.52 g HCNS (0.0936 mole) and 0.20 g "H_2SO_4" absorbed 16.9 g SO_2 (0.264 mole) to give 93.8 g (82.5 cm^3, density, 1.14 g cm^{-3}). The mole ratio, SO_2/HCNS (not given by the author) is 2.821; mole fraction, x_{SO_2} is 0.738.

At 283.15 K, 63.2g (57 cm^3, density, 1.11 g cm^{-3}) of solution containing 23.8 g HCNS, 0.403 (given by the author as 0.402) mole, absorbed 27.1 g SO_2 (0.423 mole) to give 90.3 g (72.5 cm^3, density, 1.247 g cm^{-3}) of final solution. The mole ratio, SO_2/HCNS was given as 1.054, the compiler gives 1.050, mole fraction, x_{SO_2} = 0.512.

The total pressure appeared to be an unspecified barometric pressure.

NOTE: The author gave the concentration of the hydrocyanic acid as 46.30% in the second determination; but the concentration should be given as 37.66%.

AUXILIARY INFORMATION

METHOD/APPARATUS/PROCEDURE:

Sulfur dioxide was passed into a known amount of the acid solution in a measuring cylinder. The final solution was weighed, and its volume determined.

SOURCE AND PURITY OF MATERIALS:

1. The sulfur dioxide was stated to be pure.

3. Prepared from barium thiocyanate.

ESTIMATED ERROR:

REFERENCES:

COMPONENTS:	ORIGINAL MEASUREMENTS:
1. Sulfur dioxide; SO_2; [7446-09-5] 2. Water; H_2O; [7732-18-5] 3. Ammonium chloride; NH_4Cl; [12125-02-9]	Fox, C.J.J. *Z. Phys. Chem.* 1902, *41*, 458-482
VARIABLES:	PREPARED BY:
Concentration of salt	W. Gerrard.

EXPERIMENTAL VALUES:

T/K	"Normality"	3	2.5	2	1.5	1	0.5
298	Ostwald coefficient, *L* (*L* for water given as 32.76).	42.78	41.37	39.76	38.06	36.37	34.58

The following arithmetic was shown by the original author:

a = cm^3 SO_2 (273 K, 760 mmHg) in 1 cm^3 of "solution presumably meaning the salt solution before absorption of gas.

x = cm^3 of iodine solution containing s equivalents per dm^3; 0.001 mole I_2 taken to be equivalent to 0.001 mole SO_2.

$1/2\ sx = a/22.4$, for one cm^3 of the saturated solution. For c cm^3 of the solution taken for titration, at T/K and b mmHg pressure, $a = (11.2\ sx)/c$ (T/K / 273 K). The coefficient L was taken to be equal to a (760/b). (See evaluation).

NOTE by compiler: based on the published density of sulfur dioxide, 2.9266 g dm^{-3} at 273.15 K and 1 atm, the molar volume of SO_2 is 21.87 dm^3 and not 22.4 dm^3 under these conditions; therefore all the L values are about 2% too large.

AUXILIARY INFORMATION

METHOD/APPARATUS/PROCEDURE:	SOURCE AND PURITY OF MATERIALS:
The absorption vessel was fitted with electrodes for the measurement of conductance, data not herein recorded. Sulfur dioxide was passed through the solution to saturation at barometric pressure, the apparatus being held in a thermostat, controlled at 25.05 to 25.1°C, or 35 to 35.05°C. A volume of the saturated solution was withdrawn by pipet (0.9181 to 1.436 cm^3), and the amount of SO_2 determined by iodine-thiosulfate titration.	1. Sulfur dioxide, 99.8% SO_2, 0.2% water, commercial. 2. Presumably conductivity grade. 3. Salt of Kahlbaum grade, re-crystallised and analytically attested. ESTIMATED ERROR: REFERENCES:

COMPONENTS:	ORIGINAL MEASUREMENTS:
1. Sulfur dioxide; SO_2; [7446-09-5] 2. Water; H_2O; [7732-18-5] 3. Ammonium bromide; NH_4Br; [12124-97-9]	Fox, C.J.J. *Z. Phys. Chem.* 1902, *41*, 458-482.
VARIABLES:	PREPARED BY:
Concentration of salt	W. Gerrard.

EXPERIMENTAL VALUES:

T/K	"Normality"	3	2.5	2	1.5	1	0.5
298	Ostwald coefficient, *L* (*L* for water given as 32.76)	52.25	49.17	46.06	42.78	39.46	36.28

The following arithmetic was shown by the original author:

a = cm^3 SO_2 (273 K, 760 mmHg) in 1 cm^3 of "solution", presumably meaning the salt solution before absorption of gas.

x cm^3 of iodine solution containing s equivalents per dm^3; 0.001 mole I_2 taken to be equivalent to 0.001 mole SO_2.

$1/2\ sx = a/22.4$, for one cm^3 of the saturated solution. For c cm^3 of the solution taken for titration, at T/K and b mmHg pressure, $a = (11.2\ sx)/c$ (T/K / 273 K). The coefficient L was taken to be equal to a $(760/b)$. (See evaluation).

NOTE by compiler: based on the published density of sulfur dioxide, 2.9266 g dm^{-3} at 273.15 K and 1 atm, the molar volume of SO_2 is 21.87 dm^3 and not 22.4 dm^3 under these conditions; therefore all the L values are about 2% too large.

AUXILIARY INFORMATION

METHOD/APPARATUS/PROCEDURE:

The absorption vessel was fitted with electrodes for the measurement of conductance, data not herein recorded. Sulfur dioxide was passed through the solution to saturation at barometric pressure, the apparatus being held in a thermostat, controlled at 25.05 to 25.1°C, or 35 to 35.05°C. A volume of the saturated solution was withdrawn by pipet (0.9181 to 1.436 cm^3), and the amount of SO_2 determined by iodine thiosulfate titration.

SOURCE AND PURITY OF MATERIALS:

1. Sulfur dioxide, 99.8% SO_2, 0.2% water, commercial.
2. Presumably conductivity grade.
3. Salt of Kahlbaum grade, recrystallized and analytically attested.

ESTIMATED ERROR:

REFERENCES:

COMPONENTS:

1. Sulfur dioxide; SO_2; [7446-09-5]
2. Water; H_2O; [7732-18-5]
3. Ammonium sulfate; $(NH_4)_2SO_4$; [7783-20-2]

ORIGINAL MEASUREMENTS:

Fox, C.J.J.

Z. Phys. Chem. 1902, *41*, 458-482.

VARIABLES:

Temperature, concentration of salt.

PREPARED BY:

W. Gerrard

EXPERIMENTAL VALUES: Equivalent taken by the author as 1/2 $(NH_4)_2SO_4$

T/K	"Normality"	3	2.5	2	1.5	1	0.5
298	Ostwald coefficient, *L*	35.96	35.47	34.95	34.34	33.82	33.35
	(*L* for water given as 32.76)						
308	Ostwald coefficient, *L*	24.60	24.23	23.93	23.49	23.14	22.91
	(*L* for water given as 22.43)						

The following arithmetic was shown by the original author:

a = cm^3 SO_2 (273 K, 760 mmHg) in 1 cm^3 of "solution," presumably meaning the salt solution before absorption of gas.

x = cm^3 of iodine solution containing s equivalents per dm^3 ; 0.001 mole I_2 taken to be equivalent to 0.001 mole SO_2.

$1/2\ sx = a/22.4$, for one cm^3 of the saturated solution. For c cm^3 of the solution taken for titration, at T/K and b mmHg pressure, $a = (11.2\ sx)/c$ (T/K / 273 K). The coefficient L was taken to be equal to a (760/b). (See evaluation)

NOTE by compiler: based on the published density of sulfur dioxide, 2.9266 g dm^{-3} at 273.15 K and 1 atm, the molar volume of SO_2 is 21.87 dm^3 and not 22.4 dm^3 under these conditions; therefore all the L values are about 2% too large.

AUXILIARY INFORMATION

METHOD/APPARATUS/PROCEDURE:

The absorption vessel was fitted with electrodes for the measurement of conductance, data not herein recorded. Sulfur dioxide was passed through the solution to saturation at barometric pressure, the apparatus being held in a thermostat, controlled at 25.05 to 25.1°C, or 35 to 35.05°C. A volume of the saturated solution was withdrawn by pipet (0.9181 to 1.436 cm^3), and the amount of SO_2 determined by iodine-thiosulfate titration.

SOURCE AND PURITY OF MATERIALS:

1. Sulfur dioxide, 99.8% SO_2 , 0.2% water, commercial.
2. Presumably conductivity grade.
3. Salt of Kahlbaum grade, recrystallised and analytically attested.

ESTIMATED ERROR:

REFERENCES:

COMPONENTS:	ORIGINAL MEASUREMENTS:
1. Sulfur dioxide; SO_2; [7446-09-5] 2. Water; H_2O; [7732-18-5] 3. Ammonium bisulfite; NH_4HSO_3; [10192-30-0]	Otuka, Y. *J. Soc. Chem. Ind.,Japan. Suppl.* 1939, *42*, 205B-20 B
VARIABLES:	PREPARED BY:
Temperature, pressure	W. Gerrard

EXPERIMENTAL VALUES: 760 mmHg = 1 atm = 101.325 kPa

T/K	P_{Total}/atm	P_{SO_2}/atm	Mole SO_2/dm^3 solution	H
373	1.73	0.75	0.1857	0.248
	1.98	1.00	0.2518	0.252
378	2.03	0.87	0.1819	0.209
	2.29	1.13	0.2478	0.219
383	2.36	0.98	0.1788	0.183
	2.66	1.28	0.2434	0.190
388	2.73	1.10	0.1753	0.159
	3.03	1.40	0.2399	0.171
393	3.13	1.21	0.1724	0.142
	3.47	1.56	0.2354	0.151
398	3.56	1.32	0.1693	0.128
	3.95	1.72	0.2311	0.134
403	4.06	1.45	0.1659	0.114
	4.48	1.88	0.2269	0.121
408	4.61	1.59	0.1624	0.102
	5.03	2.02	0.2233	0.110
413	5.19	1.70	0.1598	0.094
	5.61	2.14	0.2203	0.103
418	5.80	1.80	0.1574	0.087
	6.22	2.23	0.2180	0.098
423	6.42	1.85	0.1562	0.084
	6.83	2.27	0.2170	0.096

"Solution" appears to mean final solution. The author defined the solubility as $H = (H_2SO_3)/p_{SO_2}$, the pressure being in atm. (H_2SO_3) evidently means mole SO_2/dm^3 of solution. The concentration of NH_4HSO_3 was given as 0.3000 mole/dm^3. NOTE: The author refers to "pure sulfurous acid, H_2SO_3."

METHOD/APPARATUS/PROCEDURE:	SOURCE AND PURITY OF MATERIALS:
The total pressure, P_{Total}, was determined by means of a glass spring manometer. It was stated that the partial pressures were approximated from the dimensions of the apparatus, but details were not given.	Not stated. ESTIMATED ERROR: REFERENCES:

COMPONENTS:

1. Sulfur dioxide; SO_2; [7446-09-5]
2. Water; H_2O; [7732-18-5]
3. Ammonium bisulfite; NH_4HSO_3; [10192-30-0]

ORIGINAL MEASUREMENTS:

Marriner, D. E.; Whitney, R. P. *Paper Trade J.* 1948, *126* (Tappi section), 52-54.

VARIABLES:

Temperature, pressure

PREPARED BY:

W. Gerrard

EXPERIMENTAL VALUES:

760 mmHg = 1 atm = 101.325 kPa.

The liquid concentrations are expressed in g SO_2 per 100 g of water. The Total (T) SO_2 is all the SO_2 absorbed by the original aqueous solution of ammonia. The "Combined" (C) SO_2 was defined as that required to convert all "the lime to $CaSO_3$." This presumably means the amount to convert all the ammonia into $(NH_4)SO_3$. The Free SO_2 (F) is the difference T - C.

T/K	p_{SO_2}/mmHg	SO_2(T)	SO_2(C)	T/K	p_{SO_2}/mmHg	SO_2(T)	SO_2(C)
277.65	124	4.4	0.51	283.15	123	3.7	0.54
	367	10.3	0.47		333	7.9	0.60
	560	15.2	0.49		567	12.4	0.55
	655	17.9	0.50		696	14.7	0.56
	119	5.1	0.97		125	4.6	1.01
	354	11.3	0.95		313	8.3	0.94
	569	16.5	0.99		507	12.3	1.04
	162	7.1	1.40		691	15.4	0.95
	388	12.6	1.50		139	5.7	1.44
	580	17.6	1.40		355	9.9	1.44
	686	20.2	1.35		528	13.2	1.19
	117	6.8	1.93		660	15.9	1.33
	350	12.5	1.93		129	6.6	1.97
	548	17.2	1.97		331	10.3	1.85

(cont.)

AUXILIARY INFORMATION

METHOD/APPARATUS/PROCEDURE:

A mixture of sulfur dioxide and nitrogen was passed into aqueous ammonia in a glass cell until equilibrium was reached. Sulfur dioxide in the liquid and in the exit gas was estimated by a chemical tiration – iodine-thiosulfate, see White *et al.* (1). Temperature was controlled to within ± 0.5 °C.

For water at 283.15 K the SO_2 (T) (p_{SO_2}) were: 2.1 (86); 6.9 (320); 13.9 (678 mmHg).

Total and combined sulfur dioxide were estimated by a modification of the Palmrose technique (2).

SOURCE AND PURITY OF MATERIALS:

1. Sulfur dioxide: refrigeration grade.
2. Water: may be taken as pure.
3. Chemically pure ammonia solution was used.

Nitrogen from cylinder was mixed with the sulfur dioxide to obtain mixtures of determined sulfur dioxide content to estimate the p_{SO_2}.

ESTIMATED ERROR:

REFERENCES:

1. White, C. K. Jr.; Vivian, J. E.; Whitney, R. P. *Ann. Meet. Tech. Ass. Pulp and Paper Industry*, Feb. 1948.
2. Palmrose, G. V. *Paper Trade J.* 1935, *100*, 38.

COMPONENTS:	ORIGINAL MEASUREMENTS:
1. Sulfur dioxide; SO_2; [7446-09-5] 2. Water; H_2O; [7732-18-5] 3. Ammonium bisulfite; NH_4HSO_3; [10192-30-0]	Marriner, D. E.; Whitney, R. P. *Paper Trade J.* 1948, *126*, (Tappi section), 52-54.

EXPERIMENTAL VALUES:

T/K	p_{SO_2}/mmHg	SO_2(T)	SO_2(C)	T/K	p_{SO_2}/mmHg	SO_2(T)	SO_2(C)
283.15	518	13.9	2.00	298.15	119	2.5	0.51
	663	16.8	1.67		329	4.9	0.52
291.15	117	2.9	0.57		565	7.3	0.57
	562	9.3	0.56		706	9.2	0.58
	736	11.6	0.49		116	3.3	0.90
	132	3.7	1.00		335	5.7	0.98
	360	6.7	0.81		532	8.1	1.07
	551	9.8	1.00		696	9.6	0.90
	(?)00	2.7	1.35		117	4.1	1.40
	115	4.4	1.35		343	6.6	1.36
	364	7.7	1.36		544	8.8	1.42
	552	10.5	1.38		692	10.5	1.40
	706	12.7	1.34		(?)00	3.8	1.95
	175	6.2	1.93		116	5.2	1.96
	420	9.8	1.93		365	7.9	1.98
	610	13.0	2.00		558	10.1	2.04
					710	11.7	2.04

COMPONENTS:

1. Sulfur dioxide; SO_2; [7446-09-5]
2. Water; H_2O; [7732-18-5]
3. Ammonium bisulfite; NH_4SO_3H; [7803-63-6]

ORIGINAL MEASUREMENTS:

Domansky, R.; Rendos, F.

Chem. Zvesti

1957, *11*, 453-460.

VARIABLES:

PREPARED BY:

W. Gerrard

EXPERIMENTAL VALUES:

760 mmHg = 1 atm = 101.325 kPa

The values of SO_2/g (T, total), (C, combined as NH_4SO_3H), (F, free = T - C) were given in the original table (as shown in the table here) were for SO_2 absorbed by 100 g water at p_{SO_2} = 600 mmHg. The treatment seems to imply that there is a linear change of the weight of SO_2 absorbed with pressure; therefore it may be taken that weight of SO_2 absorbed for p_{SO_2} = 1 atm may be approximately calculated by multiplying the stated value by 1.267.

"Concentration of ammonia"*		0%	0.5%	1.0%	1.5%	2.0%
			SO_2/g per 100 g water			
T/K						
288.15	T	10.3	11.20	12.80	14.65	17.90
	C	-	1.90	3.80	5.70	7.60
	F	10.3	9.30	9.05	8.95	10.3
293.15	T	8.50	9.70	11.55	13.20	15.65
	C	-	1.90	3.80	5.70	7.60
	F	8.50	7.80	7.75	7.50	8.05
298.15	T	7.5	8.5	10.25	11.95	14.40
	C	-	1.90	3.80	5.70	7.60
	F	7.5	6.60	6.45	6.25	6.75
303.15	T	6.1	7.50	8.90	11.05	13.50
	C	-	1.90	3.80	5.70	7.60
	F	6.1	5.60	5.10	5.35	5.90

* appears to be wt-%.

(cont.)

AUXILIARY INFORMATION

METHOD/APPARATUS/PROCEDURE:

An absorption vessel was fitted with a stirrer, a manometer, one limb bent to pass just through the vessel cover, and the other limb open to the atmosphere, a tapped inlet tube, and a tapped outlet tube, each just passing through the cover of the vessel, for the passage of SO_2, and a thermometer. The absorption vessel was immersed in a thermostat. A stoppered sampling tube also passed through the cover and ended well below the surface of the solution in the vessel. The weight of SO_2 in a withdrawn sample was determined by an iodine-thiosulfate titration.

SOURCE AND PURITY OF MATERIALS:

It is probable that the components were of acceptable purity for the purpose of the measurements.

ESTIMATED ERROR:

REFERENCES:

COMPONENTS:	ORIGINAL MEASUREMENTS:
1. Sulfur dioxide; SO_2; [7446-09-5]	Domansky, R.; Rendos, F.
2. Water; H_2O; [7732-18-5]	*Chem. Zvesti*
3. Ammonium bisulfite; NH_4SO_3H; [7803-63-6]	1957, *11*, 453-460.

EXPERIMENTAL VALUES:

Four small diagrams were given, each for one of the stated concentrations of ammonia, and each showing plots of p_{SO_2} (mmHg) *vs.* SO_2/100 g of ammonia solution for each of the four temperatures stated. All the lines appear to be linear, and the value of g SO_2 for 600 mmHg corresponds closely with the appropriate value of SO_2/g (T) in the table. In each diagram, the four lines terminate on the base line (p_{SO_2} = 0) at the following values of g SO_2/100 g ammonia solution: 1.9 (0.5% ammonia); 3.8 (1.0% NH_3); 5.7 (1.5% NH_3); and 7.6 (2.0% NH_3), and these values are recorded in the table as SO_2/g (C), *i.e.*, combined as NH_4SO_3H.

COMPONENTS:

1. Sulfur dioxide; SO_2; [7446-09-5]
2. Water; H_2O; [7732-18-5]
3. Ammonium nitrate; NH_4NO_3; [6484-52-2]

ORIGINAL MEASUREMENTS:

Fox, C.J.J.

Z. Phys. Chem. 1902, *41*, 458-482

VARIABLES:

Temperature, concentration of salt

PREPARED BY:

W. Gerrard.

EXPERIMENTAL VALUES:

T/K	"Normality"	3	2.5	2	1.5	1	0.5
298	Ostwald coefficient, *L*	39.14	38.01	37.27	36.28	35.07	33.96
	(*L* for water given as 32.76)						
308	Ostwald coefficient, *L*	27.43	26.66	25.57	24.78	24.23	23.35
	(*L* for water given as 22.43)						

The following arithmetic was shown by the original author:

a = cm^3 SO_2 (273 K, 760 mmHg) in 1 cm^3 of "solution," presumably meaning the salt solution before absorption of gas.

x = cm^3 of iodine solution containing s equivalents per dm^3; 0.001 mole I_2 taken to be equivalent to 0.001 mole SO_2.

$1/2\ sx = a/22.4$, for one cm^3 of the saturated solution. For c cm^3 of the solution taken for titration, at T/K and b mmHg pressure, $a = (11.2\ sx)/c$ (T/K / 273 K). The coefficient L was taken to be equal to a (760/b). (See evaluation)

NOTE by compiler: based on the published density of sulfur dioxide, 2.9266 g dm^{-3} at 273.15 K and 1 atm, the molar volume of SO_2 is 21.87 dm^3 and not 22.4 dm^3 under these conditions; therefore all the L values are about 2% too large.

AUXILIARY INFORMATION

METHOD/APPARATUS/PROCEDURE:

The absorption vessel was fitted with electrodes for the measurement of conductance, data not herein recorded. Sulfur dioxide was passed through the solution to saturation at barometric pressure, the apparatus being held in a thermostat, controlled at 25.05 to 25.1°C, or 35 to 35.05°C. A volume of the saturated solution was withdrawn by pipet (0.9181 to 1.436 cm^3), and the amount of SO_2 determined by iodine-thiosulfate titration.

SOURCE AND PURITY OF MATERIALS:

1. Sulfur dioxide, 99.8% SO_2, 0.2% water, commercial.
2. Presumably conductivity grade.
3. Salt of Kahlbaum grade, re-crystallised and analytically attested.

ESTIMATED ERROR:

REFERENCES:

COMPONENTS:	ORIGINAL MEASUREMENTS:
1. Sulfur dioxide; SO_2; [7446-09-5] 2. Water; H_2O; [7732-18-5] 3. Ammonium thiocyanate; NH_4CNS; [1762-95-4]	Fox, C.J.J. *Z. Phys. Chem.* 1902, *41*, 458-482.
VARIABLES:	PREPARED BY:
Concentration of salt	W. Gerrard.

EXPERIMENTAL VALUES:

T/K	"Normality"	3	2.5	2	1.5	1	0.5
298	Ostwald coefficient, L	61.46	57.01	52.26	47.26	42.74	37.78

(L for water given as 32.76)

The following arithm tic was shown by the original author:

a = cm^3 SO_2 (273 K, 760 mmHg) in 1 cm^3 of "solution," presumably meaning the salt solution before absorption of gas.

x = cm^3 of iodine solution containing s equivalents per dm^3; 0.001 mole I_2 taken to be equivalent to 0.001 mole SO_2.

$1/2\ sx = a/22.4$, for one cm^3 of the saturated solution. For c cm^3 of the solution taken for titration, at T/K and b mmHg pressure, $a = (11.2\ sx)/c$ (T/K / 273 K). The coefficient L was taken to be equal to a $(760/b)$. (See evaluation).

NOTE: by compiler: based on the published density of sulfur dioxide, 2.9266 g dm^{-3} at 273.15 K and 1 atm, the molar volume of SO_2 is 21.87 dm^3 and not 22.4 dm^3 under these conditions; therefore all the L values are about 2% too large.

AUXILIARY INFORMATION

METHOD/APPARATUS/PROCEDURE:

The absorption vessel was fitted with electrodes for the measurement of conductance, data not herein recorded. Sulfur dioxide was passed through the solution to saturation at barometric pressure, the apparatus being held in a thermostat, controlled at 25.05 to 25.1°C, or 35 to 35.05°C. A volume of the saturated solution was withdrawn by pipet (0.9181 to 1.436 cm^3), and the amount of SO_2 determined by iodine-thiosulfate titration.

SOURCE AND PURITY OF MATERIALS:

1. Sulfur dioxide, 99.8% SO_2, 0.2% water, commercial.
2. Presumably conductivity grade.
3. Salt of Kahlbaum grade, re-crystallised and analytically attested.

ESTIMATED ERROR:

REFERENCES:

COMPONENTS:

1. Sulfur dioxide; SO_2;]7446-09-5]
2. Water; H_2O; [7732-18-5]
3. Thiocyanic acid, ammonium salt
   Ammonium thiocyanate; NH_4CNS; [1762-95-4]

ORIGINAL MEASUREMENTS:

Hansen, C. J.

Ber. 1933, *66B*, 447-450.

VARIABLES:

Temperature

PREPARED BY:

W. Gerrard

EXPERIMENTAL VALUES:

T/K	Conc. of soln. g salt/ g soln.	Wt. of soln.† /g	Wt. of soln. /g	Vol. of soln.† /cm^3	Vol. of soln. /cm^3	Wt. of SO_2 absorbed /g	Mole ratio SO_2/NH_4CNS
290.15	0.5	60	86.0	53	68	26.01	1.028
286.65	0.5	60	91.0	53	70.5	31.0	1.226
281.85	0.5	60	97.9	53	73.5	37.9	1.499
276.15	0.5	60	109.6	53	80.5	49.6	1.962

The total pressure appears to have been unspecified barometric pressure.

† Before adsorption of sulfur dioxide.

AUXILIARY INFORMATION

METHOD/APPARATUS/PROCEDURE:

A known amount of the salt solution was treated in a measuring cylinder with pure sulfur dioxide. The final solution was weighed, and its volume measured, and the weight of sulfur dioxide absorbed was thereby determined. It appears that the first temperature was 17 °C, and then the passage of SO_2 was continued to get results for the lower temperatures.

SOURCE AND PURITY OF MATERIALS:

1. Pure SO_2 was stated to be used.
2. and 3. Not specified; but may be taken of satisfactory purity.

ESTIMATED ERROR:

REFERENCES:

COMPONENTS:

1. Sulfur dioxide; SO_2; [7446-09-5]
2. Water; H_2O; [7732-18-5]
3. "Basic aluminium sulfate" (approx. $Al_2(SO_4)_3 + Al_2O_3$)

ORIGINAL MEASUREMENTS:

Applebey, M.P.

J. Soc. Chem. Ind. (Chem. Ind.) Trans. 1937, *56*, 139-146.

VARIABLES:

PREPARED BY:

W. Gerrard.

EXPERIMENTAL VALUES:

760 mmHg = 1 atm = 101.325 kPa

Data were presented only as a small diagram showing plots of (%SO_2 by volume in the gas phase") vs ("g SO_2 by 100 ml of absorbent solution"). The compiler has interpreted the %SO_2 by volume as the partial pressure, p_{SO_2}, of SO_2 in atm, and has read the following data from the plots.

No. of solution.	T/K	p_{SO_2}/atm	g SO_2/100 ml
1	293	0.01	4.7
		0.02	5.5
		0.03	6.0
		0.04	6.3
		0.05	6.7
		0.06	7.0
		0.07	7.2
		0.08	7.4
		0.09	7.6
		0.10	7.8
2	293	0.01	6.1
		0.02	7.1
		0.03	7.7
		0.04	8.1
		0.05	8.4
		0.06	8.8
		0.07	8.95

AUXILIARY INFORMATION

METHOD/APPARATUS/PROCEDURE:

None were given. The purpose of the measurements was in connection with the recovery of sulfur from smelter gases.

SOURCE AND PURITY OF MATERIALS:

None given.

ESTIMATED ERROR:

REFERENCES:

COMPONENTS:

1. Sulfur dioxide; SO_2; [7446-09-5]
2. Water; H_2O; [7732-18-5]
3. "Basic aluminium sulfate"(approx. $Al_2(SO_4)_3 + Al_2O_3$

ORIGINAL MEASUREMENTS:

Applebey, M.P.

J. Soc. Chem. Ind. (Chem. Ind.) Trans. 1937, *56*, 139-146.

EXPERIMENTAL VALUES:

No. of solution	T/K	p_{SO_2}/atm	g SO_2/100 ml
2	293	0.08	9.2
		0.09	9.4
		0.10	9 6
2	298	0.01	5.6
		0.02	6.5
		0.03	7.05
		0.04	7.45
		0.05	7.85
		0.06	8.05
		0.07	8.3
		0.08	8.6
		0.09	8.8
		0.10	8.95
2	303	0.01	5.0
		0.02	5.9
		0.03	6.5
		0.04	6.9
		0.05	7.2
		0.06	7.45
		0.07	7.7
		0.08	7.9
		0.09	8.05
		0.10	8.2
3	293	0.01	6.9
		0.02	8.3
		0.03	9.05
		0.04	9.55
		0.05	9.85
		0.06	10.1
		0.07	10.35
		0.08	10.65
		0.09	10.8
		0.10	11.05

The solution was described as "basic aluminium sulphate solution," the composition being defined as follows:

Solution No.	"Alumina" g/100 ml			"Basicity"
	Free	Combined	Total	
1	3.44	6.61	10.05	34.2
2	4.00	5.82	9.82	40.8
3	4.55	5.32	9.87	46.1

COMPONENTS:

1. Sulfur dioxide; SO_2; [7446-09-5]
2. Water; H_2O; [7732-18-5]
3. Calcium sulfite; $CaSO_3$; [10257-55-3]

ORIGINAL MEASUREMENTS:

White, C.K., Jr.; Vivian, J.E.; Whitney, R.P.

Paper Trade Journal 1948, *126*, (20), 46-48.

VARIABLES:

PREPARED BY:

W. Gerrard.

EXPERIMENTAL VALUES:

760 mmHg = 1 atm = 101.325 kPa

T/K	Wt. of SO_2(C) /g	Wt. of SO_2(T) /g	Partial Pressure p_{SO_2} /mmHg	Mole fraction $*x_{SO_2}$
283.15	0	1.98	75	0.0056
		6.70	311	0.0185
		7.54	353	0.0208
		10.53	510	0.0288
		15.45	751	0.0416
	0.65	2.10	50	0.00408
		2.23	55	0.00442
		3.54	118	0.00806
		3.81	126	0.00881
		5.25	204	0.0128
		7.53	324	0.0190
		9.05	405	0.0231
		15.80	757	0.0409
	1.20	2.69	19	0.00417
		3.61	67	0.00673
		4.60	122	0.00947
		7.35	276	0.0170
		7.94	304	0.0186
		12.51	564	0.0308
		12.85	583	0.0315
		16.03	748	0.0400

AUXILIARY INFORMATION

METHOD/APPARATUS/PROCEDURE:

A mixture of nitrogen and sulfur dioxide was passed into a cell containing water and calcium hydroxide until all the calcium hydroxide had disappeared. The final temperature was controlled to within ±0.05°C. Samples of the solution and effluent gas were taken in evacuated weighing bulbs. Total sulfur dioxide in the liquid was determined by a modified Palmrose technique (1) entailing an iodine-thiosulfate titration. From the weight of the calcium hydroxide added the combined sulfur dioxide could be determined. The pressure of the sulfur dioxide was calculated from the analysis, utilizing the measured total pressure, and correcting for deviations from the perfect gas laws.

SOURCE AND PURITY OF MATERIALS:

(1) The sulfur dioxide was of refrigeration grade.

(2) Water was presumably of satisfactory purity.

(3) C.P. calcium hydroxide was used.

ESTIMATED ERROR:

REFERENCES:

(1) Palmrose, G.V.; *Paper Trade J.* 1935, *100*, No. 3, 38.

See also Whitney, R.P.; Vivian, J.E. *Ind. Eng. Chem.* 1941, *33*, 741.

COMPONENTS:

1. Sulfur dioxide; SO_2; [7446-09-5]
2. Water; H_2O; [7732-18-5]
3. Calcium sulfite; $CaSO_3$; [10257-55-3]

ORIGINAL MEASUREMENTS:

White, C.K., Jr.; Vivian, J.E.; Whitney, R.P.

Paper Trade Journal, 1948, *126*, (20), 46-48.

EXPERIMENTAL VALUES:

T/K	Wt. of SO_2(C) /g	Wt. of SO_2(T) /g	Partial Pressure p_{SO_2} /mmHg	Mole fraction $*x_{SO_2}$
283.15	2.10	5.80	94	0.0103
		8.33	246	0.0169
		12.24	479	0.0277
		17.10	763	0.0405
291.15	0	4.03	250	0.0112
		6.21	403	0.0172
		11.20	743	0.0305
	0.65	1.74	37	0.00306
		3.21	140	0.00714
		7.84	477	0.0198
		11.79	749	0.0304
	1.20	2.86	38	0.00465
		4.39	153	0.00889
		7.89	415	0.0185
		12.36	748	0.0304
	2.10	5.03	71	0.00817
		5.89	146	0.0105
		6.87	233	0.0132
		10.62	541	0.0234
		13.45	759	0.0309
298.15	0	2.26	169	0.00632
		5.34	444	0.0148
		8.64	724	0.0237
	0.65	2.06	76	0.00395
		3.05	167	0.00670
		5.40	385	0.0132
		9.32	742	0.0238
	1.20	2.90	53	0.00476
		3.82	144	0.00731
		5.91	347	0.0128
		10.01	741	0.0242
	2.10	5.05	98	0.00823
		5.70	168	0.0100
		8.35	448	0.0173
		11.26	739	0.0251

* Calculated by compiler.

Weights of SO_2 are for (c) "combined as $CaSO_3$" and (T) total per 100 g of water at p_{SO_2}.

"Free SO_2" = (T-C)/g. Mole fraction is for free SO_2.

COMPONENTS:	ORIGINAL MEASUREMENTS:
1. Sulfur dioxide; SO_2; [7446-09-5] 2. Water; H_2O; [7732-18-5] 3. Calcium bisulfite; $Ca(HSO_3)_2$; [13780-03-5]	Otuka, Y. *J. Soc. Chem. Ind. Japan. Suppl.* 1939, *42*, 205B-207B.
VARIABLES:	PREPARED BY:
Temperature, pressure	W. Gerrard

EXPERIMENTAL VALUES:

760 mmHg = 1 atm = 101.325 kPa

T/K	p_{Total}/atm	p_{SO_2}/atm	Mole SO_2/dm^3 solution	*H*
373	2.90	1.93	0.4192	0.217
	3.27	2.30	0.5189	0.226
	3.63	2.67	0.6099	0.228
378	3.28	2.12	0.4130	0.195
	3.68	2.53	0.5118	0.202
	4.07	2.92	0.6023	0.206
383	3.69	2.32	0.4071	0.175
	4.13	2.76	0.5048	0.183
	4.54	3.18	0.5949	0.187
388	4.14	2.52	0.4012	0.159
	4.61	2.99	0.4980	0.166
	5.05	3.44	0.5878	0.171
393	4.52	2.62	0.3982	0.152
	5.12	3.22	0.4916	0.153
	5.59	3.70	0.5809	0.157
398	5.03	2.81	0.3927	0.140
	5.67	3.46	0.4849	0.140
	6.17	3.97	0.5739	0.145

AUXILIARY INFORMATION

METHOD/APPARATUS/PROCEDURE:

The total vapor pressure, p_{Total}, was determined by means of a glass spring manometer. It was stated that the partial pressures were approximated from the dimensions of the apparatus, but details were not given.

SOURCE AND PURITY OF MATERIALS:

Not stated.

ESTIMATED ERROR:

REFERENCES:

COMPONENTS:	ORIGINAL MEASUREMENTS:
1. Sulfur dioxide; SO_2; [7446-09-5] 2. Water; H_2O; [7732-18-5] 3. Calcium bisulfite; $Ca(HSO_3)_2$; [13780-03-5]	Otuka, Y. *J. Soc. Chem. Ind. Japan. Suppl.* 1939, *42*, 205B-207B

EXPERIMENTAL VALUES: 760 mmHg = 1 atm = 101.325 kPa

T/K	p_{Total}/atm	p_{SO_2}/atm	Mole SO_2/dm^3 solution	H
403	5.58	2.99	0.3877	0.130
	6.25	3.67	0.4792	0.130
	6.79	4.22	0.5674	0.134
408	6.16	3.16	0.3831	0.121
	6.86	3.88	0.4735	0.122
	7.45	4.48	0.5607	0.125
413	6.77	3.31	0.3792	0.114
	7.51	4.06	0.4688	0.115
	8.15	4.72	0.5548	0.117
418	7.43	3.46	0.3752	0.108
	8.21	4.26	0.4636	0.109
	8.90	4.94	0.5488	0.110
423	8.13	3.59	0.3720	0.103
	8.96	4.44	0.4592	0.103
	9.69	5.19	0.5438	0.105

"Solution" appears to mean the final solution.

The author defined the "solubility" as $H = (H_2SO_3)/p_{SO_2}$, p_{SO_2} in atm. (H_2SO_3) evidently means moles SO_2/dm^3 of solution.

The concentration of HSO_3 was given as 0.1476 mol dm^{-3} for the first two values of H at each temperature and 0.1481 for the third value of H at each temperature.

NOTE: The author refers to "pure sulfurous acid, H_2SO_3."

COMPONENTS:

1. Sulfur dioxide; SO_2; [7446-09-5]
2. Water; H_2O; [7732-18-5]
3. Calcium bisulfite from CaO + water + SO_2.

ORIGINAL MEASUREMENTS:

Conrad, F.H.; Beuschlein, W.L.

J. Am. Chem. Soc. 1934, *56*, 2554-2562.

VARIABLES:

PREPARED BY:

W. Gerrard.

EXPERIMENTAL VALUES:

T/K	Total Pressure /mmHg	g SO_2/100 g water Total	Combined	Free	Free SO_2 *Mole ratio	*Mole fraction
288.15	22	1.83	0.89	0.94	0.0026	0.0026
	49	3.99	1.85	2.14	0.0060	0.0060
	115	7.47	3.12	4.35	0.0122	0.0121
	154	8.72	3.51	5.21	0.0147	0.0144
	263	11.00	4.23	6.77	0.0190	0.0187
	408	13.76	4.78	8.98	0.0253	0.0246
	517	15.92	5.47	10.45	0.0294	0.0285
	761	19.25	5.95	13.30	0.0374	0.0361
298.15	40	1.91	0.91	1.00	0.0028	0.0028
	52	2.81	1.33	1.48	0.00416	0.00414
	136	5.55	2.43	3.12	0.00877	0.00870
	254	8.37	3.14	5.23	0.0147	0.0145
	461	11.52	4.06	7.46	0.0210	0.0205
	594	13.28	4.59	8.69	0.0244	0.0238
	756	15.38	5.12	10.26	0.0289	0.0281
	763	15.16	4.88	10.30	0.0290	0.0282
308.15	761	12.02	4.03	7.99	0.0225	0.0220

* Calculated by compiler.

NOTE: A solid phase, referred to as calcium oxide was present. The liquid phase was looked upon as a saturated solution of calcium (cont'd)

AUXILIARY INFORMATION

METHOD/APPARATUS/PROCEDURE:

Sulfur dioxide was passed into a measuring pipet, and then into a flask containing "calcium oxide" and water. The flask was shaken until a constant total pressure was achieved. The total pressure was measured by a manometer containing mercury. The system was allowed to settle, and a sample of the clear liquid, above the solid phase, was withdrawn into a weighed pipet for chemical titration by the iodine-thiosulfate procedure. The temperature of the thermostat was regulated to within 0.05°C. The total, free and combined sulfur dioxide were determined by the method of Hohn (see Ref. 2).

SOURCE AND PURITY OF MATERIALS:

1. Sulfur dioxide was of "refrigeration grade." and was passed through concentrated sulfuric acid before absorption.
2. Water was freshly distilled.
3. Calcium oxide was obtained by the ignition of "C.P. analysed" calcium oxalate monohydrate.

ESTIMATED ERROR:

REFERENCES:

1. Technical Association of Pulp and Paper Institute Standards, "*Analysis of Bisulfite Cooking Liquor*", T604m, 1931, December, 24.
2. Birchard, W.H. *Paper Industry*, 1926, *8*, 793.

COMPONENTS:	ORIGINAL MEASUREMENTS:
1. Sulfur dioxide; SO_2; [7446-09-5] 2. Water; H_2O; [7732-18-5] 3. Calcium bisulfite from CaO + water + SO_2.	Conrad, F.H.; Beuschlein, W.L. *J. Am. Chem. Soc.* 1934, *56*, 2554-2562.

EXPERIMENTAL VALUES: (Cont'd)

bisulfite, $Ca(SO_3H)_2$. "Free SO_2" is defined by the Technical Association of Pulp and Paper Institute (1) "as the actual free SO_2 plus half of the SO_2 in the bisulfite of calcium and is more properly called the 'available SO_2' as it indicates SO_2 in excess of the amount necessary to form monosulfite". "The combined SO_2" is that required to form the monosulfite, $CaSO_3$; the total SO_2 is the sum of these. The total, free and combined SO_2 were estimated by the method of Hohn, as described by Birchard (2).

COMPONENTS:

1. Sulfur dioxide; SO_2; [7446-09-5]
2. Water; H_2O; [7732-18-5]
3. Calcium bisulfite; $Ca(HSO_3)_2$; [13780-03-5]

or Magnesium bisulfite; $Mg(HSO_3)_2$; [13774-25-9]

ORIGINAL MEASUREMENTS:

Chodakov, K. V. *Bumaznaja Promyslennost* 1937 (No. 9), 33-38.

VARIABLES:

Temperature, pressure, concentration

PREPARED BY:

W. Gerrard

EXPERIMENTAL VALUES:

1 atm = 101.325 kPa

*Calculated by compiler: mole fraction, x_{SO_2}, is based on the supposition that the liquid is 100 cm^3 of water. The weights are given as g in 100 cm^3 of final solution. C SO_2 is combined SO_2 equivalent to $Ca(HSO_3)_2$: F SO_2 is free SO_2, from Total (T)-C: p_{SO_2} is the pressure in atm, from the % by volume in the gas phase.

T/K	Base /g	p_{SO_2} /atm	SO_2,T /g	SO_2,F /g	SO_2 mole	x_{SO_2}*
297.15	CaO 1.40 g	0.089	4.26	1.06	0.0166	0.0030
		0.264	5.66	2.46	0.0384	0.0069
		0.441	7.01	3.81	0.0595	0.0106
		0.616	8.48	5.29	0.0827	0.0147
		0.662	8.48	5.28	0.0825	0.0146
		0.850	10.16	6.96	0.109	0.0192
297.15	Dolomite equivalent to CaO, 1.40 g	0.063	3.98	0.78	0.0122	0.0022
		0.052	4.00	0.80	0.0125	0.0023
		0.231	5.31	2.11	0.0330	0.0059
		0.278	5.79	2.59	0.0405	0.0072
		0.525	7.60	4.40	0.0688	0.0122
		0.700	9.02	5.82	0.0909	0.0161
		0.820	9.90	6.70	0.105	0.0185

(cont.)

AUXILIARY INFORMATION

METHOD/APPARATUS/PROCEDURE:

Sulfur dioxide in the liquid and in the vapor phase was determined by an iodine titration.
One half of SO_2 present as bisulfite was determined by titration with alkali.

SOURCE AND PURITY OF MATERIALS:

May be taken as of satisfactory purity.

ESTIMATED ERROR:

REFERENCES:

COMPONENTS:

1. Sulfur dioxide; SO_2; [7446-09-5]
2. Water; H_2O; [7732-18-5]
3. Calcium bisulfite; $Ca(HSO_3)_2$; [13780-03-5]
or Magnesium bisulfite; $Mg(HSO_3)_2$; [13774-25-9]

ORIGINAL MEASUREMENTS:

Chodakov, K. V. *Bumaznaja Promyslennost* 1937 (No. 9), 33-38.

EXPERIMENTAL VALUES:

T/K	Base /g	p_{SO_2} /atm	SO_2,T /g	SO_2,F /g	SO_2 mole	x_{SO_2} *
297.15	MgO equivalent to CaO 1.45 g	0.091	4.16	0.85	0.0133	0.0024
		0.092	4.35	1.04	0.0163	0.0029
		0.316	5.98	2.67	0.0417	0.0075
		0.451	7.10	3.79	0.0592	0.0106
		0.483	7.20	3.89	0.0608	0.0108
		0.645	8.64	5.33	0.0833	0.0148
		0.770	9.38	6.07	0.0948	0.0168
		0.882	10.47	7.16	0.112	0.0197
		0.888	10.56	7.25	0.113	0.0200
	Wt. of CaO/g					
297.15	0.5	0.05	1.79	0.65		
		0.14	2.70	1.56		
		0.34	4.32	3.18		
		0.48	5.60	4.46		
		0.69	7.29	6.15		
		0.87	8.64	7.50		
	1.0	0.07	3.07	0.78		
		0.24	4.64	2.35		
		0.48	6.40	4.11		
		0.49	6.62	4.33		
		0.78	8.90	6.61		
		0.84	9.05	6.76		
	1.4	0.06	3.98	0.78		
		0.05	4.00	0.80		
		0.23	5.31	2.11		
		0.28	5.79	2.59		
		0.53	7.60	4.40		
		0.70	9.03	5.83		
		0.82	9.90	6.70		
	2.0	0.065	5.28	0.71		
		0.17	6.00	1.43		
		0.25	6.66	2.09		
		0.47	8.30	3.73		
		0.56	8.99	4.42		
		0.82	10.90	6.33		
		0.84	11.07	6.50		
313.15	1.0	0.16	3.50	1.21		
		0.31	4.13	1.84		
		0.38	4.67	2.38		
		0.60	5.63	3.34		
		0.62	5.70	3.41		
		0.87	6.88	4.59		
333.15	1.0	0.08	3.07	0.78		
		0.19	3.42	1.13		
		0.39	4.00	1.71		
		0.75	5.12	2.83		
353.15	1.0	0.15	3.07	0.78		
		0.22	3.32	1.03		
		0.29	3.52	1.23		
		0.34	3.68	1.39		
		0.47	3.84	1.55		
279.15	1.0	0.055	3.71	1.42		
		0.07	3.90	1.61		
		0.08	4.20	1.91		
		0.17	5.63	3.34		

(cont.)

COMPONENTS:

1. Sulfur dioxide; SO_2; [7446-09-5]
2. Water; H_2O; [7732-18-5]
3. Calcium bisulfite; $Ca(HSO_3)_2$; [13780-03-5]
or Magnesium bisulfite; $Mg(HSO_3)_2$; [13774-25-9]

ORIGINAL MEASUREMENTS:

Chodakov, K. V. *Bumaznaja Promyslennost* 1937 (No. 9), 33-38.

EXPERIMENTAL VALUES:

T/K	Base /g	SO_2 /atm	SO_2,T /g	SO_2,F /g	SO_2 mole	$x^*_{SO_2}$
	Wt. of CaO					
279.15	1.0	0.39	8.64	6.35		
		0.39	8.75	6.46		
		0.65	11.80	9.51		
		0.62	12.00	9.71		
		0.83	15.00	12.71		
		0.91	16.00	13.71		
	1.4	0.025	4.44	1.24		
		0.02	4.48	1.28		
		0.06	5.18	1.98		
		0.155	6.56	3.36		
		0.27	8.29	5.09		
		0.44	10.60	7.40		
		0.67	13.44	10.24		
		0.72	14.70	11.50		
		0.78	15.30	12.10		
		0.855	16.20	13.00		
313.15	1.4	0.06	3.65	0.45		
		0.25	4.58	1.38		
		0.47	5.65	2.45		
		0.84	7.36	4.16		
		0.88	7.52	4.32		
333.15	1.4	0.09	3.78	0.58		
		0.29	4.54	1.34		
		0.64	5.44	2.24		
		0.79	5.92	2.72		
353.15	1.4	0.17	3.84	0.64		
		0.11	3.48	0.28		
		0.28	4.16	0.96		
		0.48	4.48	1.28		

COMPONENTS:

1 Sulfur dioxide; SO_2; [7446-09-5]
2. Water; H_2O; [7732-18-5]
3. Thiocyanic acid, calcium salt, (Calcium thiocyanate); $Ca(CNS)_2$; [2092-16-2]

ORIGINAL MEASUREMENTS:

Hansen, C.J.

Ber, 1933, *66B*,447-450

VARIABLES:

PREPARED BY:

W. Gerrard.

EXPERIMENTAL VALUES:

At 280.15 K, 81.65 g of the original solution containing 22.07 g of calcium thiocyanate (volume = 70 cm^3, density = 1.167 g cm^{-3}, moles, as $Ca(CNS)_2$ = 0.141, but given by the author as 0.2828. based on one half of $Ca(CNS)_2$) absorbed 21.25 g SO_2, 0.332 mole, to give 102.9 g, and 82.3 cm^3 of solution (final), density = 1.25 g cm^{-3}. The mole ratio $SO_2/Ca(CNS)_2$ is therefore = 0.332/ 0.141 = 2.355, mole fraction, x_{SO_2}, = 0.702. The author gave the mole ratio as 1.175, based on one half of the molecular formula, $Ca(CNS)_2$. The pressure was presumably the unspecified barometric pressure.

AUXILIARY INFORMATION

METHOD/APPARATUS/PROCEDURE:

A known amount of salt solution was treated in a measuring cylinder with sulfur dioxide. The final solution was weighed and its volume determined.

SOURCE AND PURITY OF MATERIALS:

1. Sulfur dioxide was stated to be pure.

2. and 3. May be assumed to be of satisfactory purity.

ESTIMATED ERROR:

REFERENCES:

COMPONENTS:

1. Sulfur dioxide; SO_2; [7446-09-5]
2. Water; H_2O; [7732-18-5]
3. Magnesium bisulfite from MgO + water + SO_2

ORIGINAL MEASUREMENTS:

Conrad, F.H.; Brice, D.B.

J. Am. Chem. Soc. 1948, *70*, 2179-2182.

VARIABLES:

PREPARED BY:

W. Gerrard.

EXPERIMENTAL VALUES:

T/K	Total pressure /mmHg	g SO_2/100 g water Total	Combined	Free	Free SO_2 *Mole ratio	*Mole fraction
288.15	75.5	30.4	15.7	14.7	0.0413	0.0397
	87.5	28.05	14.75	13.3	0.0374	0.0361
	110.7	29.65	15.05	14.6	0.0411	0.0395
	156.8	32.9	16.9	16.0	0.0450	0.0431
	162.3	31.4	16.55	14.85	0.0418	0.0401
	245.5	39.9	21.0	18.9	0.0532	0.0505
	419.3	46.9	23.55	23.35	0.0657	0.0616
	448.7	44.8	22.4	22.4	0.0630	0.0593
	726.8	54.6	27.8	26.8	0.0754	0.0701
	730.3	53.3	26.75	26.55	0.0747	0.0695
298.15	130.2	28.2	14.2	14.0	0.0394	0.0379
	154.4	28.0	14.85	13.15	0.0370	0.0365
	213.2	35.6	18.2	17.4	0.0489	0.0467
	264.3	35.2	17.9	17.3	0.0487	0.0464
	288.0	37.5	18.75	18.75	0.0527	0.0500
	326.5	40.7	20.35	20.35	0.0572	0.0541
	410.9	42.1	21.3	20.8	0.0585	0.0553
	560.7	50.2	25.4	24.9	0.0700	0.0654
	638.4	51.6	26.3	25.3	0.0712	0.0665
	726.8	50.7	25.85	24.85	0.0699	0.0653
	728.4	52.0	26.2	25.8	0.0726	0.0677
308.15	728.7	50.1	25.7	25.8	0.0726	0.0677

* calculated by compiler.

AUXILIARY INFORMATION

METHOD/APPARATUS/PROCEDURE:

As described by Conrad and Beuschlein (1).

NOTE: For definition of "total", "combined" and "free" sulfur dioxide see Ref. (1).

Sulfur dioxide was passed into a suspension of magnesium oxide in water. The solution was deemed to be a saturated solution of sulfur dioxide in a saturated solution of magnesium bisulfite. A solid phase was present, and a sample of the clear liquid phase was removed for chemical titrations.

SOURCE AND PURITY OF MATERIALS:

(1) Sulfur dioxide was of refrigeration grade, see ref. (1)

(2) Water was freshly distilled, see ref. (1).

(3) Magnesium oxide was "C.P. analyzed."

ESTIMATED ERROR:

REFERENCES:

1. Conrad, F.H.; Beuschlein, W.L. *J. Am. Chem. Soc.* 1934, *56*, 2554.

COMPONENTS:

1. Sulfur dioxide; SO_2; [7746-09-5]
2. Water; H_2O; [7732-18-5]
3. Cadmium chloride; $CdCl_2$; [10043-52-4]

ORIGINAL MEASUREMENTS:

Fox, C.J.J.

Z. Phys. Chem. 1902, *41*, 458-482.

VARIABLES:

Temperature, concentration of salt

PREPARED BY:

W. Gerrard

EXPERIMENTAL VALUES:

Equivalent taken by authors as $\frac{1}{2}CdCl_2$

T/K	"Normality"	3	2.5	2	1.5	1	0.5
298	Ostwald coefficient, L	26.06	27.09	28.16	29.46	30.55	31.66
	(L for water given as	32.76)					
308	Ostwald coefficient, L	18.68	19.23	20.02	20.55	21.23	21.73
	(L for water given as	22.43)					

The following arithmetic was shown by the original author:

a = cm^3 SO_2 (273 K, 760 mmHg) in 1 cm^3 of "solution," presumably meaning the salt solution before absorption of gas.

x = cm^3 of iodine solution containing s equivalents per dm^3; 0.001 mole I_2 taken to be equivalent to 0.001 mole SO_2.

$1/2\ sx = a/22.4$, for one cm^3 of the saturated solution. For c cm^3 of the solution taken for titration, at T/K and b mmHg pressure, $a = (11.2\ sx)/c$ (T/K / 273 K). The coefficient L was taken to be equal to a $(760/b)$. (See evaluation).

NOTE by compiler: based on the published density of sulfur dioxide, 2.9266 g dm^{-3} at 273.15 K and 1 atm, the molar volume of SO_2 is 21.87 dm^3 and not 22.4 dm^3 under these conditions; therefore all the L values are about 2% too large.

AUXILIARY INFORMATION

METHOD/APPARATUS/PROCEDURE:

The absorption vessel was fitted with electrodes for the measurement of conductance, data not herein recorded. Sulfur dioxide was passed through the solution to saturation at barometric pressure, the apparatus being held in a thermostat, controlled at 25.05 to 25.1°C, or 35 to 35.05°C. A volume of the saturated solution was withdrawn by pipet (0.9181 to 1.436 cm^3), and the amount of SO_2 determined by iodine-thiosulfate titration.

SOURCE AND PURITY OF MATERIALS:

1. Sulfur dioxide, 99.8% SO_2, 0.2% water, commercial.
2. Presumably conductivity grade.
3. Salt of Kahlbaum grade, recrystallised and analytically attested.

ESTIMATED ERROR:

REFERENCES:

COMPONENTS:

1. Sulfur dioxide; SO_2; [7446-09-5]
2. Water; H_2O; [7732-18-5]
3. Cadmium bromide; $CdBr_2$; [7789-42-6]

ORIGINAL MEASUREMENTS:

Fox, C.J.J.

Z. Phys. Chem. 1902, *41*, 458-482.

VARIABLES:

Temperature, concentration of salt.

PREPARED BY:

W. Gerrard.

EXPERIMENTAL VALUES: Equivalent taken by the author as 1/2 $CdBr_2$

T/K	"Normality"	3	2.5	2	1.5	1	0.5
298	Ostwald coefficient, L	27.46	28.15	29.27	30.17	31.01	31.91
	(L for water given as 32.76)						
308	Ostwald coefficient, L	19.17	19.70	20.60	20.81	31.46	21.88
	(L for water given as 22.43)						

The following arithmetic was shown by the original author:

a = cm^3 SO_2 (273 K, 760 mmHg) in 1 cm^3 of "solution," presumably meaning the salt solution before absorption of gas.

x = cm^3 of iodine solution containing s equivalents per dm^3; 0.001 mole I_2 taken to be equivalent to 0.001 mole SO_2.

1/2 $sx = a/22.4$, for one cm^3 of the saturated solution. For c cm^3 of the solution taken for titration, at T/K and b mmHg pressure, $a = (11.2\ sx)/c$ (T/K / 273 K). The coefficient L was taken to be equal to a (760/b). (See evaluation).

NOTE by compiler: based on the published density of sulfur dioxide, 2.9266 g dm^{-3} at 273.15 K and 1 atm, the molar volume of SO_2 is 21.87 dm^3 and not 22.4 dm^3 under these conditions; therefore all the L values are about 2% too large.

AUXILIARY INFORMATION

METHOD/APPARATUS/PROCEDURE:

The absorption vessel was fitted with electrodes for the measurement of conductance, data not herein recorded. Sulfur dioxide was passed through the solution to saturation at barometric pressure, the apparatus being held in a thermostat, controlled at 25.05 to 25.1°C, or 35 to 35.05°C. A volume of the saturated solution was withdrawn by pipet (0.9181 to 1.436 cm^3), and the amount of SO_2 determined by iodine-thiosulfate titration.

SOURCE AND PURITY OF MATERIALS:

1. Sulfur dioxide, 99.8% SO_2, 0.2% water, commercial.
2. Presumably conductivity grade.
3. Salt of Kahlbaum grade, re-crystallised and analytically attested.

ESTIMATED ERROR:

REFERENCES:

COMPONENTS:

1. Sulfur dioxide; SO_2; [7446-09-5]
2. Water; H_2O; [7732-18-5]
3. Cadmium iodide; CdI_2; [7790-80-9]

ORIGINAL MEASUREMENTS:

Fox, C.J.J.

Z. Phys. Chem. 1902, *41*, 458-482.

VARIABLES:

Temperature, concentration of salt

PREPARED BY:

W. Gerrard

EXPERIMENTAL VALUES:

Equivalent taken by the author as 1/2 CdI_2

T/K	"Normality"	3	2.5	2	1.5	1	0.5
298	Ostwald coefficient, L	35.77	34.98	34.74	34.16	33.76	33.27
	(L for water given as 32.76)						
308	Ostwald coefficient, L	24.30	23.99	23.71	23.38	23.06	22.75
	(L for water given as 22.43)						

The following arithmetic was shown by the original author:

a = cm^3 SO_2 (273 K, 760 mmHg) in 1 cm^3 of "solution," presumably meaning the salt solution before absorption of gas.

x = cm^3 of iodine solution containing s equivalents per dm^3; 0.001 mole I_2 taken to be equivalent to 0.001 mole SO_2.

$1/2\ sx = a/22.4$, for one cm^3 of the saturated solution. For c cm^3 of the solution taken for titration, at T/K and b mmHg pressure, $a = (11.2\ sx)/c$ (T/K / 273 K). The coefficient L was taken to be equal to a $(760/b)$. (See evaluation).

NOTE by compiler: based on the published density of sulfur dioxide, 2.9266 g dm^{-3} at 273.15 K and 1 atm, the molar volume of SO_2 is 21.87 dm^3 and not 22.4 dm^3 under these conditions; therefore all the L values are about 2% too large.

AUXILIARY INFORMATION

METHOD/APPARATUS/PROCEDURE:

The absorption vessel was fitted with electrodes for the measurement of conductance, data not herein recorded. Sulfur dioxide was passed through the solution to saturation at barometric pressure the apparatus being held in a thermostat, controlled at 25.05 to 25.1°C or 35 to 35.05°C. A volume of the saturated solution was withdrawn by pipet (0.9181 to 1.436 cm^3), and the amount of SO_2 determined by iodine-thiosulfate titration.

SOURCE AND PURITY OF MATERIALS:

1. Sulfur dioxide, 99.8% SO_2, 0.2% water, commercial.
2. Presumably conductivity grade.
3. Salt of Kahlbaum grade, recrystallised and analytically attested.

ESTIMATED ERROR:

REFERENCES:

COMPONENTS:

1. Sulfur dioxide; SO_2; [7446-09-5]
2. Water; H_2O; [7732-18-5]
3. Cadmium sulfate; $CdSO_4$; [10124-36-4]

ORIGINAL MEASUREMENTS:

Fox, C.J.J.

Z. Phys. Chem. 1902, *41*, 458-482.

VARIABLES:

Temperature, concentration of salt

PREPARED BY:

W. Gerrard.

EXPERIMENTAL VALUES: Equivalent given by the author as 1/2 $CdSO_4$

T/K	"Normality"	3	2.5	2	1.5	1	0.5
298	Ostwald coefficient, L	23.76	25.14	26.58	28.24	29.71	31.11
	(L for water given as	32.76)					
308	Ostwald coefficient, L	16.25	17.41	18.31	19.42	20.43	21.45
	(L for water given as	22.43)					

The following arithmetic was shown by the original author:

a = cm^3 SO_2 (273 K, 760 mmHg) in 1 cm^3 of "solution," presumably meaning the salt solution before absorption of gas.

x = cm^3 of iodine solution containing s equivalents per dm^3; 0.001 mole I_2 taken to be equivalent to 0.001 mole SO_2.

$1/2\ sx = a/22.4$, for one cm^3 of the saturated solution. For c cm^3 of the solution taken for titration, at T/K and b mmHg pressure, $a = (11.2\ sx)/c$ (T/K / 273 K). The coefficient L was taken to be equal to a (760/b). (See evaluation)

NOTE by compiler: based on the published density of sulfur dioxide, 2.9266 g dm^{-3} at 273.15 K and 1 atm, the molar volume of SO_2 is 21.87 dm^3 and not 22.4 dm^3 under these conditions; therefore all the L values are about 2% too large.

AUXILIARY INFORMATION

METHOD/APPARATUS/PROCEDURE:

The absorption vessel was fitted with electrodes for the measurement of conductance, data not herein recorded Sulfur dioxide was passed through the solution to saturation at barometric pressure, the apparatus being held in a thermostat, controlled at 25.05 to 25.1°C or 35 to 35.05°C. A volume of the saturated solution was withdrawn by pipet (0.9181 to 1.436 cm^3), and the amount of SO_2 determined by iodine-thiosulfate titration.

SOURCE AND PURITY OF MATERIALS:

1. Sulfur dioxide, 99.8% SO_2, 0.2% water, commercial.
2. Presumably conductivity grade.
3. Salt of Kahlbaum grade, re-crystallised and analytically attested.

ESTIMATED ERROR:

REFERENCES:

COMPONENTS:

1. Sulfur dioxide; SO_2; [7446-09-5]
2. Water; H_2O; [7732-18-5]
3. Zinc sulfate; $ZnSO_4$; [7733-02-0]

ORIGINAL MEASUREMENTS:

Jager (Jaeger), L.

Chem. Prumysl. 1957, *7*, 601-602.

VARIABLES:

Temperature, concentration of salt

PREPARED BY:

W. Gerrard.

EXPERIMENTAL VALUES:

760 mmHg = 1 atm = 101.325 kPa

T/K	g $ZnSO_4$/dm^3	g SO_2/dm^3
293	88.1	84.2
	119.0	79.6
	156.0	74.3
	194.0	69.2
	232.5	63.5
	284.0	55.1
	320.5	48.1
303	28.75	67.8
	59.5	65.6
	58.9	65.4
	87.5	62.8
	125.7	59.9
	114.0	58.9
	156.0	55.1
	156.0	55.1
	199.2	50.6
	194.8	49.4
	231.0	46.2
	229.0	45.8
	286.0	41.3
	316.0	37.5
313	32.6	50.6
	30.3	49.9
	54.2	49.0
	56.7	49.6
	90.4	47.1

AUXILIARY INFORMATION

METHOD APPARATUS/PROCEDURE:

An ount of sulfur dioxide absorbed at the total pressure of about 1 atm was determined by iodimetric titration. The temperature was controlled to within ±0.1°C.

SOURCE AND PURITY OF MATERIALS:

1. Sulfur dioxide may be taken as of satisfactory purity.
2. As for (1).
3. As for (1). Content in solution was determined by chelatometric titration.

ESTIMATED ERROR:

REFERENCES:

COMPONENTS	ORIGINAL MEASUREMENTS
1. Sulfur dioxide; SO_2; [7446-09-5] 2. Water; H_2O; [7732-18-5] 3. Zinc sulfate; $ZnSO_4$; [7733-02-0]	Jager (Jaeger), L. *Chem. Prumysl.* <u>1957</u>, *7*, 601-602.

EXPERIMENTAL VALUES: 760 mmHg = 1 atm = 101.325 kPa

T/K	g $ZnSO_4$/dm^3	g SO_2/dm^3
313	159.5	42.3
	157.8	41.6
	237.0	35.8
	279.0	32.5
	282.0	32.7
	322.0	29.4
323	28.1	37.4
	26.8	36.9
	65.1	36.8
	66.2	37.2
	98.2	35.5
	98.2	35.5
	160.5	31.7
	161.2	31.9
	198.2	30.6
	284.5	25.6
	284.5	25.6
	321.5	22.9
343	31.7	21.9
	31.7	21.9
	62.9	22.1
	160.0	20.7
	160.5	20.6
	188.5	19.5
	188.4	19.5
	258.0	17.6
	331.0	15.2
	331.0	15.2
363	60.0	9.08
	60.1	9.08
	140.3	9.92
	140.2	9.92
	220.5	9.78
	220.0	9.77
	295.0	8.37
	352.5	7.67
373	70.1	3.46
	124.0	2.69
	124.1	2.69
	245.0	3.71
	293.5	3.46

<u>NOTE:</u> g SO_2/dm^3 presumably means g SO_2 absorbed by 1 dm^3 of the original solution of the salt.

COMPONENTS:

1. Sulfur dioxide; SO_2; [7446-09-5]
2. Water; H_2O; [7732-18-5]
3. Zinc bisulfite; $Zn(HSO_3)_2$; [15457-98-4]
4. Zinc sulfate; $ZnSO_4$; [7733-02-0]

ORIGINAL MEASUREMENTS:

Peisakhov, I. L.; Karmazina, V. D.
Zh. Priklad. Khim. 1959, *32*, 70-78.

VARIABLES:

Temperature, concentration of components 3 and 4

PREPARED BY:

W. Gerrard

EXPERIMENTAL VALUES:

760 mmHg = 1 atm = 101.325 kPa

SO_2(T) = total sulfur dioxide

SO_2(F) = free sulfur dioxide

ρ = density of solution

Expt.	Conc. of $Zn(HSO_3)_2$ /g dm^{-3}	Conc. of $ZnSO_4$ /g dm^{-3}	SO_2(T) /g dm^{-3}	SO_2(F) /g dm^{-3}	ρ /g cm^{-3}	T/K	p_{SO_2} /mmHg
1	5.31	4.52	2.98	0	1.006	293	0.057
						313	0.126
						323	0.160
						333	0.241
		$\log p_{SO_2}$/mmHg = 3.94 - 1520/(T/K)					
2	10.2	3.52	5.84	0	1.009	293	0.149
						313	0.393
						323	0.597
						333	0.954
		$\log p_{SO_2}$/mmHg = 4.88 - 1960/(T/K)					
3	13.9	8.1	7.97	0.19	1.016	293	0.465
						313	0.679

(cont.)

AUXILIARY INFORMATION

METHOD/APPARATUS/PROCEDURE:

Nitrogen from a cylinder was passed through a rheometer, and then through five consecutive absorption flasks containing the solution, the procedure being that of Kuzminykh and Kuznetsova (1). Presumably the sulfur dioxide solution was prepared separately. The absorption flasks were held in a thermostat. The pressure, p_{SO_2} was measured by the determination of the volume content of the gas emerging from the last flask, iodine titrations being used. Likewise the sulfur dioxide content of the solutions in the last three flasks was determined. To obtain a pulp of solid zinc sulfite, crystals of the latter were added to each flask. The rate of passage of nitrogen was 0.5 liter/minute.

SOURCE AND PURITY OF MATERIALS:

Not stated.

ESTIMATED ERROR:

REFERENCES:

1. Kuzminykh, I. N.; Kuznetsova, A. G.
Zh. Priklad. Khim. 1954, *27*, 816.

COMPONENTS:

1. Sulfur dioxide; SO_2; [7446-09-5]
2. Water; H_2O; [7732-18-5]
3. Zinc bisulfite; $Zn(HSO_3)_2$; [15457-98-4]
4. Zinc sulfate; $ZnSO_4$; [7733-02-0]

ORIGINAL MEASUREMENTS:

Peisakhov, I. L.; Karmazina, V. D.
Zh. Priklad. Khim. 1959, *32*, 70-78.

EXPERIMENTAL VALUES:

Expt.	Conc. of $Zn(HSO_3)_2$ /g dm^{-3}	Conc. of $ZnSO_4$ /g dm^{-3}	SO_2(T) /g dm^{-3}	SO_2(F) /g dm^{-3}	ρ /g cm^{-3}	T/K	p_{SO_2} /mmHg
3	13.9	8.1	7.97	0.19	1.016	323	1.29
						333	2.00
		$\log p_{SO_2}$/mmHg = 6.43 - 2040/(T/K)					
4	13.9	4.44	7.83	0	1.012	293	0.302
						313	0.767
						323	1.26
						333	2.02
		$\log p_{SO_2}$/mmHg = 6.23 - 2040/(T/K)					
5	24.2	3.71	14.0	0.26	1.018	293	1.68
						313	3.34
						323	4.61
						333	8.77
		$\log p_{SO_2}$/mmHg = 7.28 - 2110/(T/K)					
6	25.6	6.6	15.0	0.58	1.025	293	1.67
						313	3.74
						323	5.70
						333	7.62
		$\log p_{SO_2}$/mmHg = 5.77 - 1620/(T/K)					
7	39.3	11.4	23.65	1.54	1.041	293	3.8
						313	8.52
						323	12.2
						333	18.5
		$\log p_{SO_2}$/mmHg = 6.04 - 1600/(T/K)					
8	41.4	9.35	25.0	1.73	1.04	293	3.6
						313	9.34
						323	14.57
						333	19.73
		$\log p_{SO_2}$/mmHg = 6.92 - 1865/(T/K)					
9	5.56	24.8	3.36	0.23	1.026	293	0.206
						313	0.319
						323	0.356
						333	0.50
		$\log p_{SO_2}$/mmHg = 2.22 - 880/(T/K)					
10	41.3	26.8	23.7	0.45	1.051	293	6.70
						313	13.2
						323	16.8
						333	21.5
		$\log p_{SO_2}$/mmHg = 5.06 - 1240/(T/K)					

(cont.)

COMPONENTS:	ORIGINAL MEASUREMENTS:
1. Sulfur dioxide; SO_2; [7446-09-5] 2. Water; H_2O; [7732-18-5] 3. Zinc bisulfite; $Zn(HSO_3)_2$; [15457-98-4] 4. Zinc sulfate; $ZnSO_4$; [7733-02-0]	Peisakhov, I. L.; Karmazina, V. D. *Zh. Priklad. Khim.* 1959, *32*, 70-78.

EXPERIMENTAL VALUES:

Expt.	Conc. of $Zn(HSO_3)_2$ /g dm^{-3}	Conc. of $ZnSO_4$ /g dm^{-3}	SO_2(T) /g dm^{-3}	SO_2(F) /g dm^{-3}	ρ /g cm^{-3}	T/K	p_{SO_2} /mmHg
11	3.86	55.0	2.21	0.32	1.055	293	0.07
						313	0.134
						323	0.175
						333	0.259
		$\log p_{SO_2}$/mmHg = 1.85 - 850/(T/K)					
12	6.14	53.2	3.5	0.05	1.057	293	0.363
						313	0.723
						323	0.843
						333	1.035
		$\log p_{SO_2}$/mmHg = 3.5 - 1150/(T/K)					
13	42.3	49.6	25.0	1.15	1.080	293	8.44
						313	15.7
						323	20.5
						333	27.5
		$\log p_{SO_2}$/mmHg = 5.38 - 1300/(T/K)					
14	3.64	88.2	2.63	0.58	1.088	293	0.256
						313	0.289
						323	0.30
						333	0.395
15	3.64	88.2	2.63	0.58	1.088	293	0.188
						313	0.258
						323	0.311
						333	0.468
14 and 15		$\log p_{SO_2}$/mmHg = 2.09 - 830/(T/K)					
16	10.9	82.6	6.9	1.21	1.090	293	0.948
						313	1.78
						323	2.52
						333	3.58
		$\log p_{SO_2}$/mmHg = 4.76 - 1400/(T/K)					
17	22.9	82.0	13.7	1.02	1.10	293	2.68
						313	5.46
						323	6.96
						333	9.41
		$\log p_{SO_2}$/mmHg = 4.92 - 1320/(T/K)					
18	35.4	77.5	22.2	2.3	1.106	293	9.35
						313	16.77
						323	18.73
						333	21.3
19	35.4	77.5	22.2	2.3	1.106	293	8.64
						313	14.78
						323	17.6
						333	22.3
18 and 19		$\log p_{SO_2}$/mmHg = 3.88 - 850/(T/K)					

(cont.)

COMPONENTS:

1. Sulfur dioxide; SO_2; [7446-09-5]
2. Water; H_2O; [7732-18-5]
3. Zinc bisulfite; $Zn(HSO_3)_2$; [15457-98-4]
4. Zinc sulfate; $ZnSO_4$; [7733-02-0]

ORIGINAL MEASUREMENTS:

Peisakhov, I. L.; Karmazina, V. D.

Zh. Priklad. Khim. 1959, *32*, 70-78.

EXPERIMENTAL VALUES:

Expt.	Conc. of $Zn(HSO_3)_2$ /g dm^{-3}	Conc. of $ZnSO_4$ /g dm^{-3}	SO_2(T) /g dm^{-3}	SO_2(F) /g dm^{-3}	ρ /g cm^{-3}	T/K	p_{SO_2} /mmHg
20	3.64	104	2.06	0	1.088	293	0.187
						313	0.299
						323	0.329
						333	0.472
		log p_{SO_2}/mmHg = 2.68 - 1000/(T/K)					
21	5.67	103	3.44	0.24	1.099	293	0.373
						313	0.726
						323	0.933
						333	1.25
		log p_{SO_2}/mmHg = 4.01 - 1300/(T/K)					
22	42.7	101.4	24.5	0.45	1.122	293	7.83
						313	17.0
						323	22.5
						333	32.5
		log p_{SO_2}/mmHg = 6.16 - 1550/(T/K)					
23	3.28	194	3.01	1.65	1.825	293	0.274
						313	0.573
						323	0.740
						333	1.08
		log p_{SO_2}/mmHg = 4.38 - 1450/(T/K)					
24	38.1	198	21.5	0	1.216	293	5.93
						313	14.94
						323	20.53
						333	28.7
		log p_{SO_2}/mmHg = 6.66 - 1700/(T/K)					

The smoothing equations given were calculated by the authors. Care should be taken in using these equations as there are significant discrepancies between values calculated from the equations and the experimental values. The equations for experiments nos. 2, 4, 9 and 24 appear to be in error.

COMPONENTS:

1. Sulfur dioxide; SO_2; [7446-09-5]
2. Water; H_2O; [7732-18-5]
3. Sodium chloride; NaCl; [7647-14-5]

ORIGINAL MEASUREMENTS:

Fox, C.J.J.

Z. Phys. Chem. 1902, *41*, 458-482.

VARIABLES:

Concentration of salt.

PREPARED BY:

W. Gerrard

EXPERIMENTAL VALUES:

T/K	"Normality"	3	2,5	2	1.5	1	0.5
298	Ostwald coefficient, L	31.36	31.51	31.76	31.96	32.25	32.46

(L for water given as 32.76)

The following arithmetic was shown by the original author:

a = cm^3 SO_2 (273 K, 760 mmHg) in 1 cm^3 of "solution," presumably meaning the salt solution before absorption of gas.

x = cm^3 of iodine solution containing s equivalents per dm^3; 0.001 mole I_2 taken to be equivalent to 0.001 mole SO_2.

$1/2\ sx = a/22.4$, for one cm^3 of the saturated solution. For c cm^3 of the solution taken for titration, at T/K and b mmHg pressure, $a = (11.2\ sx)/c$ (T/K / 273 K). The coefficient L was taken to be equal to a (760/b). (See evaluation).

NOTE by compiler: based on the published density of sulfur dioxide, 2.9266 g dm^{-3} at 273.15 K and 1 atm, the molar volume of SO_2 is 21.87 dm and not 22.4 dm^3 under these conditions; therefore all the L values are about 2% too large.

AUXILIARY INFORMATION

METHOD/APPARATUS/PROCEDURE:

The absorption vessel was fitted with electrodes for the measurement of conductance, data not herein recorded. Sulfur dioxide was passed through the solution to saturation at barometric pressure, the apparatus being held in a thermostat, controlled at 25.05 to 25.1°C, or 35 to 35.05°C. A volume of the saturated solution was withdrawn by pipet (0.9181 to 1.436 cm^3), and the amount of SO_2 determined by iodine-thiosulfate titration.

SOURCE AND PURITY OF MATERIALS:

(1) Sulfur dioxide, 99.8 SO_2, 0.2% water, commercial.

(2) Presumably conductivity grade.

(3) Salt of Kahlbaum grade, recrystallised and analytically attested.

ESTIMATED ERROR:

REFERENCES:

COMPONENTS:	ORIGINAL MEASUREMENTS:
1. Sulfur dioxide; SO_2; [7446-09-5] 2. Water; H_2O; [7732-18-5] 3. Sodium bromide; NaBr; [7647-15-6]	Fox, C.J.J. *Z. Phys. Chem.* 1902, *41*, 458-482
VARIABLES:	PREPARED BY:
Concentration of salt	W. Gerrard

EXPERIMENTAL VALUES:

T/K	"Normality"	3	2.5	2	1.5	1	0.5
298	Ostwald coefficient, L (L for water given as 32.76)	37.74	36.84	36.26	35.27	34.54	33.76

The following arithmetic was shown by the original author:

a = cm^3 SO_2 (273 K, 760 mmHg) in 1 cm^3 of "solution," presumably meaning the salt solution before absorption of gas.

x = cm^3 of iodine solution containing s equivalents per dm^3 ; 0.001 mole I_2 taken to be equivalent to 0.001 mole SO_2.

1/2 $sx = a/22.4$, for one cm^3 of the saturated solution. For c cm^3 of the solution taken for titration, at T/K and b mmHg pressure, $a = (11.2\ sx)/c\ (T/K\ /\ 273\ K)$. The coefficient L was taken to be equal to $a\ (760/b)$. (See evaluation).

NOTE by compiler: based on the published density of sulfur dioxide, 2.9266 g dm^{-3} at 273.15 K and 1 atm, the molar volume of SO_2 is 21.87 dm^3 and not 22.4 dm^3 under these conditions; therefore all the L values are about 2% too large.

AUXILIARY INFORMATION

METHOD/APPARATUS/PROCEDURE:	SOURCE AND PURITY OF MATERIALS:
The absorption vessel was fitted with electrodes for the measurement of conductance, data not herein recorded. Sulfur dioxide was passed through the solution to saturation at barometric pressure, the apparatus being held in a thermostat, controlled at 25.05 to 25.1°C, or 35 to 35.05°C. A volume of the saturated solution was withdrawn by pipet (0.9181 to 1.436 cm^3), and the amount of SO_2 determined by iodine-thiosulfate titration.	1. Sulfur dioxide, 99.8% SO_2, 0.2% water, commercial. 2. Presumably conductivity grade. 3. Salt of Kahlbaum grade, recrystallised and analytically attested.
	ESTIMATED ERROR:
	REFERENCES:

COMPONENTS:

1. Sulfur dioxide; SO_2; [7446-09-5]
2. Water; H_2O; [7732-18-5]
3. Sodium sulfate; Na_2SO_4; [7757-82-6]

ORIGINAL MEASUREMENTS:

Fox, C.J.J.

Z. Phys. Chem. 1902, *41*, 458-482.

VARIABLES:

Temperature, concentration of salt.

PREPARED BY:

W. Gerrard

EXPERIMENTAL VALUES: Equivalent taken by the author as 1/2 Na_2SO_4

T/K	"Normality"	3	2.5	2	1.5	1	0.5
298	Ostwald coefficient, L	28.44	28.66	29.51	30.45	31.14	31.96
	(L for water given as 32.76)						
308	Ostwald coefficient, L	19.27	19.79	20.20	20.81	21.35	21.88
	(L for water given as 22.43)						

The following arithmetic was shown by the original author:

a = cm^3 SO_2 (273 K, 760 mmHg) in 1 cm^3 of "solution," presumably meaning the salt solution before absorption of gas.

x = cm^3 of iodine solution containing s equivalents per dm^3 ; 0.001 mole I_2 taken to be equivalent to 0.001 mole SO_2 .

$1/2\ sx = a/22.4$, for one cm^3 of the saturated solution. For c cm^3 of the solution taken for titration, at T/K and b mmHg pressure, $a = (11.2\ sx)/c$ (T/K / 273 K). The coefficient L was taken to be equal to a (760/b) (See evaluation).

NOTE by compiler: based on the published density of sulfur dioxide, 2.9266 g dm^{-3} at 273.15 K and 1 atm, the molar volume of SO_2 is 21.87 dm^3 and not 22.4 dm^3 under these conditions; therefore all the L values are about 2% too large.

AUXILIARY INFORMATION

METHOD/APPARATUS/PROCEDURE:

The absorption vessel was fitted with electrodes for the measurement of conductance, data not herein recorded. Sulfur dioxide was passed through the solution to saturation at barometric pressure, the apparatus being held in a thermostat, controlled at 25.05 to 25.1°C, or 35 to 35.05°C. A volume of the saturated solution was withdrawn by pipet (0.9181 to 1.436 cm^3), and the amount of SO_2 determined by iodine-thiosulfate titration.

SOURCE AND PURITY OF MATERIALS:

1. Sulfur dioxide, 99.8% SO_2, 0.2% water, commercial.
2. Presumably conductivity grade.
3. Salt of Kahlbaum grade, recrystallised and analytically attested.

ESTIMATED ERROR:

REFERENCES:

COMPONENTS:

1. Sulfur dioxide; SO_2; [7446-09-5]
2. Water; H_2O; [7732-18-5]
3. Sodium sulfate; Na_2SO_4; [7757-82-6]

ORIGINAL MEASUREMENTS:

Hudson, J.C.
J. Chem. Soc. 1925, *127*, 1332-1347

VARIABLES:

Temperature, concentration of salt

PREPARED BY:

W. Gerrard.

EXPERIMENTAL VALUES: 760 mmHg = 1 atm = 101.325 kPa

T/K								
293.15	g salt/100 g water	0	1.94	4.93	7.95	10.00	15.00	20.00
	g SO_2 100 g water	10.64	10.66	10.57	10.36	10.07	9.66	9.17
	*Mole ratio	0.0299	0.0300	0.0297	0.0291	0.0283	0.0272	0.0258
	*x_{SO_2}	0.0290	0.0291	0.0289	0.0283	0.0275	0.0265	0.0252
303.15	g salt/100 g water	0	2.04	3.99	5.47	7.83	10.00	20.01
	g SO_2/100 g water	7.608	7.709	7.776	7.773	7.716	7.586	7.006
	*Mole ratio	0.0214	0.0217	0.0219	0.0219	0.0217	0.0213	0.0197
	*x_{SO_2}	0.0209	0.0212	0.0214	0.0214	0.0212	0.0209	0.0193
313.15	g salt/100 g water	0	2.49	5.04	7.95	10.00	14.94	20.01
	g SO_2/100 g water	5.585	5.812	5.910	5.927	5.882	5.802	5.630
	*Mole ratio	0.0157	0.0163	0.0166	0.0167	0.0165	0.0163	0.0158
	*x_{SO_2}	0.0155	0.0160	0.0163	0.0164	0.0162	0.0160	0.0156
323.15	g salt/100 g water	0	2.49	5.04	10.04	14.94	19.99	
	g SO_2/100 g water	4.214	4.508	4.678	4.728	4.730	4.648	
	*Mole ratio	0.01185	0.0127	0.0132	0.0133	0.0133	0.0131	
	*x_{SO_2}	0.0117	0.0125	0.0130	0.0131	0.0131	0.0129	

AUXILIARY INFORMATION

METHOD/APPARATUS/PROCEDURE:

Sulfur dioxide was bubbled through a weighed amount of salt solution in a tube immersed in a thermostat. The pressure was measured on a manometer. The saturated solution was weighed and the SO_2 content was determined by an iodine titration. It was stated that the calculated value of the partial pressure of the gas over the solution was within 10 mmHg of 760 mmHg; the final weight of gas was given as for 760 mmHg "on the basis of Henry's law." A tentative allowance was made for the lowering of the vapor pressure of water by the dissolved salt and gas.

* Calculated by compiler

SOURCE AND PURITY OF MATERIALS:

(1) Sulfur dioxide was from a cylinder; the purity was stated to be high; the chief impurity being water, less than 0.4%.

(2) and (3) may be taken as of analytical purity.

ESTIMATED ERROR:

REFERENCES:

COMPONENTS:	ORIGINAL MEASUREMENTS:
1. Sulfur dioxide; SO_2; [7446-09-5] 2. Water; H_2O; [7732-18-5] 3. Sodium bisulfate; $NaHSO_4$; [7681-38-1]	Kobe, K.A.; Kenton, F.H. *Ind. Eng. Chem.,Anal. Edit.* 1938, *10*, 76-77.
VARIABLES:	PREPARED BY: W. Gerrard.

EXPERIMENTAL VALUES:

760 mmHg = 1 atm = 101.325 kPa

T/K = 298.15 p_{SO_2} = 760 mmHg.

The aqueous solution was made up by dissolving 200 g anhydrous sodium sulfate in 800 g of distilled water, and adding 40 cm of "concentrated, 36*N* sulfuric acid."

Volume of solution (original) / cm^3	Volume of gas absorbed /cm^3	Volume of gas dissolved in one volume of solution	Bunsen coefficient α
9.54	131.2	13.6	12.5
24.54	392.2		

NOTE: The absorption coefficient, 13.6, is based on the volume of the final solution; the so-called Bunsen coefficient, α, was obtained by 13.6 x (273/298) = 12.46. However, the Bunsen α is usually referred to the volume of the original liquid, and not the final volume. The following results are supplied by compiler.

131.2/9.54 = 13.75 : α = 12.60
329.2/24.54 = 13.41 : α = 12.29

AUXILIARY INFORMATION

METHOD/APPARATUS/PROCEDURE:	SOURCE AND PURITY OF MATERIALS:
A mine-air buret was used. See Ref. (1).	1. Not specified. 2. and 3. Analytical grade compounds were used.
	ESTIMATED ERROR:
	REFERENCES: 1. Kobe, K.A.; Williams, J.S. *Ind.Eng.Chem.Anal.Ed.* 1935, *7*, 37.

COMPONENTS:

1. Sulfur dioxide; SO_2; [7446-09-5]
2. Water; H_2O; [7732-18-5]
3. Sodium bisulfite; $NaHSO_3$; [7631-90-5]

ORIGINAL MEASUREMENTS:

Otuka, Y.

J. Soc. Chem. Ind. Japan., Suppl. 1939, *42*, 205B-209B

VARIABLES:

Temperature, pressure

PREPARED BY:

W. Gerrard

EXPERIMENTAL VALUES: 760 mmHg = 1 atm = 101.325 kPa

T/K	p_{Total}/atm	p_{SO_2}/atm	Moles SO_2/dm^3 solution	H
373	1.58	0.60	0.1594	0.266
	1.94	0.97	0.2670	0.275
378	1.86	0.69	0.1563	0.226
	2.27	1.11	0.2618	0.236
383	2.18	0.80	0.1531	0.191
	2.62	1.25	0.2567	0.205
388	2.55	0.91	0.1498	0.165
	2.99	1.36	0.2529	0.186
393	2.94	1.02	0.1468	0.144
	3.43	1.52	0.2479	0.163
398	3.37	1.13	0.1438	0.127
	3.90	1.67	0.2432	0.146
403	3.83	1.22	0.1413	0.116
	4.41	1.81	0.2387	0.132
408	4.35	1.33	0.1384	0.104
	4.98	1.98	0.2336	0.118
413	4.96	1.47	0.1347	0.092
	5.56	2.09	0.2304	0.110
418	5.55	1.54	0.1331	0.086
	6.15	2.17	0.2282	0.105
423	6.14	1.56	0.1328	0.085
	6.75	2.20	0.2274	0.103

"Solution" appears to mean the final solution. The author defined the "solubility" as $H = (H_2SO_3)/p_{SO_2}$, the pressure being in atm. (H_2SO_3) evidently means moles SO_2/dm^3 of solution. The concentration of $NaHSO_3$ was given as 0.3000 mol/dm^{-3}. NOTE: The author refers to "pure sulfurous acid, H_2SO_3."

METHOD/APPARATUS/PROCEDURE:

The total vapor pressure, p_{Total}, was determined by means of a glass spring manometer. It was stated that the partial pressures were approximated from the dimensions of the apparatus, but details were not given.

SOURCE AND PURITY OF MATERIALS:

Not stated.

ESTIMATED ERROR:

REFERENCES:

COMPONENTS:

1. Sulfur dioxide; SO_2; [7446-09-5]
2. Water; H_2O; [7732-18-5]
3. Sodium bisulfite; $NaHSO_3$; [7631-90-5]

ORIGINAL MEASUREMENTS:

Kubelka, V.

Chem. Zvesti 1959, *13*, 430-435.

VARIABLES:

Temperature, concentration

PREPARED BY:

W. Gerrard

EXPERIMENTAL VALUES:

760 mmHg = 1 atm = 101.325 kPa

Solubility was expressed as g SO_2/100 g of solution at 760 mmHg.

"Normality" of $NaHSO_3$	SO_2,	T/K 293	303	313	333	353	363
0.2	Total	10.71	7.80	5.69	3.53	1.68	1.44
	Combined	1.26	1.26	1.26	1.26	1.26	1.26
	Free	9.45	6.54	4.43	2.27	0.42	0.18
0.3	T	11.31		6.38	4.17		2.33
	C	2.10		2.10	2.10		2.10
	F	9.21		4.28	2.07		0.23
0.5	T	11.85		6.89	5.14		3.43
	C	3.13		3.13	3.13		3.13
	F	8.72		3.76	2.01		0.30
1.0	T	14.27		9.66	7.62		6.98
	C	6.12		6.12	6.12		6.12
	F	8.15		3.54	1.50		0.86
2.0	T	18.70		15.20	13.27		12.43
	C	11.72		11.72	11.72		11.72
	F	7.98		3.48	1.55		0.71

For each normality a small diagram was given to show that for a stated temperature the plot of p_{SO_2} vs. g, SO_2(Total)/100 g solution was linear, although the numerical data were not given. These lines were presumably used to indicate the g SO_2(Total) for 1 atm.

AUXILIARY INFORMATION

METHOD/APPARATUS/PROCEDURE:

The method of Domansky and Rendos (1) was used. The sulfur dioxide was determined by an iodine titration.

SOURCE AND PURITY OF MATERIALS:

1. May be taken of satisfactory purity.

2 and 3. Sodium carbonate was used as a basis of the sodium bisulfite.

ESTIMATED ERROR:

REFERENCES:

1. Domansky, R.; Rendos, F. *Chem. Zvesti* 1957, *11*, 453.

COMPONENTS:	ORIGINAL MEASUREMENTS:
1. Sulfur dioxide; SO_2; [7446-09-5] 2. Water; H_2O; [7732-18-5] 3. Sodium thiocyanate; NaCNS; [540-72-7]	Fox, C.J.J. *Z. Phys. Chem.* 1902, *41*, 458-482.
VARIABLES: Concentration of salt.	PREPARED BY: W. Gerrard

EXPERIMENTAL VALUES:

T/K	"Normality"	3	2.5	2	1.5	1	0.5
298	Ostwald coefficient, L	48.34	45.86	43.37	40.78	38.24	35.44

(L for water given as 32.76)

The following arithmetic was shown by the original author:

a = cm^3 SO_2 (273 K, 760 mmHg) in 1 cm^3 of "solution," presumably meaning the salt solution before absorption of gas.

x = cm^3 of iodine solution containing s equivalents per dm^3; 0.001 mole I_2 taken to be equivalent to 0.001 mole SO_2.

$1/2\ sx = a/22.4$, for one cm^3 of the saturated solution. For c cm^3 of the solution taken for titration, at T/K and b mmHg pressure, $a = (11.2\ sx)/c$ (T/K / 273 K). The coefficient L was taken to be equal to a $(760/b)$. (See evaluation).

NOTE by compiler: based on the published density of sulfur dioxide, 2.9266 g dm^{-3} at 273.15 K and 1 atm, the molar volume of SO_2 is 21.87 dm^3 and not 22.4 dm^3 under these conditions; therefore all the L values are about 2% too large.

AUXILIARY INFORMATION

METHOD/APPARATUS/PROCEDURE:

The absorption vessel was fitted with electrodes for the measurement of conductance, data not herein recorded. Sulfur dioxide was passed through the solution to saturation at barometric pressure, the apparatus being held in a thermostat, controlled at 25.05 to 25.1°C, or 35 to 35.05°C. A volume of the saturated solution was withdrawn by pipet (0.9181 to 1.436 cm^3), and the amount of SO_2 determined by iodine-thiosulfate titration.

SOURCE AND PURITY OF MATERIALS:

1. Sulfur dioxide, 99.8% SO_2, 0.2% water commercial.

2. Presumably conductivity grade.

3. Salt of Kahlbaum grade, re-crystallised and analytically attested.

ESTIMATED ERROR:

REFERENCES:

COMPONENTS:

1. Sulfur dioxide; SO_2; [7446-09-5]
2. Water; H_2O; [7732-18-5]
3. Sodium thiocyanate; NaCNS; [540-72-7]

ORIGINAL MEASUREMENTS:

Hansen, C.J.

Ber. 1933, *66B*, 447-450.

VARIABLES:

PREPARED BY:

W. Gerrard.

EXPERIMENTAL VALUES:

At 282.15 K, 106.3 g of the original solution containing 28.75 g NaCNS (27% by weight, d = 1.13 g cm^{-3}), i.e. 0.355 mole, and 94 cm^3, absorbed 29.9 g SO_2 , presumably at a total pressure equal to the unspecified barometric pressure. The number of moles of SO_2 was stated to be 0.477, whereas, the correct value is 0.467. The mole ratio, SO_2/NaCNS was correctly stated to be 1.32. The mole fraction, x_{SO_2}, calculated by compiler is 0.569

AUXILIARY INFORMATION

METHOD/APPARATUS/PROCEDURE:

A known amount of salt solution was treated in measuring cylinder with sulfur dioxide. The final solution was weighed, and its volume determined.

SOURCE AND PURITY OF MATERIALS:

1. Pure sulfur dioxide was stated to be used.

2. and 3. Satisfactory purity may be assumed.

ESTIMATED ERROR:

REFERENCES:

COMPONENTS:

1. Sulfur dioxide; SO_2; [7446-09-5]
2. Water; H_2O; [7732-18-5]
3. Potassium chloride; KCl; [7447-40-7]

ORIGINAL MEASUREMENTS:

Fox, C.J.J.

Z. Phys. Chem. 1902, *41*, 458-482

VARIABLES:

Temperature, concentration of salt

PREPARED BY:

W. Gerrard

EXPERIMENTAL VALUES:

T/K	"Normality"	3	2.5	2	1.5	1	0.5
298	Ostwald coefficient, L	42.27	40.96	39.32	37.76	36.05	34.42
	(L for water given as 32.76)						
308	Ostwald coefficient, L	30.02	28.93	27.94	26.54	25.15	23.74
	(L for water given as 22.43)						

The following arithmetic was shown by the original author:

a = cm^3 SO_2 (273 K, 760 mmHg) in 1 cm^3 of "solution," presumably meaning the salt solution before absorption of gas.

x = cm^3 of iodine solution containing s equivalents per dm^3; 0.001 mole I_2 taken to be equivalent to 0.001 mole SO_2.

$1/2\ sx = a/22.4$, for one cm^3 of the saturated solution. For c cm^3 of the solution taken for titration, at T/K and b mmHg pressure, $a = (11.2\ sx)/c$ (T/K / 273 K). The coefficient L was taken to be equal to a $(760/b)$. (See evaluation).

NOTE by compiler: based on the published density of sulfur dioxide, 2.9266 g dm^{-3} at 273.15 K and 1 atm, the molar volume of SO_2 is 21.87 dm^3 and not 22.4 dm^3 under these conditions; therefore all the L values are about 2% too large.

AUXILIARY INFORMATION

METHOD/APPARATUS/PROCEDURE:

The absorption vessel was fitted with electrodes for the measurement of conductance, data not herein recorded. Sulfur dioxide was passed through the solution to saturation at barometric pressure, the apparatus being held in a thermostat, controlled at 25.05 to 25.1°C, or 35 to 35.05°C. A volume of the saturated solution was withdrawn by pipet (0.9181 to 1.436 cm^3), and the amount of SO_2 determined by iodine-thiosulfate titration.

SOURCE AND PURITY OF MATERIALS:

1. Sulfur dioxide, 99.8% SO_2, 0.2% water, commercial.
2. Presumably conductivity grade.
3. Salt of Kahlbaum grade, recrystallised and analytically attested.

ESTIMATED ERROR:

REFERENCES:

COMPONENTS:

1. Sulfur dioxide; SO_2; [7446-09-5]
2. Water; H_2O; [7732-18-5]
3. Potassium chloride; KCl [7447-40-7]

ORIGINAL MEASUREMENTS:

Hudson, J.C.

J. Chem. Soc. 1925, *127*, 1332-1347

VARIABLES:

Temperature, concentration of salt

PREPARED BY:

W. Gerrard

EXPERIMENTAL VALUES: 1 atm = 760 mmHg = 101.325 kPa

T/K	P_{Total} /mmHg	P_{SO_2} /mmHg	y_{KCl} /g	S_1 /g	S /g	*Mole ratio SO_2/H_2O	*Mole fraction 1 atm, x_{SO_2}
283.15			0		15.39	0.0433	0.0415
	780.3	772.0	5.55	17.32	17.05	0.0479	0.0458
	780.3	772.2	10.22	18.88	18.58	0.0522	0.0497
	766.2	758.3	15.85	20.23	20.28	0.0570	0.0539
	766.2	758.4	20.78	21.92	21.96	0.0618	0.0582
	768.2	761.4	29.90	25.45	25.40	0.0714	0.0667
288.15			0		12.66	0.0356	0.0344
	772.2	760.8	10.39	15.30	15.32	0.0431	0.0413
	771.9	761.0	19.91	17.67	17.64	0.0496	0.0473
	772.2	761.9	30.71	20.64	20.59	0.0579	0.0547
293.15			0		10.64	0.0299	0.0290
	773.9	757.9	5.14	11.60	11.62	0.0327	0.0317
	774.6	758.9	10.30	12.50	12.52	0.0352	0.0340
	778.0	761.7	15.61	13.58	13.55	0.0381	0.0367
	774.0	759.4	24.98	15.37	15.38	0.0433	0.0415
	779.0	764.8	30.38	16.76	16.65	0.0468	0.0447
303.05			0		7.58	0.0213	0.0209
	788.6	759.3	5.19	8.204	8.21	0.0231	0.0226
	789.4	760.7	10.11	8.810	8.80	0.0247	0.0241
	784.0	756.7	20.82	10.00	10.05	0.0283	0.0275
	782.8	756.8	29.86	11.21	11.26	0.0317	0.0307

AUXILIARY INFORMATION

METHOD/APPARATUS/PROCEDURE:

Sulfur dioxide was bubbled through the weighed amount of salt solution in a tube immersed in a thermostat, and attached to a manometer assembly for the measurement of the total pressure at the saturation stage. After saturation, the tube was weighed and the sulfur dioxide content determined by oxidation with hydrogen peroxide in sodium hydroxide solution, and the gravimetric estimation of sulfate as barium sulfate.

For the estimation of the vapor pressure of water over the solution, see end of the complete table.

SOURCE AND PURITY OF MATERIALS:

(1) Sulfur dioxide was from a cylinder; the purity was stated to be high, the chief impurity being water, less than 0.4%.

(2) and (3) may be taken as analytically pure.

ESTIMATED ERROR:

REFERENCES:

COMPONENTS:

1. Sulfur dioxide; SO_2; [7446-09-5]
2. Water; H_2O; [7732-18-5]
3. Potassium chloride; KCl; [7447-40-7]

ORIGINAL MEASUREMENTS:

Hudson, J.C.

J. Chem. Soc. 1925, *127*, 1332-1347.

EXPERIMENTAL VALUES: 1 atm = 760 mmHg = 101.325 kPa

T/K	p_{Total} /mmHg	p_{SO_2} /mmHg	y_{KCl} /g	S_1 /g	S /g	*Mole ratio SO_2/H_2O	*Mole fraction 1 atm, x_{SO_2}
313.15			0		5.54	0.0156	0.0154
	816.7	764.3	5.17	5.994	5.96	0.0168	0.0165
	818.1	766.9	10.20	6.380	6.32	0.0178	0.0175
	810.4	759.2	10.40	6.380	6.39	0.0180	0.0177
	818.1	768.1	15.44	6.869	6.675	0.0188	0.0185
	810.4	761.6	20.14	7.086	7.07	0.0199	0.0195
	803.7	756.0	24.83	7.415	7.45	0.0210	0.0206
	803.7	757.0	28.95	7.721	7.75	0.0218	0.0213
321.30			0		4.39	0.0123	0.0122
	841.8	761.7	5.06	4.712	4.70	0.0132	0.0130
	841.8	763.6	10.14	4.968	4.945	0.0139	0.0137
	826.9	750.7	15.82	5.180	5.24	0.0147	0.0145
	839.8	765.7	21.11	5.521	5.48	0.0154	0.0152
	826.9	754.0	25.09	5.703	5.74	0.0161	0.0159
	839.8	769.7	32.72	6.258	6.18	0.0174	0.0171
333.15			0		3.25	0.0091	0.0090
	889	746	5.30	3.384	3.45	0.0097	0.0096
	888	748	10.67	3.468	3.52	0.0099	0.0098
	896	763.5	21.12	3.943	3.92	0.0110	0.0109
	890.5	764	30.98	4.258	4.24	0.0119	0.0118
343.15			0		2.61	0.0073	0.00725
	969	747.5	8.74	2.756	2.80	0.0079	0.0078
	960.5	752	21.68	3.016	3.05	0.0086	0.0085
	960	757	30.73	3.251	3.26	0.0092	0.0091
353.15			0		2.13	0.0060	0.0060
	958	620	8.66	1.850	2.27	0.0064	0.0063
	961	643	21.35	2.043	2.415	0.0068	0.0067
	961	656	29.68	2.179	2.52	0.0071	0.0070
363.15			0		1.81	0.0051	0.0051
	967.5	468	9.23	1.142	1.855	0.0052	0.0052
	958	482	19.49	1.248	1.97	0.0055	0.0055
	965.5	520	32.75	1.445	2.11	0.0059	0.0059

* Calculated by compiler.

p_{Total} = total pressure.

p_{SO_2} = partial pressure of SO_2.

y_{KCl} = g of KCl in 100 g of water.

"Solubility" S_1 = g of SO_2 per 100 g of water at p_{SO_2}

S = g of SO_2 per 100 g of water at 760 mmHg

$S = S_1 \times 760/p_{SO_2}$

Allowance for the vapor pressure of water, p_w. It was assumed that lowering of the vapor pressure of water was equal to the sum of the depressions that would be caused by SO_2 and salt separately. Data for for the salt were taken from the literature; the lowering due to the gas was estimated by Raoult's law on the assumption that water is bimolecular. The correction was deemed to be small, 0.1% at 283 K, 1.25% at 363 K. Uncertainty was attributed to the combination of sulfur dioxide with water and with the salt, see the evaluation.

COMPONENTS:

1. Sulfur dioxide; SO_2; [7446-09-5]
2. Water; H_2O; [7732-18-5]
3. Potassium bromide; KBr; [7758-02-3]

ORIGINAL MEASUREMENTS:

Fox, C.J.J.

Z. Phys. Chem. 1902, *41*, 458-482.

VARIABLES:

Temperature, concentration of salt

PREPARED BY:

W. Gerrard

EXPERIMENTAL VALUES:

T/K	"Normality"	3	2.5	2	1.5	1	0.5
298	Ostwald coefficient, L	52.26	48.87	44.96	42.41	39.11	35.94
	(L for water given as 32.76)						
308	Ostwald coefficient, L	36.14	34.12	31.93	29.64	27.49	24.83
	(L for water given as 22.43)						

The following arithmetic was shown by the original author:

a = cm^3 SO_2 (273 K, 760 mmHg) in 1 cm^3 of "solution," presumably meaning the salt solution before absorption of gas.

x = cm^3 of iodine solution containing s equivalents per dm^3; 0.001 mole I_2 taken to be equivalent to 0.001 mole SO_2.

$1/2\ sx = a/22.4$, for one cm^3 of the saturated solution. For c cm^3 of the solution taken for titration, at T/K and b mmHg pressure, $a = (11.2\ sx)/c$ (T/K / 273 K). The coefficient L was taken to be equal to a $(760/b)$. (See evaluation).

NOTE by compiler: based on the published density of sulfur dioxide, 2.9266 g dm^{-3} at 273.15 K and 1 atm, the molar volume of SO_2 is 21.87 dm^3 and not 22.4 dm^3 under these conditions; therefore all the L values are about 2% too large.

AUXILIARY INFORMATION

METHOD/APPARATUS/PROCEDURE:

The absorption vessel was fitted with electrodes for the measurement of conductance, data not herein recorded. Sulfur dioxide was passed through the solution to saturation at barometric pressure, the apparatus being held in a thermostat, controlled at 25.05 to 25.1°C or 35 to 35.05°C. A volume of the saturated solution was withdrawn by pipet (0.9181 to 1.436 cm^3), and the amount of SO_2 determined by iodine-thiosulfate titration.

SOURCE AND PURITY OF MATERIALS:

1. Sulfur dioxide, 99.8% SO_2, 0.2% water commercial.
2. Presumably conductivity grade.
3. Salt of Kahlbaum grade, recrystallised and analytically attested.

ESTIMATED ERROR:

REFERENCES:

COMPONENTS:

1. Sulfur dioxide; SO_2; [7446-09-5]
2. Water; H_2O; [7732-18-5]
3. Potassium iodide; KI; [7681-11-0]

ORIGINAL MEASUREMENTS:

Fox, C.J.J.

Z. Phys. Chem. 1902, *41*, 458-482.

VARIABLES:

Temperature, concentration of salt.

PREPARED BY:

W. Gerrard.

EXPERIMENTAL VALUES:

T/K	"Normality"	3	2.5	2	1.5	1	0.5
298	Ostwald coefficient, *L*	68.36	62.63	56.75	50.58	44.76	38.66
	(*L* for water given as	32.76)					
308	Ostwald coefficient, *L*	45.43	41.87	38.04	34.64	30.25	26.30
	(*L* for water given as	22.43)					

The following arithmetic was shown by the original author:

a = cm^3 SO_2 (273 K, 760 mmHg) in 1 cm^3 of "solution," presumably meaning the salt solution before absorption of gas.

x = cm^3 of iodine solution containing s equivalents per dm^3; 0.001 mole I_2 taken to be equivalent to 0.001 mole SO_2.

$1/2\ sx = a/22.4$, for one cm^3 of the saturated solution. For c cm^3 of the solution taken for titration, at T/K and b mmHg pressure, $a = (11.2\ sx)/c$ (T/K / 273 K). The coefficient L was taken to be equal to a $(760/b)$. (See evaluation).

NOTE by compiler: based on the published density of sulfur dioxide, 2.9266 g dm^{-3} at 273.15K and 1 atm, the molar volume of SO_2 is 21.87 dm^3 and not 22.4 dm^3 under these conditions; therefore all the L values are about 2% too large.

AUXILIARY INFORMATION

METHOD/APPARATUS/PROCEDURE:

The absorption vessel was fitted with electrodes for the measurement of conductance, data not herein recorded. Sulfur dioxide was passed through the solution to saturation at barometric pressure, the apparatus being held in a thermostat, controlled at 25.05 to 25.1°C, or 35 to 35.05°C. A volume of the saturated solution was withdrawn by pipet (0.9181 to 1.436 cm^3), and the amount of SO_2 determined by iodine-thiosulfate titration.

SOURCE AND PURITY OF MATERIALS:

1. Sulfur dioxide, 99.8% SO_2, 0.2% water, commercial.
2. Presumably conductivity grade.
3. Salt of Kahlbaum grade, re-crystallised and analytically attested.

ESTIMATED ERROR:

REFERENCES:

COMPONENTS:

1. Sulfur dioxide; SO_2; [7446-09-5]
2. Water; H_2O; [7732-18-5]
3. Potassium sulfate; K_2SO_4; [7778-80-5]

ORIGINAL MEASUREMENTS:

Fox, C.J.J.

Z. Phys. Chem. 1902, *41*, 458-482.

VARIABLES:

Concentration of salt

PREPARED BY:

W. Gerrard.

EXPERIMENTAL VALUES: Equivalent taken by the author as 1/2 K_2SO_4

T/K	"Normality"	1	0.5
298	Ostwald coefficient, *L* (*L* for water given as 32.76).	33.61	33.20

The following arithmetic was shown by the original author:

a = cm^3 SO_2 (273 K, 760 mmHg) in 1 cm^3 of "solution," presumably meaning the salt solution before absorption of gas.

x = cm^3 of iodine solution containing s equivalents per dm^3; 0.001 mole I_2 taken to be equivalent to 0.001 mole SO_2.

$1/2\ sx = a/22.4$, for one cm^3 of the saturated solution. For c cm^3 of the solution taken for titration, at T/K and b mmHg pressure, $a = (11.2\ sx)/c$ (T/K / 273 K). The coefficient L was taken to be equal to a (760/b). (See evaluation).

NOTE by compiler: based on the published density of sulfur dioxide, 2.9266 g dm^{-3} at 273.15K and 1 atm, the molar volume of SO_2 is 21.87 dm^3 and not 22.4 dm^3 under these conditions; therefore all the L values are about 2% too large.

AUXILIARY INFORMATION

METHOD/APPARATUS/PROCEDURE:

The absorption vessel was fitted with electrodes for the measurement of conductance, data not herein recorded. Sulfur dioxide was passed through the solution to saturation at barometric pressure, the apparatus being held in a thermostat, controlled at 25.05 to 25.1°C, or 35 to 35.05°C. A volume of the saturated solution was withdrawn by pipet (0.9181 to 1.436 cm^3), and the amount of SO_2 determined by iodine-thiosulfate titration.

SOURCE AND PURITY OF MATERIALS:

1. Sulfur dioxide, 99.8% SO_2, 0.2% water, commercial.
2. Presumably conductivity grade.
3. Salt of Kahlbaum grade, recrystallised and analytically attested.

ESTIMATED ERROR:

REFERENCES:

COMPONENTS:	ORIGINAL MEASUREMENTS:
1. Sulfur dioxide; SO_2; [7446-09-5] 2. Water; H_2O; [7732-18-5] 3. Potassium nitrate; KNO_3; [7757-79-1]	Fox, C.J.J. *Z. Phys. Chem.* 1902, *41*, 458-482.

VARIABLES:	PREPARED BY:
Temperature, concentration of salt	W. Gerrard.

EXPERIMENTAL VALUES:

T/K	"Normality"	3	2.5	2	1.5	1	0.5
298	Ostwald coefficient, L	38.52	37.57	36.66	35.77	34.79	33.80
	(L for water given as	32.76)					
308	Ostwald coefficient, L	27.33	26.54	25.72	24.79	24.03	23.27
	(L for water given as	22.43)					

The following arithmetic was shown by the original author:

a = cm^3 SO_2 (273 K, 760 mmHg) in 1 cm^3 of "solution," presumably meaning the salt solution before absorption of gas.

x = cm^3 of iodine solution containing s equivalents per dm^3; 0.001 mole I_2 taken to be equivalent to 0.001 mole SO_2.

1/2 $sx = a/22.4$, for one cm^3 of the saturated solution. For c cm^3 of the solution taken for titration, at T/K and b mmHg pressure, $a = (11.2\ sx)/c$ (T/K /273 K). The coefficient L was taken to be equal to a (760/b). (See evaluation.)

NOTE by compiler: based on the published density of sulfur dioxide, 2.9266 g dm^{-3} at 273.15 K and 1 atm, the molar volume of SO_2 is 21.87 dm^3 and not 22.4 dm^3 under these conditions; therefore all the L values are about 2% too large.

AUXILIARY INFORMATION

METHOD/APPARATUS/PROCEDURE:

The absorption vessel was fitted with electrodes for the measurement of conductance, data not herein recorded. Sulfur dioxide was passed through the solution to saturation at barometric pressure, the apparatus being held in a thermostat, controlled at 25.05 to 25.1°C, or 35 to 35.05°C. A volume of the saturated solution was withdrawn by pipet (0.9181 to 1.436 cm^3), and the amount of SO_2 determined by iodine-thiosulfate titration.

SOURCE AND PURITY OF MATERIALS:

1. Sulfur dioxide, 99.8% SO_2, 0.2% water, commercial.
2. Presumably conductivity grade.
3. Salt of Kahlbaum grade, recrystallised and analytically attested.

ESTIMATED ERROR:

REFERENCES:

COMPONENTS:

1. Sulfur dioxide; SO_2; [7446-09-5]
2. Water; H_2O; [7732-18-5]
3. Potassium thiocyanate; KCNS; [333-20-0]

ORIGINAL MEASUREMENTS:

Fox, C.J.J.

Z. Phys. Chem. 1902, *41*, 458-482.

VARIABLES:

Temperature, concentration of salt.

PREPARED BY:

W. Gerrard

EXPERIMENTAL VALUES:

T/K	"Normality"	3.0	2.5	2.0	1.5	1.0	0.5
298	Ostwald coefficient, *L*	61.26	55.87	51.86	47.02	42.38	37.57
	(*L* for water given as	32.76)					
308	Ostwald coefficient, *L*	42.94	38.13	35.05	32.03	28.79	25.63
	(*L* for water given as	22.43)					

The following arithmetic was shown by the original author:

a = cm^3 SO_2 (273 K, 760 mmHg) in 1 cm^3 of "solution," presumably meaning the salt solution before absorption of gas.

x = cm^3 of iodine solution containing s equivalents per dm^3; 0.001 mole I_2 taken to be equivalent to 0.001 mole SO_2 .

$1/2\ sx = a/22.4$, for one cm^3 of the saturated solution. For c cm^3 of the solution taken for titration, at T/K and b mmHg pressure, $a = (11.2\ sx)/c$ (T/K / 273 K). The coefficient L was taken to be equal to a (760/b). (See evaluation).

NOTE by compiler: based on the published density of sulfur dioxide, 2.9266 g dm^{-3} at 273.15 K and 1 atm, the molar volume of SO_2 is 21.87 dm^3 and not 22.4 dm^3 under these conditions; therefore all the L values are about 2% too large.

AUXILIARY INFORMATION

METHOD/APPARATUS/PROCEDURE:

The absorption vessel was fitted with electrodes for the measurement of conductance, data not herein recorded. Sulfur dioxide was passed through the solution to saturation at barometric pressure, the apparatus being held in a thermostat, controlled at 25.05 to 25.1°C, or 35 to 35.05°C. A volume of the saturated solution was withdrawn by pipet (0.9181 to 1.436 cm^3), and the amount of SO_2 determined by iodine-thiosulfate titration.

SOURCE AND PURITY OF MATERIALS:

1. Sulfur dioxide, 99.8% SO_2, 0.2% water, commercial.
2. Presumably conductivity grade.
3. Salt of Kahlbaum grade, recrystallised and analytically attested.

ESTIMATED ERROR:

REFERENCES:

COMPONENTS:	ORIGINAL MEASUREMENTS:
1. Sulfur dioxide; SO_2; [7446-09-5] 2. Water; H_2O; [7732-18-5] 3. Potassium thiocyanate; KCNS; [333-20-0]	Hansen, C.J. *Ber*, 1933, *66B*, 447-450.
VARIABLES:	PREPARED BY: W. Gerrard

EXPERIMENTAL VALUES:

At 281.15 K, 50 g of the original solution of potassium thiocyanate (25 g, 0.258 moles) absorbed 32.2 g SO_2 to give 82.2 g of final solution of 58 cm^3 (density, * 1.42 g cm^{-3}. The mole ratio, SO_2/KCNS was given as 0.503/0.258 = 1.95; mole fraction, x_{SO_2}, calculated by compiler is 0.661.

*The density was given as 1.48.

At 285.15 K 60 g of the original solution containing 50 g of KCNS (0.515 mole) absorbed 52.5 g of SO_2 (0.820 mole) to give 112.5 g (68 cm^3, density 1.65 g cm^{-3}) of final solution. The mole ratio, SO_2/KCNS was given as 0.820/0.515 = 1.59. The corresponding mole fraction, x_{SO_2}, = 0.614.

The total pressure appears to have been the unspecified barometric pressure.

AUXILIARY INFORMATION

METHOD/APPARATUS/PROCEDURE:	SOURCE AND PURITY OF MATERIALS:
A known amount of the salt solution was treated in a measuring cylinder with sulfur dioxide. The final solution was weighed, and its volume was determined.	1. Pure SO_2 was stated to be used. 2. and 3. Satisfactory purity may be assumed.
	ESTIMATED ERROR:
	REFERENCES:

COMPONENTS:

1. Sulfur dioxide; SO_2; [7446-09-5]
2. Water; H_2O; [7732-18-5]
3. Trichloromethane (Chloroform); $CHCl_3$; [75-43-4]

ORIGINAL MEASUREMENTS:

McCrae, J.; Wilson, W. E.

Z. anorg. Chem. 1903, *35*, 11-15.

VARIABLES:

PREPARED BY:

W. Gerrard

EXPERIMENTAL VALUES:

T/K = 293.15

Water phase g SO_2/dm^3 (soln.) c_1	*Equiv. SO_2/dm^3 (soln.) e_1	x_{SO_2} ** (water)	Chloroform phase g SO_2/dm^3 (soln.) c_2	*Equiv. SO_2/dm^3 (soln.) e_2	x_{SO_2} ** ($CHCl_3$)	e_1/e_2
1.738	0.05426	0.000489	1.123	0.03505	0.00140	1.55
1.753	0.05473		1.122	0.03503		1.56
2.326	0.07263		1.704	0.05320		1.37
2.346	0.07324		1.703	0.05320		1.38
2.628	0.08206		1.897	0.05924		1.38
3.039	0.09488		2.395	0.07478		1.27
3.058	0.09548		2.385	0.07447		1.28
3.686	0.1151		3.063	0.09563		1.20
3.735	0.1166		3.062	0.09561		1.22
4.226	0.1319		3.626	0.1132		1.17
5.269	0.1645		4.798	0.1498		1.10
5.372	0.1678		4.813	0.1503		1.12
6.588	0.2057		6.183	0.1930		1.07
31.92	0.9968	0.0089	33.84	1.056	0.0407	0.94
33.26	1.038	0.0093	37.25	1.163	0.0445	0.89

* As 1/2 SO_2.

** The mole fraction, x_{SO_2}, water, is based on 1 dm^3 of water, *i.e.*, on the assumption that there was no change in volume when SO_2 dissolved. Likewise for $CHCl_3$.

AUXILIARY INFORMATION

METHOD/APPARATUS/PROCEDURE:

A freshly prepared aqueous solution (50 cm^3) of sulfur dioxide was shaken thoroughly with chloroform (50 cm^3) in a separatory funnel in a thermostat at 293 K. The amount of SO_2 absorbed at equilibrium in a measured volume of each liquid phase was determined by titration (iodine-thiosulfate).

NOTE: Pressure was not mentioned. The results merely provide so-called distribution coefficients as ratios of mole/l in the aqueous phase/ mole/l in the chloroform phase for a range of concentrations in the aqueous phase.

SOURCE AND PURITY OF MATERIALS:

Not specified.

ESTIMATED ERROR:

REFERENCES:

COMPONENTS:

1. Sulfur dioxide; SO_2; [7446-09-5]
2. Trichloromethane (Chloroform); $CHCl_3$; [75-43-4]
3. Water; H_2O; [7732-18-5]
4. Hydrogen chloride; HCl; [7647-01-0]

ORIGINAL MEASUREMENTS:

McCrae, J.; Wilson, W. E.

Z. anorg. Chem. 1903, *35*, 11-15.

VARIABLES:

PREPARED BY:

W. Gerrard

EXPERIMENTAL VALUES:

T/K = 293.15   * As 1/2 SO_2

	Aqueous phase		Chloroform phase		
N HCl	g SO_2/dm^3 (soln.) c_1	*Equiv. SO_2/dm^3 (soln.) e_1	g SO_2/dm^3 (soln.) c_2	*Equiv. SO_2/dm^3 (soln.) e_2	Distribution Coefficient e_1/e_2
0.05	1.86	0.0581	1.46	0.0456	1.28
0.05	3.076	0.0960	2.830	0.0884	1.08
0.05	4.277	0.1336	4.07	0.1271	1.04
0.05	5.340	0.1667	5.420	0.1692	0.96
0.1	1.25	0.0390	1.410	0.0440	0.88
0.1	1.324	0.0413	1.416	0.0442	0.93
0.1	2.78	0.0868	3.080	0.0962	0.90
0.1	3.86	0.1199	4.08	0.1275	0.94
0.1	5.161	0.1612	5.715	0.1784	0.90
0.2	1.268	0.0396	1.509	0.0471	0.84
0.2	1.914	0.0597	2.274	0.0710	0.84
0.2	2.464	0.0769	3.040	0.0949	0.81
0.2	3.967	0.1239	4.898	0.1530	0.81
0.4	1.202	0.0375	1.614	0.0504	0.79
0.4	1.894	0.0591	2.263	0.0706	0.83

AUXILIARY INFORMATION

METHOD/APPARATUS/PROCEDURE:

A freshly prepared aqueous solution of sulfur dioxide and HCl was shaken thoroughly with chloroform in a separatory funnel in a thermostat at 293.15 K. After attainment of equilibrium, the amount of sulfur dioxide in each phase was determined by iodine-thiosulfite titrations.

NOTE: Pressure was not mentioned. The results merely provide so-called distribution coefficients as ratios of mole/l in the aqueous phase divided by mole/l in the chloroform phase, for a range of concentrations in the aqueous phase.

SOURCE AND PURITY OF MATERIALS:

Not specified.

ESTIMATED ERROR:

REFERENCES:

COMPONENTS:	ORIGINAL MEASUREMENTS:
1. Sulfur Dioxide, SO_2; [7446-09-5] 2. Water; H_2O; [7732-18-5] 3. Methanol; CH_4O; [67-56-1]	Tokunaga, J. *J. Chem. Engng. Data.* 1974, *19*, 162-165.

VARIABLES:	PREPARED BY:
Temperature	C.L. Young

EXPERIMENTAL VALUES:

T/K	Mole fraction of alcohol in soln. (before absorption).	Total Pressure P/bar	Partial pressure of sulfur dioxide P/bar	$S^{\S}$	S_o^{+}
283.15	0.0000	1.024	1.012	15.66	15.66
	0.0608	1.030	1.013	16.06	16.06
	0.0813	1.032	1.013	16.92	16.91
	0.1415	1.044	1.019	20.55	20.42
	0.2066	1.043	1.015	25.04	24.98
	0.3428	1.045	1.015	43.60	43.50
	0.4161	1.053	1.017	51.75	51.53
	0.4620	1.056	1.019	56.63	56.26
	0.5862	1.049	1.010	75.96	76.17
	0.6792	1.063	1.023	91.79	90.85
	0.7490	1.058	1.016	97.44	97.07
	0.8694	1.066	1.024	112.18	110.97
	0.9263	1.059	1.016	120.38	120.02
	1.0000	1.060	1.015	131.85	131.56
293.15	0.0000	1.036	1.013	10.90	10.90
	0.0107	1.037	1.013	10.45	10.44
	0.0675	1.048	1.014	11.14	11.12
	0.1835	1.061	1.011	14.48	14.51
	0.2659	1.074	1.016	19.05	18.99
	0.3503	1.076	1.011	24.51	24.56
	0.4233	1.082	1.013	31.41	31.40
	0.5080	1.084	1.022	38.64	38.27

AUXILIARY INFORMATION

METHOD /APPARATUS/PROCEDURE:

Sulfur dioxide bubbled through aqueous solution. Saturated solution analysed by iodometric titration. Details in source. Partial pressure of sulfur dioxide calculated from known vapor pressure of water + alcohol mixture and assuming Raoult's law for the lowering of vapor pressure of aqueous alcohol solution owing to dissolved sulfur dioxide.

SOURCE AND PURITY OF MATERIALS:

1. High purity sample with purity of 99.96 mole per cent.
2. No details given.
3. Fractionated.

ESTIMATED ERROR:

$\delta T/K = \pm 0.1$; $\delta P/bar = \pm 0.001$; $\delta x_{SO_2} = \pm 1\%$. (estimated by compiler)

REFERENCES:

COMPONENTS :	ORIGINAL MEASUREMENTS:
1. Sulfur Dioxide; SO_2; [7446-09-5]	Tokunaga, J.
2. Water; H_2O; [7732-18-5]	*J. Chem. Engng. Data*, 1974, *19*, 162-165.
3. Methanol; CH_4O; [67-56-1]	

EXPERIMENTAL VALUES :

T/K	Mole fraction of alcohol in soln. (before absorption).	Total pressure P/bar	Partial pressure of sulfur dioxide P/bar	$S^{§}$	S_o^{+}
293.15	0.5798	1.095	1.019	45.87	45.58
	0.6694	1.097	1.018	53.02	52.74
	0.7638	1.096	1.014	63.57	63.44
	0.8996	1.089	1.002	75.33	76.15
	1.0000	1.102	1.012	86.11	86.19
303.15	0.0000	1.057	1.016	7.79	7.77
	0.0811	1.079	1.014	8.35	8.33
	0.1890	1.102	1.011	10.80	10.88
	0.2849	1.120	1.010	14.13	14.16
	0.4193	1.132	1.010	21.18	21.23
	0.4679	1.133	1.006	23.41	23.56
	0.6197	1.147	1.009	32.69	32.82
	0.6760	1.153	1.009	35.69	35.79
	0.7506	1.157	1.008	40.36	40.53
	0.8734	1.168	1.011	49.09	49.19
	0.9673	1.169	1.007	56.60	56.92
	1.0000	1.180	1.014	57.63	57.53
313.15	0.0000	1.086	1.013	5.76	5.76
	0.0830	1.125	1.011	6.26	6.26
	0.1402	1.153	1.016	7.22	7.20
	0.3185	1.201	1.012	11.98	11.98
	0.4041	1.221	1.015	14.88	14.85
	0.5332	1.242	1.017	20.94	20.85
	0.5822	1.240	1.007	22.21	22.33
	0.6354	1.232	1.010	24.69	24.75
	0.6740	1.262	1.016	26.33	26.23
	0.7957	1.271	1.010	31.96	32.05
	0.8400	1.288	1.011	33.56	33.62
	0.9165	1.290	1.013	37.46	37.44
	1.0000	1.301	1.013	40.82	40.80

$S^{§}$ - grams of sulfur dioxide per 100 grams of solution (solute gas free) at total pressure.

S_o^{+} - grams of sulfur dioxide per 100 grams of solution (solute gas free) at 1 atm. partial pressure.

Original gives pressure in units of mmHg

760 mmHg = 1 atm = 1.01325 bar

COMPONENTS:

(1) Sulfur dioxide, SO_2; [7466-09-5]
(2) Water, H_2O; [7732-18-5]
(3) Acetonitrile; C_2H_3N; [75-05-8]

ORIGINAL MEASUREMENTS:

Byerley, J.J.; Rempel, G.L.; Le, V.T.

J. Chem. Eng. Data. 1980, *25*, 55-56

VARIABLES:

T/K: 298.15, 323.15
Total p/kPa 101.3 (1 atm)
CH_3CN/wt% 0-100

PREPARED BY:

H.L. Clever

EXPERIMENTAL VALUES:

Liquid components CH_3CN Wt %	H_2O Wt %	298.15 K SO_2 solubility,[d] g/100g Single phase	Two phase CH_3CN rich	Two phase H_2O rich	323.15 K SO_2 solubility, g/100g single phase
0	100	8.24			3.77
5	95	8.37			-
10	90	9.03			3.95
15	85	10.0			-
20	80	b	a	a	5.08
30	70	b	59.4 (0.25)[c]	12.4 (0.74)[c]	6.85
40	60	b	59.4 (0.45)	12.4 (0.55)	8.33
50	50	b	59.4 (0.60)	12.4 (0.40)	11.09
60	40	b	59.4 (0.74)	12.4 (0.25)	13.75
70	30	b	59.4 (0.87)	12.4 (0.13)	15.60
80	20	b	a	a	18.67
90	10	73.8			23.10
95	5	78.7			-
100	0	84.6			25.64

a. Phase separation. ill defined reading to poor reproducibility.

b. Two phases.

c. Number in c) is the phase fraction. The total composition of the CH_3CN rich phase is 12.9 wt% H_2O, 37.3 wt % SO_2, and 49.8 wt % acetonitrile with density 0.975 g cm^{-3}. The water rich phase is 73.0 wt % H_2O, 11.0 wt % SO_2 and 16.0 wt % acetonitrile with density 1.005 g cm^{-3}

d. The SO_2 solubility is expressed as g SO_2/100 g of gas free solution.

AUXILIARY INFORMATION

METHOD/APPARATUS/PROCEDURE:

The mixed solvent was placed in a 50 ml water jacketted buret. Sulfur dioxide was bubbled at a rate of 60 ml m^{-1} from the bottom of the buret. A spiral condenser at the top prevented excessive loss of acetonitrile. The sulfur dioxide concentration after saturation was measured by titrating excess iodine with standard thiosulfate solution using starch as the indicator. In the cases where there was phase separation upon saturation with SO_2, the composition of both phases was determined by gas chromatography. The column, 0.91m long and 0.4 cm i.d., stainless steel was packed with 50-80 mesh Porapak Q. The carrier gas was a helium hydrogen mixture.

SOURCE AND PURITY OF MATERIALS:

1. Sulfur dioxide. Union Carbide Canada. Ltd. Anhydrous.

2. Water, distilled, carbon dioxide free.

3. Acetonitrile. J.T. Baker Co. Reagent grade. Used as received.

ESTIMATED ERROR:

$\delta T/K = 0.1$

REFERENCES:

COMPONENTS:

1. Sulfur Dioxide; SO_2; [7446-09-5]
2. Water; H_2O; [7732-18-5]
3. Ethanol; C_2H_6O; [64-17-5]

ORIGINAL MEASUREMENTS:

Tokunaga, J.

J. Chem. Engng. Data. 1974, *19*, 162-165.

VARIABLES:

Temperature

PREPARED BY:

C.L. Young

EXPERIMENTAL VALUES:

T/K	Mole fraction of alcohol in soln. (before adsorption).	Total Pressure P/bar	Partial pressure of sulfur dioxide P/bar	$S^{§}$	S_o^{+}
283.15	0.0000	1.024	1.012	15.66	15.66
	0.0628	1.029	1.012	14.44	14.45
	0.0986	1.035	1.016	14.87	14.82
	0.1722	1.036	1.014	20.18	20.15
	0.2423	1.036	1.014	26.88	26.84
	0.3325	1.037	1.013	35.76	35.73
	0.3827	1.039	1.015	39.73	39.62
	0.4728	1.038	1.014	47.06	46.98
	0.6010	1.038	1.015	57.08	56.94
	0.6650	1.043	1.019	60.83	60.46
	0.7955	1.036	1.014	65.55	65.44
	0.8730	1.034	1.012	67.60	67.61
	1.0000	1.034	1.014	70.58	70.50
293.15	0.0000	1.036	1.013	10.90	10.90
	0.0563	1.048	1.015	9.21	9.19
	0.0986	1.043	1.005	9.78	9.86
	0.2168	1.043	0.998	15.08	15.30
	0.2781	1.064	1.017	19.00	18.92
	0.3399	1.064	1.016	23.09	23.00
	0.4753	1.060	1.013	30.88	30.88
	0.5923	1.059	1.012	35.73	35.76
	0.6832	1.062	1.015	39.15	39.06

AUXILIARY INFORMATION

METHOD /APPARATUS/PROCEDURE:

Sulfur dioxide bubbled through aqueous solution. Saturated solution analysed by iodometric titration. Details in source.

The partial pressure of sulfur dioxide was calculated from the known vapor pressure of water + alcohol mixture and assuming Raoults law for the lowering of vapor pressure due to dissolution of sulfur dioxide

SOURCE AND PURITY OF MATERIALS:

1. High purity sample with purity of 99.96 mole per cent.
2. No details given.
3. Fractionated.

ESTIMATED ERROR:

$\delta T/K = \pm 0.1$; $\delta P/bar = \pm 0.001$; $\delta x_{SO_2} = \pm 1\%$. (estimated by compiler).

REFERENCES:

COMPONENTS :	ORIGINAL MEASUREMENTS :
1. Sulfur Dioxide; SO_2; [7446-09-5]	Tokunaga, J.
2. Water; H_2O; [7732-18-5]	*J. Chem. Engng. Data.* 1974, *19*, 162-165.
3. Ethanol; C_2H_6O; [64-17-5]	

EXPERIMENTAL VALUES:

T/K	Mole fraction of alcohol in soln. (before adsorption).	Total pressure P/bar	Partial pressure of sulfur dioxide P/bar	$S^§$	S_o^+
293.15	0.7927	1.059	1.014	42.37	42.33
	0.8941	1.059	1.014	44.91	44.83
	1.0000	1.059	1.016	46.95	46.81
303.15	0.0000	1.057	1.016	7.79	7.77
	0.0651	1.074	1.010	6.63	6.44
	0.1159	1.080	1.007	7.22	7.27
	0.1992	1.093	1.012	10.32	10.32
	0.3004	1.096	1.011	14.20	14.22
	0.3663	1.101	1.014	16.46	16.43
	0.4684	1.102	1.014	19.98	19.96
	0.5004	1.100	1.011	20.47	20.49
	0.6082	1.102	1.013	23.04	23.02
	0.7307	1.101	1.013	26.21	26.19
	0.8182	1.099	1.011	28.10	28.14
	0.8807	1.098	1.011	28.79	28.84
	0.9551	1.098	1.013	30.33	30.32
	1.0000	1.101	1.017	31.76	31.62
313.15	0.0000	1.086	1.013	5.76	5.76
	0.0644	1.118	1.007	5.29	5.32
	0.1035	1.135	1.012	5.99	6.09
	0.1900	1.156	1.019	7.99	7.94
	0.3000	1.157	1.011	10.01	10.02
	0.3982	1.159	1.009	12.60	12.64
	0.5111	1.159	1.004	15.07	15.19
	0.6055	1.168	1.013	17.13	17.12
	0.7294	1.167	1.011	18.30	18.32
	0.8205	1.169	1.014	20.33	20.31
	0.9152	1.167	1.013	21.69	21.68
	1.0000	1.164	1.012	23.80	23.80

$S^§$ - grams of sulfur dioxide per 100 grams of solution (solute gas free) at total pressure.

S_o^+ - grams of sulfur dioxide per 100 grams of solution (solute gas free) at 1 atm. partial pressure.

Original gives pressure in units of mmHg

760 mmHg = 1 atm = 1.01325 bar

COMPONENTS:

1. Sulfur dioxide; SO_2; [7746-09-5]
2. Water; H_2O; [7732-18-5]
3. Sulfinybismethane, (Dimethylsulfoxide); C_2H_6OS; [67-68-5]

ORIGINAL MEASUREMENTS:

Smedslund, T.H.

Finska Kemistsamfundets Medd. 1950, *59*, 40-43.

VARIABLES:

Temperature

PREPARED BY:

W. Gerrard

EXPERIMENTAL VALUES:

Dimethylsulfoxide containing "10%" water.

T/K	SO_2, Weight %	Presumably for total pressure of 1 atm.
292.15	54.3	
350.15	22.1	
408.15	6.0	
420.15	2.5	
293.15	5.2	p_{SO_2} = 11.2 mmHg
293.15	12.1	p_{SO_2} = 31 mmHg
293.15	16.1	p_{SO_2} = 46 mmHg

AUXILIARY INFORMATION

METHOD/APPARATUS/PROCEDURE:

The first set of data are for saturation at 1 atm; details are not available for the second set.

SOURCE AND PURITY OF MATERIALS:

(1) Information not available

(2) Dimethylsulfoxide was prepared by the catalytic oxidation of dimethyl sulfide in the vapor state (1). The purified product had b.p. 63°C/6 mmHg.

ESTIMATED ERROR:

REFERENCES:

1. Smedslund, T.H. *U.S. Patent*, 2,581,050, 1952

COMPONENTS:

1. Sulfur dioxide; SO_2; [7446-09-5]
2. Water; H_2O; [7732-18-5]
3. 1-Propanol; C_3H_8O; [71-23-8]

ORIGINAL MEASUREMENTS:

Tokunaga, J.

J. Chem. Engng. Data. 1974, *19*, 162-165.

VARIABLES:

Temperature

PREPARED BY:

C.L. Young

EXPERIMENTAL VALUES:

T/K	Mole fraction of alcohol in soln. (before adsorption).	Total Pressure P/bar	Partial Pressure of sulfur dioxide P/bar	$S^{\S}$	S_o^+
283.15	0.0000	1.024	1.024	15.66	15.66
	0.0124	1.030	1.017	14.59	14.52
	0.0352	1.027	1.013	14.28	14.28
	0.0651	1.026	1.011	15.51	15.53
	0.1036	1.030	1.011	19.02	19.04
	0.2000	1.028	1.012	26.56	26.58
	0.3026	1.027	1.012	33.27	33.29
	0.3960	1.027	1.013	38.36	38.35
	0.4986	1.027	1.013	42.09	42.06
	0.5927	1.027	1.014	45.54	45.46
	0.7068	1.025	1.013	47.80	47.77
	0.8160	1.023	1.012	48.96	48.97
	0.9286	1.023	1.014	49.52	49.43
	1.0000	1.020	1.013	49.63	49.60
293.15	0.0000	1.036	1.013	10.90	10.90
	0.0218	1.040	1.013	9.96	9.95
	0.0501	1.042	1.013	10.16	10.15
	0.0922	1.045	1.014	12.24	12.22
	0.1882	1.051	1.019	17.22	17.11
	0.2172	1.044	1.013	18.21	18.21
	0.2967	1.044	1.014	21.04	21.02

AUXILIARY INFORMATION

METHOD /APPARATUS/PROCEDURE:

Sulfur dioxide bubbled through aqueous solution. Saturated solution analysed by iodometric titration. Details in source.

The partial pressure of sulfur dioxide was calculated from the known vapor pressure of water + alcohol mixture and assuming Raoults law for the lowering of vapor pressure due to dissolution of sulfur dioxide.

SOURCE AND PURITY OF MATERIALS:

1. High purity sample with purity of 99.96 mole per cent.
2. No details given.
3. Fractionated.

ESTIMATED ERROR:

$\delta T/K = \pm 0.1$; $\delta P/bar = \pm 0.001$; $\delta x_{SO_2} = \pm 1\%$. (estimated by compiler).

REFERENCES:

COMPONENTS :	ORIGINAL MEASUREMENTS :
1. Sulfur dioxide; SO_2; [7446-09-5]	Tokunaga, J.
2. Water; H_2O; [7732-18-5]	*J. Chem. Engng. Data.* 1974, *19*, 162-165
3. 1-Propanol; C_3H_8O; [71-23-8]	

EXPERIMENTAL VALUES :

T/K	Mole fraction of alcohol in soln. (before adsorption).	Total pressure P/bar	Partial pressure of sulfur dioxide P/bar	$S^{\S}$	S_o^+
293.15	0.3980	1.044	1.014	24.28	24.24
	0.4846	1.043	1.014	26.55	26.50
	0.5960	1.041	1.013	28.47	28.47
	0.6948	1.038	1.013	29.38	29.39
	0.7860	1.036	1.013	30.27	30.27
	0.8941	1.033	1.013	30.37	30.36
	1.0000	1.027	1.012	30.68	30.68
303.15	0.0000	1.057	1.017	7.79	7.77
	0.0223	1.064	1.014	7.34	7.34
	0.0516	1.069	1.014	8.04	8.03
	0.0761	1.071	1.012	8.79	8.80
	0.1964	1.073	1.014	12.78	12.77
	0.3009	1.058	1.000	15.28	15.49
	0.3976	1.071	1.014	17.67	17.66
	0.4900	1.068	1.012	18.98	18.99
	0.5984	1.066	1.012	20.43	20.45
	0.7077	1.065	1.014	21.32	21.29
	0.7732	1.061	1.013	21.53	21.53
	0.9026	1.052	1.013	21.91	21.92
	1.0000	1.044	1.013	22.05	22.06
313.15	0.0000	1.087	1.014	5.76	5.76
	0.0186	1.085	1.009	5.57	5.60
	0.0355	1.107	1.012	5.59	5.60
	0.0763	1.116	1.016	6.49	6.47
	0.1043	1.119	1.014	7.21	7.20
	0.1974	1.119	1.014	9.30	9.30
	0.2992	1.118	1.013	11.22	11.22
	0.3950	1.116	1.013	12.39	12.39
	0.4994	1.116	1.015	13.60	13.58
	0.6068	1.112	1.014	14.53	14.52
	0.7108	1.106	1.014	15.15	15.14
	0.7824	1.097	1.010	15.54	15.59
	0.9112	1.088	1.015	15.69	15.66
	1.0000	1.071	1.013	15.95	15.95

$S^{\S}$ - grams of sulfur dioxide per 100 grams of solution (solute gas free) at total pressure.

S_o^+ - grams of sulfur dioxide per 100 grams of solution (solute gas free) at 1 atm. partial pressure.

Original gives pressure in units of mmHg

760 mmHg = 1 atm = 1.01325 bar

COMPONENTS:

1. Sulfur dioxide; SO_2; [7446-09-5]

2. Alkanes

Colin L. Young,
School of Chemistry,
University of Melbourne,
Parkville, Victoria 3052,
Australia.

November 1980

CRITICAL EVALUATION:

Although these systems have been investigated by several workers, there are very serious discrepancies between solubilities reported by different authors. The most extensive studies are those of Makranczy *et al.* (1) and Sololev *et al.* (2) and values of their mole fraction solubilities at 298.15 K and 101.3 kPa are contrasted below.

Alkane	x_{SO_2} (Sololev *et al.*)	x_{SO_2} (Makranczy *et al.*)
Octane	0.0319	0.0586
Dodecane	0.0378	0.0662
Hexadecane	-	0.0745
Heptadecane	0.0372	-

The evaluator is of the opinion that the values of Sololev *et al.* (2) are probably in error whereas those of Makranczy *et al.* (1) are moderately reliable. The value of Gerrard (3) for decane at 293.15 K and 273.15 K extrapolate to 298.15 K to give a value of about 35% higher than the value of Makranczy *et al.* (1). The mole fraction solubilities for hexadecane of Tremper and Prausnitz (4), Lenoir *et al.* (5), Sano (6) and Makranczy *et al.* (1) agree within about 15% whereas the value of Sololev *et al.* (2) is approximately have of the average value of the other workers.

Therefore the values of Sololev *et al.* (2) and Gerrard (3) are classified as doubtful whereas those of Makranczy *et al.* (1), Lenoir *et al.* (5), Sano (6) and Tremper and Prausnitz (4) are classified as tentative.

The data of Makranczy *et al.* for the alkanes can be represented as a function of carbon number, C_n, by the equation

$$\ln x_{SO_2} = 0.02933C_n - 3.068$$

well within the experimental error of 3 per cent. A plot of $\ln x_{SO_2}$ against C_n is given on the following page.

The value of Lenoir *et al.* (5) was determined from gas chromatographic retention volume data and is effectively at infinite dilution of sulfur dioxide. The extrapolation to 101.3 kPa pressure assumes Henry's law holds up to that pressure and consequently involves considerable approximation.

The data presented by Komissarov *et al.* (7) on the system sulfur dioxide + tetradecane appears to be greatly in error, the mole fraction solubility calculated from their data being about double the value of Makranczy *et al.* (1).

References

1. Makranczy, J.; Megyerg-Balog, K.; Rusz, L.; Patyi, L. *Hung. J. Ind. Chem.* 1976, *4*, 269.

2. Sololev, I. A.; Kukarin, V. A.; Dzhagatspanyan, R. V.; Kosorotov, V. I.; Zogarets, P. A.; Popov, A. I. *Khim. Prom.* 1970, *46*, 668.

3. Gerrard, W. *Solubility of Gases and Liquids*, Plenum Press, New York, 1976, p.239.

4. Tremper, K. K.; Prausnitz, J. M. *J. Chem. Engng. Data* 1976, *21*, 295.

5. Lenoir, J. Y.; Renault, P.; Renon, H. *J. Chem. Engng. Data* 1971, *16*, 340.

6. Sano, H. *Nippon Kagaku Zasshi* 1968, *89*, 362.

7. Komissarov, V. D.; Saitova, M. A.; Timirova, R. G. *Khim. Prom.* 1973, *49*, 738.

COMPONENTS:	EVALUATOR:
1. Sulfur dioxide; SO_2; [7446-09-5] 2. Alkanes	Colin L. Young, School of Chemistry, University of Melbourne, Parkville, Victoria 3052, Australia. November 1980

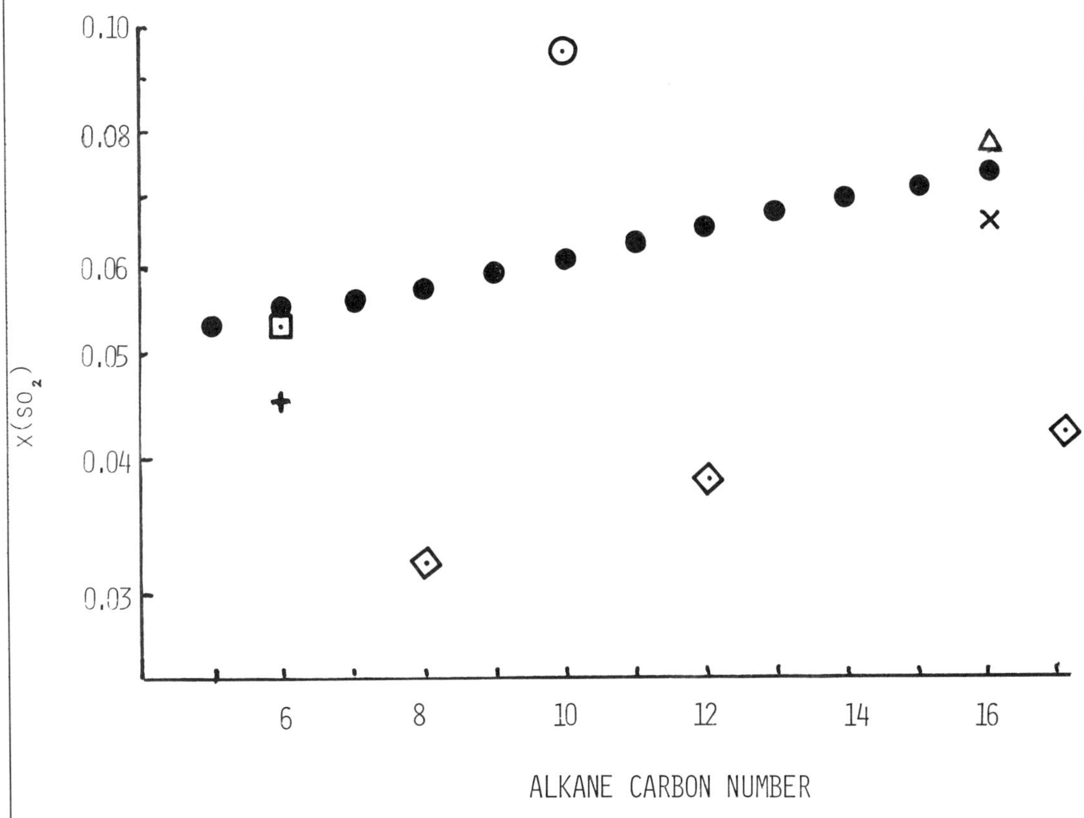

Figure 1. Comparison of mole fraction solubilities for sulfur dioxide in alkanes.

- ● Makranczy, Megyery-Balog, Rusz and Patyi (1)
- ⊙ Gerard (3)
- ◈ Sololev, Kukarin, Dzhagatspanyan, Kosorotov, Zogarets and Popov (2)
- ⊡ Patyi, Furmer, Makranczy, Sadilenko, Stepanova and Berengarten
- △ Lenoir, Renault, Renon (5)
- × Tremper and Prausnitz (4)
- \+ Sano (6)

COMPONENTS:

1. Sulfur Dioxide; SO_2; [7446-09-5];
2. Pentane; C_5H_{12}; [109-66-0]

ORIGINAL MEASUREMENTS:

Makranczy, J.; Megyery-Balog, K.; Rusz, L.; Patyi, L.

Hung. J. Ind. Chem. 1976, *4(2)*, 269-280.

VARIABLES:

T/K: 298.15 - 313.15

P/kPa: 101.325 (1 atm)

PREPARED BY:

S. A. Johnson/ H. L. Clever

EXPERIMENTAL VALUES:

T/K	Mole fraction x_{SO_2}	Bunsen coefficient α	Ostwald coefficient L
298.15	0.0543	10.880	11.876
313.15	0.0402	7.76	8.90

Mole fractions and Bunsen coefficients were calculated by the compiler.

AUXILIARY INFORMATION

METHOD /APPARATUS/PROCEDURE:

Apparently the volumetric apparatus described in ref.(1) was modified for use at temperatures above 0° C. The apparatus was designed to be operated at a partial pressure of sulfur dioxide of 760 torr.

SOURCE AND PURITY OF MATERIALS:

Analytical grade reagents of Hungarian and foreign origin were used (both liquids and gases). No further information.

ESTIMATED ERROR:

$\delta x_{SO_2}/x_{SO_2}$ = ±0.03.

REFERENCES:

1. Bodor, E.; Bor, Gy.; Mohai, B.; Sipos, G. *Veszprémi Vegyipari Egyetem Köslemények* 1957, *1*, 55; *CA 55*, 3175h.

COMPONENTS:

1. Sulfur dioxide; SO_2; [7446-09-5]
2. Hexane; C_6H_{14}; [110-54-3]

ORIGINAL MEASUREMENTS:

Sano, H.

Nippon Kagaku Zasshi

1968, *89*, 362-368.

VARIABLES:

PREPARED BY:

C. L. Young

EXPERIMENTAL VALUES:

$T/°C = 30$ $T/K = 303$

Vapor pressure of hexane, $p_s^\circ = 0.240$ atm

Absorption coefficient, $S^{a,b}$	Bunsen coefficient,[b] α	Bunsen coefficient,[c] α	$\ell^{c,d}$	Mole ratio	Mole fraction[e]
6.8	8.1	8.2	6.5	0.047	0.0449

[a] Volume of sulfur dioxide absorbed by unit volume of solvent at experimental temperature at a total pressure of 1 atmosphere.

[b] Determined by static absorption method.

[c] Determined by flow method.

[d] $\alpha = \ell(1 - p_s^\circ)$ where p_s° is vapor pressure of solvent.

[e] Calculated by compiler.

AUXILIARY INFORMATION

METHOD/APPARATUS/PROCEDURE:

Two techniques were used: (a) static absorption method and (b) flow method. Static method consisted of equilibrating solvent and sulfur dioxide in a 50 ml flask and measuring the absorption volumetrically. The flow method consists of passing sulfur dioxide through a micro gas absorption tube and estimating sulfur dioxide by iodometric titration. The vapor pressure of the solvents were measured roughly and were assumed to be equal to the partial pressure of the solvent by the authors when calculating the Bunsen coefficient.

SOURCE AND PURITY OF MATERIALS:

1. Purity 99.0-99.8 mole per cent.

ESTIMATED ERROR:

$\delta T/K = \pm 0.5$; $\delta x_{SO_2} = \pm 5\%$

(estimated by compiler).

REFERENCES:

COMPONENTS:

1. Sulfur Dioxide; SO_2; [7446-09-5]

2. Hexane; C_6H_{14}; [110-54-3]

ORIGINAL MEASUREMENTS:

Makranczy, J.; Megyery-Balog, K.; Rusz, L.; Patyi, L.

Hung. J. Ind. Chem. 1976, *4(2)*, 269-280.

VARIABLES:

T/K: 298.15 - 313.15

P/kPa: 101.325 (1 atm)

PREPARED BY:

S. A. Johnson/ H. L. Clever

EXPERIMENTAL VALUES:

T/K	Mole fraction x_{SO_2}	Bunsen coefficient α	Ostwald coefficient L
298.15	0.0562	9.964	10.876
313.15	0.0408	7.00	8.02

Mole fractions and Bunsen coefficients were calculated by the compiler.

AUXILIARY INFORMATION

METHOD/APPARATUS/PROCEDURE:

Apparently the volumetric apparatus described in ref.(1) was modified for use at temperatures above 0°C. The apparatus was designed to be operated at a partial pressure of sulfur dioxide of 760 torr.

SOURCE AND PURITY OF MATERIALS:

Analytical grade reagents of Hungarian and foreign origin were used (both liquids and gases). No further information.

DATA CLASS:

ESTIMATED ERROR:

$\delta x_{SO_2}/x_{SO_2} = \pm 0.03$.

REFERENCES:

1. Bodor, E.; Bor, Gy.; Mohai, B.; Sipos, G. *Veszprémi Vegyipari Egyetem Köslemén yei* 1957, *1*, 55; *CA 55*, 3175h.

COMPONENTS:

1. Sulfur dioxide; SO_2; [7446-09-5]
2. Hexane; C_6H_{14}; [110-54-3]

ORIGINAL MEASUREMENTS:

Patyi, L.; Furmer, I. E. Makranczy, J.; Sadilenko, A. S.; Stepanova, Z. G.; Berengarten, M. G.

Zh. Prikl. Khim. 1978, *51*, 1296-1300.

VARIABLES:

PREPARED BY:

C. L. Young

EXPERIMENTAL VALUES:

T/K	α†	Mole fraction of sulfur dioxide at a partial pressure of 101.325 kPa x_{SO_2}
298.15	9.38	0.05336

† volume of gas (measured at 101.325 kPa and 273.15 K) dissolved by one volume of hexane.

AUXILIARY INFORMATION

METHOD/APPARATUS/PROCEDURE:

Volumetric method. Pressure measured when known amounts of gas were added, in increments, to a known amount of liquid in a vessel of known dimensions. Corrections were made for the partial pressure of solvent. Details in ref. (1).

SOURCE AND PURITY OF MATERIALS:

Purity better than 99 mole per cent as determined by gas chromatography.

ESTIMATED ERROR:

$\delta T/K = \pm 0.1$; $\delta\alpha = \pm 4\%$ or less.

REFERENCES:

1. Bodor, E.; Bor, G. J.; Mohai, B.; Sipos, G. *Veszpremi. Vegyip. Egy. Kosl.* 1957, *1*, 55.

COMPONENTS:	ORIGINAL MEASUREMENTS:
1. Sulfur Dioxide; SO_2; [7446-09-5] 2. Heptane; C_7H_{16}; [142-82-5]	Makranczy, J.; Megyery-Balog, K.; Rusz, L.; Patyi, L. *Hung. J. Ind. Chem.* 1976, *4(2)*, 269-280.
VARIABLES:	PREPARED BY:
T/K: 298.15 - 313.15 *P*/kPa: 101.325 (1 atm)	S. A. Johnson / H. L. Clever

EXPERIMENTAL VALUES:

T/K	Mole fraction x_{SO_2}	Bunsen coefficient α	Ostwald coefficient L
298.15	0.0567	8.969	9.790
313.15	0.0417	6.40	7.34

Mole fractions and Bunsen coefficients were calculated by the compiler.

AUXILIARY INFORMATION

METHOD/APPARATUS/PROCEDURE:

Apparently the volumetric apparatus described in ref. (1) was modified for use at temperatures above 0°C. The apparatus was designed to be operated at a partial pressure of sulfur dioxide of 760 torr.

SOURCE AND PURITY OF MATERIALS:

Analytical grade reagents of Hungarian and foreign origin were used (both liquids and gases). No further information.

ESTIMATED ERROR:

$\delta x_{SO_2}/x_{SO_2}$ = ±0.03.

REFERENCES:

1. Bodor, E.; Bor, Gy.; Mohai, B.; Sipos, G. *Veszprémi Vegyipari Egyetem Köslemények* 1957, *1*, 55; *CA 55*, 3175h.

COMPONENTS:

1. Sulfur dioxide; SO_2; [7446-09-5]
2. Octane; C_8H_{18}; [111-65-9]

ORIGINAL MEASUREMENTS:

Sobolev, I.A.; Kukarin, V.A.; Dzhagatspanyan, R.V.; Kosorotov, V.I. Zogorets, P.A.; Popov, A.I.

Khim. Prom. 1970, *46*, 668-70.

VARIABLES:

Temperature.

PREPARED BY:

C.L. Young.

EXPERIMENTAL VALUES:

T/K	Solubility,* S /mol l^{-1}	Mole fraction of sulfur dioxide in in liquid, x_{SO_2} +
283.15	0.283	0.044
288.15	0.254	0.040
293.15	0.227	0.036
298.15	0.201	0.032
303.15	0.182	0.029

\+ calculated by compiler.

* at atmospheric pressure.

$\ln S = 830/(T/K) - 6.35$ (where S is in units of mol l^{-1} atm^{-1}).

AUXILIARY INFORMATION

METHOD/APPARATUS/PROCEDURE:

Bubbler method. Samples of saturated liquid taken and added to sodium hydroxide solution and then excess alkali back titrated.

SOURCE AND PURITY OF MATERIALS:

1. Purity at least 95 mole per cent.
2. Chemically pure. Dried.

ESTIMATED ERROR:

$\delta T/K = \pm 0.1$; δ(Solubility) = $\pm 2\%$; $\delta x_{SO_2} = \pm 0.001$, (estimated by compiler).

REFERENCES:

COMPONENTS:

1. Sulfur Dioxide; SO_2; [7446-09-5]

2. Octane; C_8H_{18}; [111-65-9]

ORIGINAL MEASUREMENTS:

Makranczy, J.; Megyery-Balog, K.; Rusz, L.; Patyi, L.
Hung. J. Ind. Chem. 1976, *4(2)*, 269-280.

VARIABLES:

T/K: 298.15 - 313.15
P/kPa: 101.325 (1 atm)

PREPARED BY:

S. A. Johnson / H. L. Clever

EXPERIMENTAL VALUES:

T/K	Mole fraction x_{SO_2}	Bunsen coefficient α	Ostwald coefficient L
298.15	0.0583	8.330	9.092
313.15	0.0429	5.95	6.82

Mole fractions and Bunsen coefficients were calculated by the compiler.

AUXILIARY INFORMATION

METHOD /APPARATUS/PROCEDURE:

Apparently the volumetric apparatus described in ref.(1) was modified for use at temperatures above 0°C. The apparatus was designed to be operated at a partial pressure of sulfur dioxide of 760 torr.

SOURCE AND PURITY OF MATERIALS:

Analytical grade reagents of Hungarian and foreign origin were used (both liquids and gases). No further information.

ESTIMATED ERROR:

$\delta x_{SO_2}/x_{SO_2}$ = ±0.03.

REFERENCES:

1. Bodor, E.; Bor, Gy.; Mohai, B.; Sipos, G. *Veszprémi Vegyipari Egyetem Közlemenyei* 1957, *1*, 55; *CA 55*, 3175h.

COMPONENTS:

1. Sulfur dioxide; SO_2; [7446-09-5]
2. 2,2,4-Trimethylpentane; C_8H_{18}; [540-84-1]

ORIGINAL MEASUREMENTS:

Sobolev, I.A.; Kukarin, V.A.; Dzhagatspanyan, R.V.; Kosorotov, V.I; Zogorets, P.A.; Popov, A.I.

Khim. Prom. 1970, *46*, 668-70.

VARIABLES:

PREPARED BY:

C.L. Young

EXPERIMENTAL VALUES:

T/K	Solubility*, S /mol l^{-1}	Mole fraction of sulfur dioxide[+] x_{SO_2}
293.15	0.223	0.0355

* at atmospheric pressure

\+ calculated by compiler

AUXILIARY INFORMATION

METHOD/APPARATUS/PROCEDURE:

Bubbler method. Samples of saturated liquid taken and added to sodium hydroxide solution and then excess alkali back titrated

SOURCE AND PURITY OF MATERIALS:

1. Purity at least 95 mole per cent.
2. Chemically pure. Dried.

ESTIMATED ERROR:

$\delta T/K = \pm 0.1$; δ(Solubility) = $\pm 2\%$; $\delta x_{SO_2} = \pm 0.001$; (estimated by compiler).

REFERENCES:

COMPONENTS:

1. Sulfur dioxide; SO_2; [7446-09-5]

2. 2,2,4-Trimethylpentane; C_8H_{18}; [540-84-1]

ORIGINAL MEASUREMENTS:

Nitta, T.; Kido. O.; Katayama, T.;

J. Chem. Engng. Japan, 1976, *9*, 317-8.

VARIABLES:

Pressure

PREPARED BY:

C.L. Young

EXPERIMENTAL VALUES:

T/K	*P*/bar	Mole fraction of sulfur dioxide in liquid, x_{SO_2}
293.15	0.7021	0.0445
	1.031	0.0710
	1.516	0.1184
	2.042	0.1847
	2.521	0.2706
	2.796	0.3362
	3.170	0.6412
	3.193	0.7897
	3.200	0.8506

AUXILIARY INFORMATION

METHOD:

Static method liquid phase composition estimated by iodometry. Few details given in source. Method similar to that used in ref. (1).

SOURCE AND PURITY OF MATERIALS:

1. Seitetsu Chemical Co.sample, purity better than 99.9 mole per cent.

2. Nakarai Chemicals Ltd. sample, guaranteed grade.

ESTIMATED ERROR:

$\delta T/K = \pm 0.1$; $\delta P/\text{bar} = \pm 0.001$ (below 1.3 bar) ± 0.005 (above 1.3 bar); $\delta x_{SO_2} = 1.5\%$

REFERENCES:

Quitzich, K.; Ulbrecht, H.; Geiseler, G.

Z. Physik Chem. 1967, *234*, 33.

COMPONENTS:

1. Sulfur Dioxide; SO_2; [7446-09-5]

2. Nonane; C_9H_{20}; [111-84-2]

ORIGINAL MEASUREMENTS:

Makranczy, J.; Megyery-Balog, K.; Rusz, L.; Patyi, L.

Hung. J. Ind. Chem. 1976, *4(2)*, 269-280.

VARIABLES:

T/K: 298.15 - 313.15

P/kPa: 101.325 (1 atm)

PREPARED BY:

S. A. Johnson / H. L. Clever

EXPERIMENTAL VALUES:

T/K	Mole fraction x_{SO_2}	Bunsen coefficient α	Ostwald coefficient L
298.15	0.0602	7.840	8.558
313.15	0.0441	5.58	6.40

Mole fractions and Bunsen coefficients were calculated by the compiler.

AUXILIARY INFORMATION

METHOD /APPARATUS/PROCEDURE:

Apparently the volumetric apparatus described in ref.(1) was modified for use at temperatures above 0°C. The apparatus was designed to be operated at a partial pressure of sulfur dioxide of 760 torr.

SOURCE AND PURITY OF MATERIALS:

Analytical grade reagents of Hungarian and foreign origin were used (both liquids and gases). No further information.

ESTIMATED ERROR:

$\delta x_{SO_2}/x_{SO_2} = \pm 0.03$.

REFERENCES:

1. Bodor, E.; Bor, Gy.; Mohai, B.; Sipos, G. *Veszprémi Vegyipari Egyetem Közleményei* 1957, *1*, 55; *CA 55*, 3175h.

COMPONENTS:

1. Sulfur Dioxide; SO_2; [7446-09-5]

2. Decane; $C_{10}H_{22}$; [124-18-5]

ORIGINAL MEASUREMENTS:

Makranczy, J.; Megyery-Balog, K.; Rusz, L.; Patyi, L.
Hung. J. Ind. Chem. 1976, *4(2)*, 269-280.

VARIABLES:

T/K: 298.15 - 313.15
P/kPa: 101.325 (1 atm)

PREPARED BY:

S. A. Johnson/ H. L. Clever

EXPERIMENTAL VALUES:

T/K	Mole fraction x_{SO_2}	Bunsen coefficient α	Ostwald coefficient L
298.15	0.0620	7.430	8.110
313.15	0.0456	5.30	6.08

Mole fractions and Bunsen coefficients were calculated by the compiler.

AUXILIARY INFORMATION

METHOD /APPARATUS/PROCEDURE:

Apparently the volumetric apparatus described in ref.(1) was modified for use at temperatures above 0°C. The apparatus was designed to be operated at a partial pressure of sulfur dioxide of 760 torr.

SOURCE AND PURITY OF MATERIALS:

Analytical grade reagents of Hungarian and foreign origin were used (both liquids and gases). No further information.

ESTIMATED ERROR:

$\delta x_{SO_2}/x_{SO_2} = \pm 0.03$.

REFERENCES:

1. Bodor, E.; Bor, Gy.; Mohai, B.; Sipos, G. *Veszprémi Vegyipari Egyetem Közleményei* 1957, *1*, 55; *CA 55*, 3175h.

COMPONENTS:

1. Sulfur dioxide; SO_2; [7446-09-5]

2. Decane; $C_{10}H_{22}$; [124-18-5] or Decahydronaphthalene; (Decalin); $C_{10}H_{18}$; [91-17-8] or 1,2,3,4-Tetrahydronaphthalene; (Tetralin); $C_{10}H_{12}$; [119-64-2]

ORIGINAL MEASUREMENTS:

Gerrard, W. *"Solubility of Gases and Liquids,"* Plenum Press, New York, 1976, p.239.

VARIABLES:

PREPARED BY:

W. Gerrard.

EXPERIMENTAL VALUES:

T/K	Weight of Component (2) g	Weight of So$_2$ absorbed, g	Mole ratio SO_2 /Comp. (2)	Mole fraction x_{SO_2}
		Decane	(Total pressure, 1 atm)	
273.15	3.8122	0.3040	0.179	0.151
293.15	3.8122	1.7832	0.105	0.095
		(cis-trans) Decalin	(Total pressure, 1 atm)	
293.15	7.2121	0.2090	0.062	0.060
		Tetralin	(Total pressure, 1 atm)	
293.15	7.5100	1.0732	0.295	0.228

760 mmHg = 1 atm = 101.325 kPa

AUXILIARY INFORMATION

METHOD:

Sulfur dioxide was bubbled into a weighed amount of component (2) in a bubbler tube. The amount of gas absorbed was determined by re-weighing the tube. The total pressure was barometric, very nearly 1 atm. See ref. (1).

SOURCE AND PURITY OF MATERIALS:

1. Sulfur dioxide was the best grade contained in a small cylinder. Its high degree of purity was attested by the measurement of vapor pressure from 258-268 K.

2. The best commercial specimen was purified and attested.

ESTIMATED ERROR:

REFERENCES:

1. Gerrard, W. *J. Appl. Chem. Biotechnol.* 1972, *22*, 623.

COMPONENTS:

1. Sulfur Dioxide; SO_2; [7446-09-5]

2. Undecane; $C_{11}H_{24}$; [1120-21-4]

ORIGINAL MEASUREMENTS:

Makranczy, J.; Megyery-Balog, K.; Rusz, L.; Patyi, L.
Hung. J. Ind. Chem. 1976, *4(2)*, 269-280.

VARIABLES:

T/K: 298.15 - 313.15
P/kPa: 101.325 (1 atm)

PREPARED BY:

S. A. Johnson / H. L. Clever

EXPERIMENTAL VALUES:

T/K	Mole fraction x_{SO_2}	Bunsen coefficient α	Ostwald coefficient L
298.15	0.0647	7.170	7.826
313.15	0.0471	5.07	5.81

Mole fractions and Bunsen coefficients were calculated by the compiler.

AUXILIARY INFORMATION

METHOD/APPARATUS/PROCEDURE:

Apparently the volumetric apparatus described in ref.(1) was modified for use at temperatures above 0°C. The apparatus was designed to be operated at a partial pressure of sulfur dioxide of 760 torr.

SOURCE AND PURITY OF MATERIALS:

Analytical grade reagents of Hungarian and foreign origin were used (both liquids and gases). No further information.

ESTIMATED ERROR:

$\delta x_{SO_2}/x_{SO_2} = \pm 0.03.$

REFERENCES:

1. Bodor, E.; Bor, Gy.; Mohai, B.; Sipos, G. *Veszprêmi Vegyipari Egyetem Kösleményei* 1957, *1*, 55; *CA 55*, 3175h.

COMPONENTS:	ORIGINAL MEASUREMENTS:
1. Sulfur dioxide; SO_2; [7446-09-5] 2. Dodecane; $C_{12}H_{26}$; [112-40-3]	Sobolev, I.A.; Kukarin, V.A.; Dzhagatspanyan, R.V.; Kosorotov, V.I. Zogorets, P.A.; Popov, A.I. *Khim. Prom.* 1970, *46*, 668-670.

VARIABLES:	PREPARED BY:
Temperature	C.L. Young

EXPERIMENTAL VALUES:

T/K	Solubility*, S /mol l^{-1}	Mole fraction of+ sulfur dioxide in liquid, x_{SO_2}
283.15	0.243	0.052
288.15	0.214	0.046
293.15	0.191	0.042
298.15	0.172	0.038
303.15	0.153	0.034

\+ calculated by compiler.

* at atmospheric pressure (authors claim linear relationship between pressure and solubility up to 1 atmosphere partial pressure of SO_2)

$\ln S = 852 / (T/K) - 6.50$ (where S is in units of mol l^{-1} atm^{-1})

AUXILIARY INFORMATION

METHOD/APPARATUS/PROCEDURE:

Bubbler method. Samples of saturated liquid taken and added to sodium hydroxide solution and then excess alkali back titrated

SOURCE AND PURITY OF MATERIALS:

1. Purity at least 95 mole per cent.
2. Chemically pure. Dried.

ESTIMATED ERROR:

$\delta T/K = \pm 0.1$; δ(Solubility) = ±2%; $\delta x_{SO_2} = \pm 0.001$; (estimated by compiler).

REFERENCES:

COMPONENTS:

1. Sulfur Dioxide; SO_2; [7446-09-5]

2. Dodecane; $C_{12}H_{26}$; [112-40-3]

ORIGINAL MEASUREMENTS:

Makranczy, J.; Magyery-Balog, K.; Rusz, L.; Patyi, L.
Hung. J. Ind. Chem. 1976, *4(2)*, 269-280.

VARIABLES:

T/K: 298.15 - 313.15
P/kPa: 101.325 (1 atm)

PREPARED BY:

S. A. Johnson / H. L. Clever

EXPERIMENTAL VALUES:

T/K	Mole fraction x_{SO_2}	Bunsen coefficient α	Ostwald coefficient L
298.15	0.0662	6.820	7.444
313.15	0.0484	4.85	5.56

Mole fractions and Bunsen coefficients were calculated by the compiler.

AUXILIARY INFORMATION

METHOD /APPARATUS/PROCEDURE:

Apparently the volumetric apparatus described in ref.(1) was modified for use at temperatures above 0°C. The apparatus was designed to be operated at a partial pressure of sulfur dioxide of 760 torr.

SOURCE AND PURITY OF MATERIALS:

Analytical grade reagents of Hungarian and foreign origin were used (both liquids and gases). No further information.

ESTIMATED ERROR:

$\delta x_{SO_2}/x_{SO_2} = \pm 0.03$.

REFERENCES:

1. Bodor, E.; Bor, Gy.; Mohai, B.; Sipos, G. *Veszprémi Vegyipari Egyetem Közleményei* 1957, *1*, 55; *CA 55*, 3175h.

COMPONENTS:

1. Sulfur Dioxide; SO_2; [7446-09-5]

2. Tridecane; $C_{13}H_{28}$; [629-50-5]

ORIGINAL MEASUREMENTS:

Makranczy, J.; Megyery-Balog, K.; Rusz, L.; Patyi, L.

Hung. J. Ind. Chem. 1976, *4(2)*, 269-280.

VARIABLES:

T/K: 298.15 - 313.15

P/kPa: 101.325 (1 atm)

PREPARED BY:

S. A. Johnson/ H. L. Clever

EXPERIMENTAL VALUES:

T/K	Mole fraction x_{SO_2}	Bunsen coefficient α	Ostwald coefficient L
298.15	0.0682	6.580	7.182
313.15	0.0500	4.68	5.36

Mole fractions and Bunsen coefficients were calculated by the compiler.

AUXILIARY INFORMATION

METHOD/APPARATUS/PROCEDURE:

Apparently the volumetric apparatus described in ref.(1) was modified for use at temperatures above 0°C. The apparatus was designed to be operated at a partial pressure of sulfur dioxide of 760 torr.

SOURCE AND PURITY OF MATERIALS:

Analytical grade reagents of Hungarian and foreign origin were used (both liquids and gases). No further information.

ESTIMATED ERROR:

$\delta x_{SO_2}/x_{SO_2} = \pm 0.03.$

REFERENCES:

1. Bodor, E.; Bor, Gy.; Mohai, B.; Sipos, G. *Veszprémi Vegyipari Egyetem Közleményei* 1957, *1*, 55; *CA 55*, 3175h.

COMPONENTS:

1. Sulfur Dioxide; SO_2; [7446-09-5]

2. Tetradecane; $C_{14}H_{30}$; [629-59-4]

ORIGINAL MEASUREMENTS:

Makranczy, J.; Megyery-Balog, K.; Rusz, L.; Patyi, L.
Hung. J. Ind. Chem. 1976, *4(2)*, 269-280.

VARIABLES:

T/K: 298.15 - 313.15
P/kPa: 101.325 (1 atm)

PREPARED BY:

S. A. Johnson/ H. L. Clever

EXPERIMENTAL VALUES:

T/K	Mole fraction x_{SO_2}	Bunsen coefficient α	Ostwald coefficient L
298.15	0.0702	6.360	6.942
313.15	0.0514	4.52	5.18

Mole fractions and Bunsen coefficients were calculated by the compiler.

AUXILIARY INFORMATION

METHOD /APPARATUS/PROCEDURE:

Apparently the volumetric apparatus described in ref.(1) was modified for use at temperatures above 0°C. The apparatus was designed to be operated at a partial pressure of sulfur dioxide of 760 torr.

SOURCE AND PURITY OF MATERIALS:

Analytical grade reagents of Hungarian and foreign origin were used (both liquids and gases). No further information.

ESTIMATED ERROR:

$\delta x_{SO_2}/x_{SO_2} = \pm 0.03$.

REFERENCES:

1. Bodor, E.; Bor, Gy.; Mohai, B.; Sipos, G. *Veszprémi Vegyipari Egyetem Közleményei* 1957, *1*, 55; *CA 55*, 3175h.

COMPONENTS:

1. Sulfur Dioxide; SO_2; [7446-09-5]

2. Pentadecane; $C_{15}H_{32}$; [629-62-9]

ORIGINAL MEASUREMENTS:

Makranczy, J.; Megyery-Balog, K.; Rusz, L.; Patyi, L.

Hung. J. Ind. Chem. 1976, *4(2)*, 269-280.

VARIABLES:

T/K: 298.15 - 313.15

P/kPa: 101.325 (1 atm)

PREPARED BY:

S. A. Johnson / H. L. Clever

EXPERIMENTAL VALUES:

T/K	Mole fraction x_{SO_2}	Bunsen coefficient α	Ostwald coefficient L
298.15	0.0722	6.170	6.735
313.15	0.0531	4.40	5.04

Mole fractions and Bunsen coefficients were calculated by the compiler.

AUXILIARY INFORMATION

METHOD /APPARATUS/PROCEDURE:

Apparently the volumetric apparatus described in ref.(1) was modified for use at temperatures above 0°C. The apparatus was designed to be operated at a partial pressure of sulfur dioxide of 760 torr.

SOURCE AND PURITY OF MATERIALS:

Analytical grade reagents of Hungarian and foreign origin were used (both liquids and gases). No further information.

ESTIMATED ERROR:

$\delta x_{SO_2}/x_{SO_2} = \pm 0.03$.

REFERENCES:

1. Bodor, E.; Bor, Gy.; Mohai, B.; Sipos, G. *Veszprêmi Vegyipari Egyetem Köslemênyei* 1957, *1*, 55; *CA 55*, 3175h.

COMPONENTS:	ORIGINAL MEASUREMENTS:
1. Sulfur dioxide; SO_2; [7446-09-5] 2. Hexadecane; $C_{16}H_{34}$; [544-76-3]	Tremper, K.K.; Prausnitz, J.M. *J. Chem. Engng. Data* 1976, *21*, 295-9
VARIABLES:	PREPARED BY:
Temperature	C.L. Young

EXPERIMENTAL VALUES:

T/K	Henry's Constant[a] /atm	Mole fraction[b] of sulfur dioxide at 1 atm partial pressure, x_{SO_2}
300	15.1	0.0662
325	22.3	0.0448
350	30.8	0.0325
375	40.2	0.0249
400	50.1	0.0200
425	60.1	0.0166
450	69.8	0.0143
475	77.1	0.0130

a. Authors stated measurements were made at several pressures and values of solubility used were all within the Henry's-Law region.

b. Calculated by compiler assuming linear relationship between mole fraction and pressure.

AUXILIARY INFORMATION

METHOD/APPARATUS/PROCEDURE

Volumetric apparatus similar to that described by Dymond and Hildebrand (1). Pressure measured with a null detector and precision gauge. Details in ref. (2).

NOTE

It was stated that "while all measurements were made at low pressures, experiments were made at several pressures. The total pressures were always less than 1000 mm Hg and usually much less."

SOURCE AND PURITY OF MATERIALS:

Solvent degassed, no other details given.

ESTIMATED ERROR:

$\delta T/K = \pm 0.1$; $\delta x_{SO_2} = \pm 1\%$.

REFERENCES:

1. Dymond, J.; Hildebrand, J.H. *Ind. Eng. Chem. Fundam.* 1967, *6*, 130.
2. Cukor, P.M.; Prausnitz, J.M. *Ind. Eng. Chem. Fundam.* 1971, *10*, 638.

COMPONENTS:	ORIGINAL MEASUREMENTS:
1. Sulfur Dioxide; SO_2; [7446-09-5] 2. Hexadecane; $C_{16}H_{34}$; [544-76-3]	Makranczy, J.; Megyery-Balog, K.; Rusz, L.; Patyi, L. *Hung. J. Ind. Chem.* 1976, *4(2)*, 269-280.
VARIABLES: T/K: 298.15 - 313.15 *P*/kPa: 101.325 (1 atm)	PREPARED BY: S. A. Johnson / H. L. Clever

EXPERIMENTAL VALUES:

T/K	Mole fraction x_{SO_2}	Bunsen coefficient α	Ostwald coefficient L
298.15	0.0745	6.020	6.571
313.15	0.0545	4.27	4.90

Mole fractions and Bunsen coefficients were calculated by the compiler.

AUXILIARY INFORMATION

METHOD/APPARATUS/PROCEDURE:

Apparently the volumetric apparatus described in ref.(1) was modified for use at temperatures above 0°C. The apparatus was designed to be operated at a partial pressure of sulfur dioxide of 760 torr.

SOURCE AND PURITY OF MATERIALS:

Analytical grade reagents of Hungarian and foreign origin were used (both liquids and gases). No further information.

ESTIMATED ERROR:

$\delta x_{SO_2}/x_{SO_2} = \pm 0.03$.

REFERENCES:

1. Bodor, E.; Bor, Gy.; Mohai, B.; Sipos, G. *Veszprémi Vegyipari Egyetem Közleményei* 1957, *1*, 55; *CA 55*, 3175h.

COMPONENTS:	ORIGINAL MEASUREMENTS:
1. Sulfur dioxide, SO_2; [7446-09-5] 2. Hexadecane; $C_{16}H_{34}$; [544-76-3] or Heptadecane; $C_{17}H_{36}$; [629-78-7]	Lenoir, J-Y.; Renault, P.; Renon, H. *J. Chem. Eng. Data.* 1971, *16*, 340-342
VARIABLES:	PREPARED BY: W. Gerrard.

EXPERIMENTAL VALUES:

SOLVENT	T/K	HENRY'S CONSTANT H_{SO_2}/ATM	MOLE FRACTION AT 1 ATM *
Hexadecane	298	12.9	0.0775
Heptadecane	323	16.4	0.0610

* Calculated by compiler assuming a linear function of p_{SO_2} vs x_{SO_2}, i.e. x_{SO_2} (1 atm) = $1/H_{SO_2}$.

The authors gave "Henry's constant" as $H_{SO_2} = \lim_{x_{SO_2} \to 0} \dfrac{f_{SO_2}(P,T)}{x_{SO_2}}$

x_{SO_2} is the mole fraction. f_{SO_2} is the fugacity of SO_2 "in the equilibrium condition" of pressure and temperature. H_{SO_2} is related to experimental chromatographic parameters based on the number of moles of solvent in the column, and the absolute retention time of the solute. Helium was the carrier gas.

AUXILIARY INFORMATION

METHOD/APPARATUS/PROCEDURE:

A conventional gas-liquid chromatographic unit was used. The chromatograph was an IGC 12 M Intersmat model, equipped with a thermal conductivity detector. The temperature control was stated to be within ±0.1° C. The pressure drop in the column, was measured by a mercury manometer. Helium was the carrier gas.

There is serious uncertainty in the estimation of accuracy.

SOURCE AND PURITY OF MATERIALS:

1. Sulfur dioxide was provided by l'Air Liquide, and had a stated purity of 99.9%. It was used as provided.
2. Provided by Touzart & Matignon and Serlabo, and stated to have a minimum purity of 99%. Distillation was carried out when necessary, and the refractive index was measured.

ESTIMATED ERROR:

REFERENCES:

COMPONENTS:

1. Sulfur dioxide; SO_2; [7446-09-5]
2. Heptadecane; $C_{17}H_{36}$; [629-78-7]

ORIGINAL MEASUREMENTS:

Sobolev, I.A.; Kukarin, V.A.; Dzhagatspanyan, R.V.; Kosorotov, V.I.; Zogorets, P.A.; Popov, A.I.;

Khim, Prom. 1970, *46*, 668-70

VARIABLES:

Temperature

PREPARED BY:

C.L. Young

EXPERIMENTAL VALUES:

T/K	Solubility*, S /mol l^{-1}	Mole fraction of+ sulfur dioxide in liquid, x_{SO_2}
303.15	0.124	0.037
313.15	0.104	0.031
323.15	0.0902	0.027
333.15	0.0768	0.023

\+ calculated by compiler.

* at atmospheric pressure.

$\ln S = 565/(T/K) - 5.67$ (where S is in units of mol l^{-1} atm^{-1}).

AUXILIARY INFORMATION

METHOD/APPARATUS/PROCEDURE:

Bubbler method. Samples of saturated liquid taken and added to sodium hydroxide solution and then excess alkali back titrated.

SOURCE AND PURITY OF MATERIALS:

1. Purity at least 95 mole per cent.
2. Chemically pure. Dried.

ESTIMATED ERROR:

$\delta T/K = \pm 0.1$; δ(Solubility) = ±2%; $\delta x_{SO_2} = \pm 0.001$; (estimated by compiler).

REFERENCES:

COMPONENTS:

1. Sulfur dioxide; SO_2; [7446-09-5]
2. Cyclohexane; C_6H_{12}; [110-82-7]
   Decahydronaphthalene (*decalin*); $C_{10}H_{18}$; [91-17-8]
   1,2,3,4-Tetrahydronaphthalene (*tetralin*); $C_{10}H_{12}$; [119-64-2]

ORIGINAL MEASUREMENTS:

Sano, H.

Nippon Kogaku Zasshi

1968, *89*, 362-368.

VARIABLES:

PREPARED BY:

C. L. Young

EXPERIMENTAL VALUES:

T/°C = 30 T/K = 303

Vapor pressure of cyclohexane, p_s^o = 0.140 atm

Absorption coefficient, $S^{a,b}$	Bunsen coefficient,[b] α	Bunsen coefficient,[c,f] α	$\ell^{c,d}$	Mole ratio	Mole fraction[e]
		Cyclohexane			
8.0	8.5	8.6	7.25	0.041	0.0394
		Decahydronaphthalene (mixed isomers)			
7.0	6.4	6.2	6.2	0.045	0.0431
		1,2,3,4-Tetrahydronaphthalene			
45.6	41.4	42.0	42.0	0.253	0.202

[a] Volume of sulfur dioxide absorbed by unit volume of solvent at experimental temperature and a total pressure of 1 atmosphere.

[b] Determined by static absorption method.

[c] Determined by flow method.

[d] $\alpha = \ell(1 - p_s^o)$ where p_s^o is the vapor pressure of solvent.

[e] Calculated by compiler.

[f] Assuming vapor pressure of decahydronaphthalene and tetrahydronaphthalene are negligible.

AUXILIARY INFORMATION

METHOD/APPARATUS/PROCEDURE:

Two techniques were used: (a) static absorption method and (b) flow method. Static method consisted of equilibrating solvent and sulfur dioxide in a 50 ml flask and measuring the absorption volumetrically. The flow method consists of passing sulfur dioxide through a micro gas absorption tube and estimating sulfur dioxide by iodometric titration. The vapor pressure of the solvents were measured roughly and were assumed to be equal to the partial pressure of the solvent by the authors when calculating the Bunsen coefficient.

SOURCE AND PURITY OF MATERIALS:

1. Purity 99.0-99.8 mole per cent.

ESTIMATED ERROR:

$\delta T/K = \pm 0.5$; $\delta x_{SO_2} = \pm 5\%$

(estimated by compiler).

REFERENCES:

COMPONENTS:

1. Sulfur dioxide; SO_2; [7446-09-5]
2. Cyclohexane; C_6H_{12}; [110-82-7]

ORIGINAL MEASUREMENTS:

Patyi, L.; Furmer, I. E.; Makranczy, J.; Sadilenko, A. S.; Stepanova, Z. G.; Berengarten, M. G.

Zh. Prikl. Khim. 1978, *51*, 1296-1300.

VARIABLES:

PREPARED BY:

C. L. Young

EXPERIMENTAL VALUES:

T/K	$\alpha^{\dagger}$	Mole fraction of sulfur dioxide at a partial pressure of 101.325 kPa x_{SO_2}
298.15	10.71	0.06251

† volume of gas (measured at 101.325 kPa and 273.15 K) dissolved by one volume of cyclohexane.

AUXILIARY INFORMATION

METHOD/APPARATUS/PROCEDURE:

Volumetric method. Pressure measured when known amounts of gas were added, in increments, to a known amount of liquid in a vessel of known dimensions. Corrections were made for the partial pressure of solvent. Details in ref. (1).

SOURCE AND PURITY OF MATERIALS:

Purity better than 99 mole per cent as determined by gas chromatography.

ESTIMATED ERROR:

$\delta T/K = \pm 0.1$; $\delta\alpha = \pm 4\%$ or less.

REFERENCES:

1. Bodor, E.; Bor, G. J.; Mohai, B,; Sipos. G. *Veszpremi. Vegyip. Egy. Kozl.* 1957, *1*, 55.

COMPONENTS:	ORIGINAL MEASUREMENTS:
1. Sulfur dioxide; SO_2; [7446-09-5] 2. Cyclohexane, C_6H_{12}; [110-82-7]	Benoit, R.L.; Milanova, E. *Can. J. Chem.*, 1979, *57*, 1319-1323
VARIABLES:	PREPARED BY: C.L. Young

EXPERIMENTAL VALUES:

T/K	Henry's law constant[+] H/atm
298.15	28.3

[+] Defined in original as

$$H = \frac{\text{partial pressure of sulfur dioxide}}{\text{mole fraction of sulfur dioxide}}$$

AUXILIARY INFORMATION

METHOD/APPARATUS/PROCEDURE:

Few details given. Pressure measured with quartz spiral gauge. Partial pressure calculated assuming Raoult's law. Concentration of sulfur dioxide determined by iodometric titration. Concentration of sulfur dioxide ranged from 0.05 to 0.10 mol l^{-1}

SOURCE AND PURITY OF MATERIALS:

1. Anhydrous sample from Canadian Liquid Air, purity 99.98 mole per cent.

2. Dried over 4A molecular sieve

ESTIMATED ERROR:

$\delta T/K = \pm 0.1$; $\delta H/atm = \pm 4\%$ (estimated by compiler)

REFERENCES:

COMPONENTS:

1. Sulfur dioxide; SO_2; [7446-09-5]
2. Decahydronaphthalene, (decalin); $C_{10}H_{18}$; [91-17-8]

ORIGINAL MEASUREMENTS:

Weissenberger, G,; Hadwiger, H.

Z. angew. Chem. 1927, *40*, 734-736.

VARIABLES:

PREPARED BY:

W. Gerrard

EXPERIMENTAL VALUES:

T/K	Volume of Comp. (2)/cm^3	Volume of SO_2 absorbed/cm^3	*Mole ratio SO_2/Comp. (2)	*Mole fraction x_{SO_2}
			(pressure assumed to be barometric)	
293.15	157.5 ("Molar volume")	1181.25	0.0503	0.0479

* Calculated by compiler

AUXILIARY INFORMATION

METHOD/APPARATUS/PROCEDURE:

Used an absorption vessel of 200 cm^3 capacity, and a vessel with a two-way tap. Simply stated that the absorptiometer was like that used by Manchot, but the reference was not cited; probably ref. (1).

SOURCE AND PURITY OF MATERIALS:

No details given.

ESTIMATED ERROR:

REFERENCES:

1. Manchot, W. *Z. anorg. Chem.* 1924, *141*, 38.

COMPONENTS:

1. Sulfur dioxide, SO_2; [7446-09-5]
2. Decahydronaphthalene, (decalin); $C_{10}H_{18}$; [91-17-8]

ORIGINAL MEASUREMENTS:

Lenoir, J-Y.; Renault, P.; Renon, H.

J. Chem. Eng. Data 1971, *16*, 340-342.

VARIABLES:

PREPARED BY:

W. Gerrard

EXPERIMENTAL VALUES:

760 mmHg = 1 atm = 101.325 kPa.

T/K	Henry's constant H_{SO_2}/atm	*Mole fraction at 1 atm
298	18.8	0.0532
323	25.6	0.0391

* Calculated by compiler assuming a linear function of p_{SO_2} vs x_{SO_2}, i.e. x_{SO_2} (1 atm) = $1/H_{SO_2}$.

The authors gave "Henry's constant" as $H_{SO_2} = \lim_{x_{SO_2}\to 0} \frac{f_{SO_2}(P,T)}{x_{SO_2}}$

x_{SO_2} is the mole fraction. f_{SO_2} is the fugacity of SO_2 "in the equilibrium condition"of pressure and temperature. H_{SO_2} is related to experimental chromatographic parameters based on the number of moles of solvent in the column, and the absolute retention time of the solute. Helium was the carrier gas.

AUXILIARY INFORMATION

METHOD/APPARATUS/PROCEDURE:

A conventional gas-liquid chromatographic unit was used. The chromatograph was an IGC 12M Intersmat model equipped with a thermal conductivity detector. The temperature control was stated to be within $\pm 0.1^{\circ}C$. The pressure drop in the column was measured by a mercury manometer. Helium was the carrier gas.

There is serious uncertainty in the estimation of accuracy.

SOURCE AND PURITY OF MATERIALS:

1. Sulfur dioxide was provided by l'Air Liquide, and had a stated purity of 99.9%. It was used as provided.

2. Provided by Touzart & Matignon and Serlabo, and stated to have a minimum purity of 99%. Distillation was carried out when necessary, and the refractive index was measured.

ESTIMATED ERROR:

REFERENCES:

COMPONENTS:	EVALUATOR:
1. Sulfur dioxide; SO_2; [7446-09-5] 2. Aromatic hydrocarbons and Hydronaphthalenes	Colin L. Young, School of Chemistry, University of Melbourne, Parkville, Victoria 3052, Australia. November 1980

CRITICAL EVALUATION:

Benzene; C_6H_6; [71-43-2]

This system has been studied by six groups. Lloyd's (1) data for benzene are in the form of concentration of sulfur dioxide in units of grams per litre of solution for a total pressure equal to the barometric pressure. The calculated mole fraction is based on the assumption that the volume of solution after absorption is the same as that before absorption. Calculation of the mole fraction at a partial pressure of sulfur dioxide of 101.3 kPa (1 atm) entails further uncertainty as the partial pressure of benzene becomes larger and increasingly uncertain for the higher temperatures.

Ipatiev and Monroe (2) made measurements at pressures between 120 kPa and 250 kPa for temperatures between 298 K and 300 K. The most extensive data for this system are those of Horiuti (3) who studied the system at temperatures between 283.15 K and 333.15 K. The table below compares the mole fraction solubility from the six groups extrapolated, where necessary, to 298.15 K and a partial pressure of 101.3 kPa. The data of Horiuti (3) were used to give an approximate estimate of the temperature dependence of the solubility for the purpose of this extrapolation. The data of Benoit and Milanova (4) were determined at low pressure and the mole fraction at a partial pressure of 101.3 kPa was obtained assuming Henry's law is obeyed. There is reasonable agreement between the data of Lloyd (1), Horiuti (3), Benoit and Milanova (4), Sano (6) and Ipatiev and Monroe (2). The result of Patyi *et al.* (5) is about 10 per cent smaller and is classified as doubtful.

Mole fraction solubility at 298 K and 101.3 kPa partial pressure

Patyi *et al.* (5)	0.2020
Benoit and Milanova (4)	0.2217
Lloyd (1)	0.21
Horiuti (3)	0.23
Ipatiev and Monroe (2)	0.23
Sano (6)	0.21

Although Horiuti (3) and Ipatiev and Monroe (2) data agree at 298 K and 101.3 kPa when the complete mole fraction data at 298 K are plotted against pressure, the plot of Horiuti, which covers the pressure range up to 135 kPa total pressure, has a slightly different slope from that obtained by plotting Ipatiev and Monroe's data. Therefore no data are recommended but the data of Horiuti (3), Ipatiev and Monroe (2), Lloyd (1), Sano (6), and Benoit and Milanova (4) are all classified as tentative.

Methylbenzene; C_7H_8; [108-88-3]

This system has been studied in detail by Lorimer *et al.* (7) at 227.6 K, 237.4 K and 249.8 K. In this temperature range sulfur dioxide is a liquid at atmospheric pressure. (The normal boiling point is

COMPONENTS:	EVALUATOR:
1. Sulfur dioxide; SO_2; [7446-09-5] 2. Aromatic hydrocarbons and Hydronaphthalenes	Colin L. Young, School of Chemistry, University of Melbourne, Parkville, Victoria 3052, Australia. November 1980

CRITICAL EVALUATION:

∿ 263 K.) Lloyd's data (1) cover the temperature range 293.15 K to 333.15 K. Since conversion of the mass of sulfur dioxide per unit volume of solution to a mole fraction value required an approximation for the density of solution, Lloyd's data must be assessed with caution. Gerrard (8) has determined the solubility of sulfur dioxide at 273.15 K and a total pressure of 101.3 kPa. The mole fraction solubility value of Sano (6) is somewhat (∿ 10 per cent) larger than the value of Lloyd at the same temperature. However, in view of the different range of temperature covered it is impossible to make any *precise* comparisons between the work of the four groups. The results of Lorimer *et al.* (7) appear to be self-consistent and are probably the most accurate and therefore are classified as tentative. The data of Sano (6), Gerrard (8) and Lloyd (1) are probably less reliable.

1,3-Dimethylbenzene; C_8H_{10}; [108-38-3]

This system has also been studied by Lorimer *et al.* (7) at 227.6 K, 237.4 K and 249.9 K and by Gerrard (9) over the temperature range 273.15 to 298.15 K. In this case there is better agreement between these two groups' results than in the case of methylbenzene and both are classified as tentative. The isolated value of Sano (6) for a mixture or xylenes appears to be slightly small compared with the value obtained by extrapolation of the values of Gerrard (9) to the same temperature.

1,3,5-Trimethylbenzene; C_9H_{12}; [108-67-8]

This system has also been investigated by Lorimer *et al.* (7) at ‿27.6 K, 237.4 K and 250.0 K and by Gerrard (9) over the temperature range 273.15 to 293.15 K. The temperature dependence of Gerrard's data indicates that the mole fraction solubility at the temperatures studied by Lorimer *et al.* (7) would be less than that given by Raoult's law, whereas the opposite was found experimentally. This could arise because of small errors in weighing in Gerrard's measurements when large amounts of gas are absorbed. The data of both groups are classified as tentative but the measurement of Gerrard at 273.15 K is doubtful.

1,1'-Methylenebis(1-methylethyl)benzene; $C_{19}H_{24}$; [25566-92-1]
1,1'-Methylenebis(methylbenzene); $C_{15}H_{16}$; [1335-47-3]

The solubility of sulfur dioxide in these two compounds has been studied by Gel'perin *et al.* (10). The temperature dependence of the solubility is more or less as would be expected for a system which shows only moderate deviations from Raoult's law. In the absence of any other evidence as to the reliability of these data, they are classified as tentative.

COMPONENTS:	EVALUATOR:
1. Sulfur dioxide; SO_2; [7446-09-5] 2. Aromatic hydrocarbons and Hydronaphthalenes	Colin L. Young School of Chemistry, University of Melbourne, Parkville, Victoria 3052, Australia. November 1980

CRITICAL EVALUATION:

Hydronaphthalenes

The mole fraction solubility of sulfur dioxide in tetrahydronaphthalene given by Gerrard (9) at 293.15 K and 101.3 kPa total pressure is 0.228 and 0.202 at 303 K and a partial pressure of 101.3 kPa according to Sano (6), where as Weissenberger and Hadwiger (11) reported that the solubility of sulfur dioxide in this compound was too small to be of consequence. Gerrard's value for decahydronaphthalene is 0.060 at 298.15 K and 101.3 kPa total pressure whereas the value based on Weissenberger and Hadwiger's primary data is 0.0479 under the same conditions. Sano (6) obtained a value at 303 K and 101.3 kPa of 0.0431. It is to be expected that the value for decahydronaphthalene should be much less than that for the tetrahydro-compound (*cf.* the solubility in cyclohexane which is much smaller than in benzene). It is therefore suggested that the data of Weissenberger and Hadwiger should be rejected. Lenoir *et al.* (12) gave solubility data at 298 K which, with some approximation, lead to a mole fraction solubility of sulfur dioxide in decahydronaphthalene of 0.053. In view of the approximation made, this must be regarded as in good agreement with Gerrard's value.

Turpentine

Gerrard's mole fraction solubility values for pinene (representing turpentine) are 0.200 at 273 K and 0.110 at 293 K for a total pressure of 101.3 kPa. Weissenberger and Hadwiger's (11) value of 0.055 for turpentine at 293 K appears to be unacceptable.

References

1. Lloyd, S. J. *J. Phys. Chem.* 1918, *22*, 300.
2. Ipatiev, V. N.; Monroe, G. S. *Ind. Eng. Chem. Anal. Ed.* 1942, *14*, 166.
3. Horiuti, J. *Sci. Pap. Inst. Phys. Chem. Res. Tokyo* 1931, *17*, 125.
4. Benoit, R. L.; Milanova, E. *Can. J. Chem.* 1979, *57*, 1319.
5. Patyi, L.; Furmer, I. E.; Makranczy, J.; Sadilenko, A. S.; Stepanova , Z. G.; Berengarten, M. G. *Zh. Prikl. Khim.* 1978, *51*, 1296.
6. Sano, H. *Nippon Kagaku Zasshi* 1968, *89*, 362.
7. Lorimer, J. W.; Smith, B. C.; Smith, G. H. *J. C. S. Faraday I* 1975, *71*, 2232.
8. Gerrard, W. *J. Appl. Chem. Biotechnol.* 1972, *22*, 623.
9. Gerrard, W. *Solubility of Gases and Liquids*, Plenum Press, New York, 1976, p.239.
10. Gel'perin, N. I.; Matveev, I. G.; Bil'shau, K. V. *Zh. Prikl. Khim.* 1958, *31*, 1323.
11. Weissenberger, G.; Hadwiger, H. Z. *Angew. Chem.* 1927, *40*, 734.
12. Lenoir, J.-Y.; Renault, P.; Renon, H. *J. Chem. Engng. Data* 1971, *16*, 340.

COMPONENTS:

1. Sulfur dioxide; SO_2; [7446-09-5]
2. Benzene; C_6H_6; [71-43-2]

ORIGINAL MEASUREMENTS:

Lloyd, S.J.

J. Phys. Chem. 1918, *22*, 300-302.

VARIABLES:

Temperature

PREPARED BY:

W. Gerrard.

EXPERIMENTAL VALUES:

T/K	Weight of SO_2 per dm^3 of solution /g.	*Speculative values *Mole ratio, SO_2/ Comp. (2)	*Mole fraction (as for 1 atm) x_{SO_2}
303.15	127.5	0.210	0.174
313.15	82.9	0.131	0.115
323.15	60.3	0.093	0.085
333.15	34.0	0.052	0.049

760 mmHg = 1 atm = 101.325 kPa.

* Calculated by compiler by assuming that the volume of solution is the same as the volume of the component (2) before absorption of gas.

AUXILIARY INFORMATION

METHOD:/APPARATUS/PROCEDURE:

Gas was bubbled through component (2) in a long graduated tube of small diameter. After equilibrium had been attained at a total pressure of about 1 atm (barometric pressure, 756-760 mmHg), the volume of the solution (about 5 cm^3) was measured, and the weight of absorbed sulfur dioxide was determined by an iodometric titration.

SOURCE AND PURITY OF MATERIALS:

(1) Prepared from concentrated sulfuric acid and a saturated aqueous solution of sodium bisulfite. Dried by sulfuric acid and phosphorus pentoxide.

(2) Distilled, dried by calcium chloride, and redistilled before use.

ESTIMATED ERROR:

REFERENCES:

COMPONENTS:	ORIGINAL MEASUREMENTS:
1. Sulfur dioxide; SO_2; [7446-09-5] 2. Benzene; C_6H_6; [71-43-2]	Horiuti, J., *Sci. Pap. Inst. Phys. Chem. Res. (Jpn.)*, 1931/32, *17*, 125-256.
VARIABLES:	PREPARED BY:
Temperature, pressure	C. L. Young

EXPERIMENTAL VALUES:

T/K	$P^{\S}$/bar	Mole fraction of sulfur dioxide in liquid, x_{SO_2}	T/K	$P^{\dagger}$/bar	Mole fraction of sulfur dioxide in liquid, x_{SO_2}	Ostwald coefficient, L
298.15	0.1248	0.0000	283.15	0.9749	0.3764	126.4
	0.3833	0.0592	288.15	0.9592	0.3107	103.2
	0.3921	0.0618	293.15	0.9390	0.2549	84.81
	0.6577	0.1247	298.15	0.9136	0.2079	70.01
	0.6730	0.1275	303.15	0.8829	0.1722	59.50
	0.8842	0.1783	308.15	0.8452	0.1405	50.37
	0.9350	0.1910	313.15	0.7999	0.1139	43.01
	1.078	0.2252	318.15	0.7457	0.0921	37.25
	1.147	0.2401	323.15	0.6806	0.0737	32.63
	1.230	0.2605	328.15	0.6062	0.0570	28.28
	1.301	0.2796	333.15	0.5172	0.0435	25.36
	1.348	0.2909				

§ total pressure.

† partial pressure, total pressure = 1 atmosphere = 1.01325×10^5 Pa.

Ostwald coefficient (as $x \to 0$, T/K = 298.15) = 60.3.

Henry's law constant, dp/dx (as $x \to 0$, T/K = 298.15) = 3454 mmHg = 4.605 bar.

AUXILIARY INFORMATION

METHOD/APPARATUS/PROCEDURE:

Apparatus consisted of a gas buret, a solvent reservoir, and an absorption pipet. Dry gas introduced into degassed solvent. System mixed using a magnetic stirrer until saturation occurred. Care was taken to prevent solvent vapor from mixing with gas in the gas buret. Volume of gas determined from gas buret reading; volume of liquid determined from height of meniscus in the absorption pipet.

SOURCE AND PURITY OF MATERIALS:

1. Sample obtained from Nippon Sulphuric Acid Manufacturing Co. Dried and fractionated several times.

2. Merck, extra pure sulfur-free sample refluxed over sodium amalgam. B. pt. 80.18 °C.

ESTIMATED ERROR:

$\delta T/K = \pm 0.1$; $\delta P/bar = \pm 0.001$;

$\delta x_{SO_2} = \pm 0.001$ (estimated by compiler)

REFERENCES:

COMPONENTS:

1. Sulfur dioxide; SO_2; [7446-09-5]
2. Benzene; C_6H_6; [71-43-2]

ORIGINAL MEASUREMENTS:

Ipatieff, V.N.; Monroe, G.S.

Ind. Eng. Chem. Anal. Edn. 1942, *14*, 166-171.

VARIABLES:

Pressure

PREPARED BY:

C.L. Young

EXPERIMENTAL VALUES:

t/°C	T/K	Total pressure p/atm	p/kPa	Solubility g SO_2/100g H_2O	Mole fraction* of sulfur dioxide, x_{SO_2}
26	299	1.20	1.22	24.7	0.231
26	299	1.75	1.77	37.6	0.314
27	300	2.40	2.43	84.8	0.508
25	298	2.65	2.69	125.5	0.605

* calculated by compiler.

AUXILIARY INFORMATION

METHOD/APPARATUS/PROCEDURE:

Rotating bomb of 3.5 dm^3 capacity. Pressure measured with a Bourdon gauge and temperature measured with thermocouple. Samples of liquid phase analysed by removing sulfur dioxide by a stream of air and absorption into potassium hydroxide solution. Benzene determined gravimetrically.

SOURCE AND PURITY OF MATERIALS:

1. No details given.
2. Baker C.P. thiophene free sample, n_D^{20} 1.5012.

ESTIMATED ERROR:

$\delta T/K = \pm 0.5$; $\delta x_{SO_2} = \pm 5\%$ (estimated by compiler).

REFERENCES:

COMPONENTS:	ORIGINAL MEASUREMENTS:
1. Sulfur dioxide; SO_2; [7446-09-5] 2. Benzene; C_6H_6; [71-43-2] Methylbenzene (*toluene*); C_7H_8; [108-88-3] Dimethylbenzene (*xylene, mixture of isomers*); C_8H_{10}; [1330-20-7]	Sano, H. *Nippon Kagaku Zasshi* 1968, *89*, 362-368.
VARIABLES:	PREPARED BY: C. L. Young

EXPERIMENTAL VALUES:

T/°C = 30   T/K = 303

Vapor pressure of benzene, p_s° = 0.117 atm

Vapor pressure of methylbenzene, p_s° = 0.037 atm

Vapor pressure of dimethylbenzene (mixture of isomers), p_s° = 0.013 atm

Absorption coefficient, $S^{a,b}$	Bunsen coefficient,[b] α	Bunsen coefficient,[c] α	$\ell^{c,d}$	Mole ratio	Mole fraction[e]
		Benzene			
52.0	53.4	54.2	45.2	0.213	0.176
		Methylbenzene			
61.8	58.0	59.2	56.2	0.276	0.216
		Dimethylbenzene			
58.05	53.5	54.1	53.8	0.294	0.227

[a] Volume of sulfur dioxide absorbed by unit volume of solvent at experimental temperature and a total pressure of 1 atmosphere.

[b] Determined by static absorption method.

[c] Determined by flow method.

[d] $\alpha = \ell(1 - p_s^\circ)$ where p_s° is the vapor pressure of solvent.

[e] Calculated by compiler.

AUXILIARY INFORMATION

METHOD/APPARATUS/PROCEDURE:

Two techniques were used: (a) static absorption method and (b) flow method. Static method consisted of equilibrating solvent and sulfur dioxide in a 50 ml flask and measuring the absorption volumetrically. The flow method consists of passing sulfur dioxide through a micro gas absorption tube and estimating sulfur dioxide by iodometric titration. The vapor pressure of the solvents were measured roughly and were assumed to be equal to the partial pressure of the solvent by the authors when calculating the Bunsen coefficient.

SOURCE AND PURITY OF MATERIALS:

1. Purity 99.0-99.8 mole per cent.

ESTIMATED ERROR:

$\delta T/K = \pm 0.5$; $\delta x_{SO_2} = \pm 5\%$ (estimated by compiler).

REFERENCES:

COMPONENTS:	ORIGINAL MEASUREMENTS:
1. Sulfur dioxide; SO_2; [7446-09-5] 2. Benzene; C_6H_6; [71-43-2]	Patyi, L.; Furmer, I. E.; Makranczy, J.; Sadilenko, A. S.; Stepanova, Z. G.; Berengarten, M. G. *Zh. Prikl. Khim.* 1978, *51*, 1296-1300.
VARIABLES:	PREPARED BY:
	C. L. Young

EXPERIMENTAL VALUES:

T/K	α†	Mole fraction of sulfur dioxide at a partial pressure of 101.325 kPa x_{SO_2}
298.15	62.05	0.2020

† volume of gas (measured at 101.325 kPa and 273.15 K) dissolved by one volume of benzene.

AUXILIARY INFORMATION

METHOD/APPARATUS/PROCEDURE:

Volumetric method. Pressure measured when known amounts of gas were added, in increments, to a known amount of liquid in a vessel of known dimensions. Corrections were made for the partial pressure of solvent. Details in ref. (1).

SOURCE AND PURITY OF MATERIALS:

Purity better than 99 mole per cent as determined by gas chromatography.

ESTIMATED ERROR:

$\delta T/K = \pm 0.1$; $\delta\alpha = \pm 4\%$ or less.

REFERENCES:

1. Bodor, E.; Bor, G. J.; Mohai, B.; Sipos, G. *Veszpremi. Vegyip. Egy. Kozl.* 1957, *1*, 55.

COMPONENTS:

1. Sulfur dioxide; SO_2; [7446-09-5]
2. Benzene; C_6H_6; [71-43-2]

ORIGINAL MEASUREMENTS:

Benoit, R.L.; Milanova, E.

Can. J. Chem., 1979, *57*, 1319-1323

VARIABLES:

PREPARED BY:

C.L. Young

EXPERIMENTAL VALUES:

T/K	Henry's law constant[+] H/atm
298.15	4.51

[+] Defined in original as

$$H = \frac{\text{partial pressure of sulfur dioxide}}{\text{mole fraction of sulfur dioxide}}$$

AUXILIARY INFORMATION

METHOD/APPARATUS/PROCEDURE:

Few details given. Pressure measured with quartz spiral gauge. Partial pressure calculated assuming Raoults law. Concentration of sulfur dioxide determined by iodometric titration. Concentration of sulfur dioxide ranged from 0.05 to 0.2 mol l^{-1}

SOURCE AND PURITY OF MATERIALS:

1. Anhydrous sample from Canadian Liquid Air, purity 99.98 mole per cent.

2. Dried over 4A molecular sieve

ESTIMATED ERROR:

$\delta T/K = \pm 0.1$; $\delta H/\text{atm} = \pm 4\%$ (estimated by compiler)

REFERENCES:

COMPONENTS:

1. Sulfur dioxide; SO_2; [7446-09-5]
2. Methylbenzene; C_7H_8; [108-88-3]

ORIGINAL MEASUREMENTS:

Lloyd, S.J.

J. Phys. Chem. 1918, *22*, 300-302.

VARIABLES:

PREPARED BY:

W. Gerrard

EXPERIMENTAL VALUES:

T/K	Weight of SO_2 per dm^3 of solution /g.	*Speculative values *Mole ratio, SO_2/Comp. (2)	*Mole fraction (as for 1 atm) x_{SO_2}
293.15	217.5	0.482	0.325
298.15	170.4	0.354	0.261
303.15	124.4	0.244	0.196
313.15	93.6	0.178	0.151
323.15	77.2	0.145	0.127
333.15	54.7	0.101	0.092

760 mmHg = 1 atm = 101.325 kPa

* Calculated by compiler by assuming that the volume of solution is the same as the volume of the component (2) before absorption of gas.

AUXILIARY INFORMATION

METHOD:/APPARATUS/PROCEDURE:

Gas was bubbled through component (2) in a long graduated tube of small diameter. After equilibrium had been attained at a total pressure of about 1 atm (barometric pressure, 756-760 mmHg), the volume of the solution (about 5 cm^3) was measured, and the weight of absorbed sulfur dioxide was determined by an iodometric titration.

SOURCE AND PURITY OF MATERIALS:

(1) Prepared from concentrated sulfuric acid and a saturated aqueous solution of sodium bisulfite. Dried by sulfuric acid and phosphorus pentoxide.

(2) Distilled, dried by calcium chloride and redistilled before use.

ESTIMATED ERROR:

REFERENCES:

COMPONENTS:	ORIGINAL MEASUREMENTS:
1. Sulfur dioxide; SO_2; [7446-09-5] 2. Methylbenzene (toluene); C_7H_8; [108-88-3]	Gerrard, W. *J. Appl. Chem. Biotechnol.* 1972, *22*, 623-650.
VARIABLES:	PREPARED BY: C. L. Young

EXPERIMENTAL VALUES:

760 mmHg = 1 atm = 101.325 kPa

T/K	Pressure, total /mmHg	Mole ratio SO_2/toluene	*Mole fraction x_{SO_2}
273.15	760	1.82	0.646

* Calculated by compiler.

AUXILIARY INFORMATION

METHOD/APPARATUS/PROCEDURE:

Sulfur dioxide was bubbled into a weighed amount of component (2) in a bubbler tube as described in detail, diagram given, in the original paper. The amount of gas absorbed at equilibrium and at the observed temperature and pressure was weighed. By means of a manometer assembly, to which the bubbler tube was attached, the weight of gas absorbed at successively lower pressure was measured. Eventually the pressure was reduced to that of the component (2). The refractive index, and the infrared spectrum of the residue showed it to be essentially pure component (2).

SOURCE AND PURITY OF MATERIALS:

1. Sulfur dioxide was the best grade contained in a small cylinder. Its high purity was attested by the measurement of vapor pressure from 258 to 268 K.

2. The best obtainable specimen was carefully purified and rigorously attested.

ESTIMATED ERROR:

REFERENCES:

COMPONENTS:

1. Sulfur dioxide; SO_2; [7446-09-5]
2. Methylbenzene, (Toluene); C_7H_8; [108-88-3]

ORIGINAL MEASUREMENTS:

Lorimer, J.W.; Smith, B.C.; Smith, G.H.

J.C.S. Faraday I, 1975, *71*, 2232-50.

VARIABLES:

Temperature, pressure

PREPARED BY:

C.L. Young

EXPERIMENTAL VALUES:

T/K	p/mmHg [a]	p/kPa [a,b]	Mole fraction of sulfur dioxide in liquid, x_{SO_2}	in vapor, y_{SO_2}
227.6	0.42	0.056	0.0000	0.000
	7.88	1.051	0.0864	0.952
	14.83	1.977	0.1949	0.979
	30.19	4.025	0.2879	0.991
	43.59	5.812	0.3834	0.995
	61.73	8.230	0.5135	0.997
	75.19	10.025	0.6031	0.998
	81.78	10.903	0.6557	0.999
	93.38	12.450	0.7623	0.999
	104.17	13.888	0.8812	1.000
	116.67	15.555	1.0000	1.000
237.4	0.83	0.111	0.0000	0.000
	14.48	1.931	0.0830	0.948
	30.55	4.073	0.1850	0.979
	56.03	7.470	0.2794	0.990
	78.34	10.444	0.3625	0.994
	110.59	14.744	0.4904	0.997
	131.91	17.587	0.5900	0.998
	143.80	19.172	0.6369	0.998
	166.33	22.176	0.7517	0.999
	186.12	24.814	0.8782	0.999
	208.43	27.788	1.0000	1.000

AUXILIARY INFORMATION

METHOD/APPARATUS/PROCEDURE:

Static vapor-liquid equilibrium apparatus. Total pressure measured using mercury manometers and cathetometer. Temperature measured using a sulfur dioxide vapor pressure thermometer. Some details in source, together with activity coefficients. Vapor compositions calculated from the total vapor pressure and liquid composition using Barker's technique (ref. 1).

SOURCE AND PURITY OF MATERIALS:

No details given

ESTIMATED ERROR:

$\delta T/K = \pm 0.1$; $\delta p/kPa = \pm 0.01$; δx_{SO_2}, δy_{SO_2} = ±1%.

REFERENCES:

1. Barker, J. A. *Aust. J. Chem.* 1953, *6*, 207.

COMPONENTS:	ORIGINAL MEASUREMENTS:
1. Sulfur dioxide; SO_2; [7746-09-5] 2. Methylbenzene, (Toluene); C_7H_8; [108-88-3]	Lorimer, J.W.; Smith, B.C.; Smith, G.H. *J.C.S. Faraday I*, 1975, *71*, 2232-50

EXPERIMENTAL VALUES:

T/K	p/mmHg [a]	p/kPa [a,b]	Mole fraction of sulfur dioxide in liquid, x_{SO_2}	in vapor, y_{SO_2}
249.8	1.85	0.247	0.0000	0.000
	28.82	3.842	0.0762	0.941
	65.48	8.730	0.1745	0.977
	107.02	14.268	0.2632	0.988
	139.74	18.630	0.3248	0.992
	202.40	26.984	0.4525	0.995
	245.28	32.701	0.5585	0.997
	265.84	35.442	0.5947	0.997
	306.92	40.919	0.7289	0.998
	363.40	48.449	0.8695	0.999
	405.40	54.049	1.0000	1.000

[a] total pressure.

[b] calculated by compiler.

COMPONENTS:

1. Sulfur dioxide; SO_2; [7446-09-5]

2. 1,3-Dimethylbenzene, (*m*-xylene); C_8H_{10}; [108-38-3]

ORIGINAL MEASUREMENTS:

Gerrard, W.

J. Appl. Chem. Biotechnol. 1972, *22*, 623-650.

VARIABLES:

Temperature

PREPARED BY:

W. Gerrard

EXPERIMENTAL VALUES:

T/K	Pressure, total /mmHg	Mole ratio SO_2/Xylene.	*Mole fraction x_{SO_2}
273.15	765	2.15	0.683
283.15	765	0.895	0.472
293.15	765	0.51	0.338
298.15	765	0.40	0.286

760 mmHg = 1 atm = 101.325 kPa.

* Calculated by compiler.

AUXILIARY INFORMATION

METHOD/APPARATUS/PROCEDURE:

Sulfur dioxide was bubbled into a weighed amount of component (2) in a bubbler tube as described in detail, diagram given, in the original paper. The amount of gas absorbed at equilibrium and at the observed temperature and pressure was weighed. By means of a manometer assembly, to which the bubbler tube was attached, the weight of gas absorbed at successively lower pressure was measured. Eventually the pressure was reduced to that of the component (2). The refractive index, and the infrared spectrum of the residue showed it to be essentially pure component (2).

SOURCE AND PURITY OF MATERIALS:

1. Sulfur dioxide was the best grade contained in a small cylinder. Its high purity was attested by the measurement of vapor pressure from 258 to 268 K.

2. The best obtainable specimen was carefully purified and rigorously attested.

ESTIMATED ERROR:

REFERENCES:

COMPONENTS:	ORIGINAL MEASUREMENTS:
1. Sulfur dioxide; SO_2; [7446-09-5] 2. 1,3-Dimethylbenzene, (*m*-xylene); C_8H_{10}; [108-38-3]	Lorimer, J.W.; Smith, B.C.; Smith, G.H. *J.C.S. Faraday I*, 1975, *71*, 2232-50
VARIABLES:	PREPARED BY:
Temperature, pressure	C.L. Young

EXPERIMENTAL VALUES:

T/K	p/mmHg[a]	p/kPa[a,b]	Mole fraction of sulfur dioxide in liquid, x_{SO_2}	in vapor, y_{SO_2}
227.6	0.22	0.029	0.0000	0.000
	7.63	1.017	0.0893	0.974
	16.04	2.138	0.2020	0.989
	23.95	3.193	0.3040	0.994
	38.22	5.096	0.3695	0.997
	52.36	6.981	0.4704	0.998
	67.18	8.957	0.5724	0.999
	80.23	10.696	0.6250	0.999
	93.14	12.418	0.7595	1.000
	105.16	14.020	0.8760	1.000
	116.67	15.555	1.0000	1.000
237.4	0.35	0.047	0.0000	0.000
	14.58	1.944	0.0854	0.978
	30.57	4.076	0.1955	0.991
	46.76	6.234	0.2962	0.995
	67.73	9.030	0.3516	0.997
	92.90	12.386	0.4525	0.998
	121.07	16.141	0.5565	0.999
	134.04	17.871	0.6046	0.999
	166.71	22.226	0.7477	1.000
	188.58	25.142	0.8725	1.000
	209.13	27.882	1.0000	1.000

AUXILIARY INFORMATION

METHOD/APPARATUS/PROCEDURE:

Static vapor-liquid equilibrium apparatus. Total pressure measured using mercury manometers and cathetometer. Temperature measured using a sulfur dioxide vapor pressure thermometer. Some details in source, together with activity coefficients. Vapor compositions calculated from the total vapor pressure and liquid composition using Barker's technique (ref. 1).

SOURCE AND PURITY OF MATERIALS:

No details given.

ESTIMATED ERROR:

$\delta T/K = \pm 0.1$; $\delta p/kPa = \pm 0.01$; $\delta x_{SO_2}, \delta y_{SO_2} = \pm 1\%$.

REFERENCES:

1. Barker, J. A. *Aust. J. Chem.* 1953, *6*, 207.

COMPONENTS:

1. Sulfur dioxide; SO_2; [7446-09-5]

2. 1,3-Dimethylbenzene, (*m*-xylene); C_8H_{10}; [108-38-3]

ORIGINAL MEASUREMENTS:

Lorimer, J.W.; Smith, B.C.; Smith, G.H.

J.C.S. Faraday I, 1975, *71*, 2232-50.

EXPERIMENTAL VALUES:

T/K	p/mmHg[a]	p/kPa[a,b]	Mole fraction of sulfur dioxide in liquid, x_{SO_2}	in vapor, y_{SO_2}
249.9	0.47	0.063	0.0000	0.000
	28.21	3.761	0.0779	0.985
	60.19	8.025	0.1826	0.994
	92.18	12.290	0.2805	0.996
	119.30	15.905	0.3188	0.997
	168.48	22.462	0.4163	0.999
	222.02	29.600	0.5233	0.999
	240.60	32.077	0.5560	0.999
	312.50	41.663	0.7195	1.000
	363.18	48.420	0.8639	1.000
	408.08	54.405	1.0000	1.000

[a] total pressure.

[b] calculated by compiler.

COMPONENTS:

1. Sulfur dioxide; SO_2; [7446-09-5]
2. 1,3,5-Trimethylbenzene, (mesitylene); C_9H_{12}; [108-67-8]

ORIGINAL MEASUREMENTS:

Lorimer, J.W.; Smith, B.C.; Smith, G.H.

J.C.S. Faraday I, 1975, *71*, 2232-50.

VARIABLES:

Temperature, pressure

PREPARED BY:

C.L. Young

EXPERIMENTAL VALUES:

T/K	p/mmHg[a]	p/kPa[a,b]	Mole fraction of sulfur dioxide in liquid, y_{SO_2}	in vapor, y_{SO_2}
227.6	0.17	0.023	0.0000	0.000
	6.53	0.871	0.1124	0.977
	14.13	1.884	0.1940	0.990
	24.29	3.238	0.2956	0.995
	38.56	5.141	0.3970	0.998
	55.12	7.349	0.4882	0.999
	73.35	9.779	0.5903	0.999
	86.89	11.584	0.6860	1.000
	100.25	13.366	0.7975	1.000
	108.69	14.491	0.9012	1.000
	116.67	15.555	1.0000	1.000
237.4	0.23	0.031	0.0000	0.000
	12.57	1.676	0.1084	0.984
	25.30	3.373	0.1880	0.993
	46.06	6.141	0.2868	0.997
	66.72	8.895	0.3781	0.998
	95.21	12.694	0.4682	0.999
	127.27	16.968	0.5725	0.999
	149.09	19.877	0.6604	1.000
	176.06	23.473	0.7845	1.000
	195.12	26.014	0.8977	1.000
	208.65	27.818	1.0000	1.000

AUXILIARY INFORMATION

METHOD/APPARATUS/PROCEDURE:

Static vapor-liquid equilibrium apparatus. Total pressure measured using mercury manometers and cathetometer. Temperature measured using a sulfur dioxide vapor pressure thermometer. Some details in source, together with activity coefficients. Vapor compositions calculated from the total vapor pressure and liquid composition using Barker's technique (ref. 1).

SOURCE AND PURITY OF MATERIALS:

No details given.

ESTIMATED ERROR:

$\delta T/K = \pm 0.1$; $\delta p/kPa = \pm 0.01$; δx_{SO_2}, $\delta y_{SO_2} = \pm 1\%$.

REFERENCES:

1. Barker, J. A. *Aust. J. Chem.* 1953, *6*, 207.

COMPONENTS

1. Sulfur dioxide; SO_2; [7446-09-5]
2. 1,3,5-Trimethylbenzene, (mesitylene); C_9H_{12}; [108-67-8]

ORIGINAL MEASUREMENTS:

Lorimer, J.W.; Smith, B.C.; Smith, G.H.

J.C.S. Faraday I, 1975, *71*, 2232-50.

EXPERIMENTAL VALUES:

T/K	p/mmHg [a]	p/kPa [a,b]	Mole fraction of sulfur dioxide in liquid, x_{SO_2}	in vapor, y_{SO_2}
250.0	0.48	0.064	0.0000	0.000
	25.29	3.372	0.1004	0.983
	50.73	6.763	0.1742	0.992
	89.13	11.883	0.2676	0.996
	119.25	15.899	0.3403	0.998
	170.08	22.675	0.4267	0.999
	231.39	30.849	0.5328	0.999
	264.15	35.217	0.5987	0.999
	334.28	44.567	0.7490	1.000
	377.45	50.323	0.8888	1.000
	410.55	54.735	1.0000	1.000

[a] total pressure.

[b] calculated by compiler.

COMPONENTS:

1. Sulfur dioxide; SO_2; [7446-09-5]
2. 1,3,5-Trimethylbenzene, (Mesitylene); C_9H_{12}; [526-73-8]

ORIGINAL MEASUREMENTS:

Gerrard, W. *"Solubility of Gases and Liquids,"* Plenum Press, New York, 1976, p.239.

VARIABLES:

Temperature

PREPARED BY:

W. Gerrard

EXPERIMENTAL VALUES:

T/K	Weight of Component(2)	Weight of SO_2 absorbed, g	Mole ratio SO_2 /Comp.(2)	Mole fraction x_{SO_2}
			(Total pressure, 1 atm)	
273.15	3.8690	3.5828	1.740	0.635
283.15	3.8690	1.9450	0.944	0.485
293.15	3.8690	1.2539	0.610	0.380

760 mmHg = 1 atm = 101.325 kPa.

AUXILIARY INFORMATION

METHOD:

Sulfur dioxide was bubbled into a weighed amount of component (2) in a bubbler tube. The amount of gas absorbed was determined by re-weighing the tube. The total pressure was barometric, very nearly 1 atm. See ref. (1).

SOURCE AND PURITY OF MATERIALS:

1. Sulfur dioxide was the best grade contained in a small cylinder. Its high degree of purity was attested by the measurement of vapor pressure from 258-268 K.
2. The best commercial specimen was purified and attested.

ESTIMATED ERROR:

REFERENCES:

1. Gerrard, W. *J. Appl. Chem. Biotechnol.* 1972, *22*, 623.

COMPONENTS:

1. Sulfur dioxide; SO_2; [7446-09-5]
2. Bis(1-methylethyl)benzene, (diisopropylbenzene); $C_{12}H_{18}$; [25321-09-9]

ORIGINAL MEASUREMENTS:

Gel'perin, N. I.; Matveev, I. G.; Bil'shau, K. V.

Zh. Prikl. Khim.

1958, *31*, 1323-1332.

VARIABLES:

PREPARED BY:

W. Gerrard

EXPERIMENTAL VALUES:

760 mmHg = 1 atm = 101.325 kPa

T/K	g SO_2/100 g Comp. (2)	*Mole ratio SO_2/Comp. (2)	*Mole fraction x_{SO_2}
277.65	26.5	0.672	0.402
298.15	10.7	0.271	0.213
315.15	7.5	0.190	0.160

* Calculated by compiler.

AUXILIARY INFORMATION

METHOD/APPARATUS/PROCEDURE:

The amount of gas absorbed "at normal pressure" was determined by weighing, to constant weight. The data expressed as g SO_2/100 g component (2) were stated to agree to within 1-2% with the data obtained by titration with 0.1 *N* potassium hydroxide solution to a phenolphthalein end-point.

SOURCE AND PURITY OF MATERIALS:

1. Simply referred to as "concentrated 100%" sulfur dioxide.

2. Fraction from a cracking plant.

ESTIMATED ERROR:

REFERENCES:

COMPONENTS:	ORIGINAL MEASUREMENTS:
1. Sulfur dioxide; SO_2; [7446-09-5] 2. 1,1'-Methylenebismethylbenzene (Ditotylmethane); $C_{15}H_{16}$; (a mixture of *o*-, *m*-, and *p*- isomers of constant composition); [1335-47-3]	Gel'perin, N. I.; Matveev, I. G.; Bil'shau, K. V. *Zh. Prikl. Khim.* 1958, *31*, 1323-1332.
VARIABLES:	PREPARED BY: W. Gerrard

EXPERIMENTAL VALUES: The original data appear to be for 1 atm.
760 mmHg = 1 atm = 101.325 kPa.

T/K	g SO_2 for 100 g of Comp.(2)	*Mole ratio SO_2/Comp.(2)	*Mole fraction x_{SO_2}
277.65	59	1.810	0.644
282.15	45	1.380	0.580
290.65	19.9	0.610	0.379
298.15	14.5	0.445	0.308
307.65	10.8	0.331	0.249
315.15	9.17	0.281	0.220
327.65	5.92	0.182	0.154
333.15	5.19	0.159	0.137
338.15	3.96	0.121	0.108
343.15	3.79	0.116	0.104
348.15	3.27	0.100	0.0912
362.65	2.06	0.0632	0.0594
403.15	1.65	0.0506	0.0482
413.15	0.95	0.0291	0.0283

Presumably to demonstrate the effect of a change in partial pressure of SO_2, a mixture of air and SO_2 of stated % SO_2 by volume was used. It appeared to be understood that $y/100$ was the partial pressure of SO_2 in atm, y being the % by volume of SO_2 in the SO_2 + air mixture. Data are given below.

AUXILIARY INFORMATION

METHOD/APPARATUS/PROCEDURE:

The amount of gas absorbed "at normal pressure" was determined by weighing, to constant weight. The data, expressed as g SO_2/100 g component (2), were stated to agree to within 1-2% with the data obtained by titration with 0.1 *N* potassium hydroxide solution to a phenolphthalein end-point.

SOURCE AND PURITY OF MATERIALS:

1. Simply referred to as "concentrated 100%" sulfur dioxide.
2. From the method of preparation ditolylmethane was a mixture of *o*-, *m*-, and *p*- isomers of constant composition. Various properties such as density, viscosity, and vapor pressure were stated.

ESTIMATED ERROR:

REFERENCES:

COMPONENTS:	ORIGINAL MEASUREMENTS:
1. Sulfur dioxide; SO_2; [7446-09-5] 2. 1,1'-Methylenebismethylbenzene (Ditotylmethane); $C_{15}H_{16}$; (a mixture of *o*-, *m*-, and *p*- isomers of constant composition); [1335-47-3]	Gel'perin, N. I.; Matveev, I. G.; Bil'shau, K. V. *Zh. Prikl. Khim.* 1958, *31*, 1323-1332.

T/K	y	Solubility /(g SO_2/ 100 g solvent)	Mole ratio SO_2/solvent *	Mole fraction x_{SO_2} *
276.15	57.75	21.9	0.672	0.402
287.65		14.2	0.436	0.303
298.95		9.14	0.280	0.219
308.75		6.46	0.198	0.165
322.15		4.22	0.129	0.115
347.15		2.32	0.0712	0.0664
276.65	21.00	9.68	0.297	0.229
287.65		5.69	0.175	0.149
298.15		3.8	0.117	0.105
308.75		3.02	0.0926	0.0848
322.75		1.97	0.0604	0.0570
347.75		0.94	0.0288	0.0280
281.65	7.08	2.95	0.0905	0.0830
287.65		1.92	0.0589	0.0556
299.15		1.34	0.0411	0.0395
309.15		0.932	0.0286	0.0278
316.95		0.716	0.0220	0.0215
329.45		0.516	0.0158	0.0156
339.15		0.408	0.0125	0.0124
348.35		0.354	0.0109	0.0107
293.15	1.7	0.385	0.0118	0.0117
298.15		0.32	0.00982	0.0972
308.15		0.180	0.00552	0.00549

* Calculated by compiler.

COMPONENTS:

1. Sulfur dioxide; SO_2; [7446-09-5]
2. (1-Methylethyl)-1,1′-methylene-bisbenzene, (Dicumyl-methane), (diisopropyl-phenylmethane); $C_{19}H_{24}$; [25566-92-1]

ORIGINAL MEASUREMENTS:

Gel'perin, N.I.; Matveev, I.G.; Bil'shau, K.V.

Zh. Prikl. Khim. 1958, *31*, 1323-1332.

VARIABLES:

PREPARED BY:

W. Gerrard.

EXPERIMENTAL VALUES: 760 mmHg = 1 atm = 101.325 kPa.

T/K	g SO_2/100 g component(2)	*Mole ratio SO_2/Comp. (2)	*Mole fraction x_{SO_2}	(Appears to be for 1 atm)
276.15	35.2	1.389	0.581	
277.65	29.2	1.152	0.535	
282.15	21.5	0.848	0.459	
290.65	13.3	0.525	0.344	
298.15	10.8	0.426	0.299	
307.65	8.25	0.326	0.246	
315.15	7.58	0.299	0.230	
327.65	3.87	0.153	0.132	
338.15	3.54	0.140	0.123	
348.15	2.62	0.103	0.0934	
362.65	1.65	0.0651	0.0611	

Presumably to demonstate the effect of a change in partial pressure of SO_2, a mixture of air and SO_2 of stated % SO_2 by volume was used.

It appeared to be understood that y/100 was the partial pressure in atm of SO_2 in these mixtures, y being the % SO_2 by volume. Data are given on continuation sheet.

AUXILIARY INFORMATION

METHOD/APPARATUS/PROCEDURE:

The amount of gas absorbed "at normal pressure" was determined by weighing, to constant weight. The data expressed as g SO_2/100 g component (2) were stated to agree to within 1-2% with the data obtained by titration with 0.1*N* potassium hydroxide solution to a phenolphthalein end-point.

SOURCE AND PURITY OF MATERIALS:

1. Simply referred to as "concentrated 100%" sulfur dioxide.

2. From the method of preparation component (2) was a mixture of *o*-, *m*- and *p*- isomers of constant composition.

ESTIMATED ERROR:

REFERENCES:

COMPONENTS:	ORIGINAL MEASUREMENTS:
1. Sulfur dioxide; SO_2; [7446-09-5] 2. (1-Methylethyl)-1,1′-methylene-bisbenzene, (Dicumylmethane), (diisopropyl-phenylmethane); $C_{19}H_{24}$; [25566-92-1]	Gel'perin, N.I.; Matveev, I.G.; Bil'shau, K.V. *Zh. Prikl. Khim.* 1958, *31*,1323-1332

EXPERIMENTAL VALUES: 760 mmHg = 1 atm = 101.325 kPa.

T/K	y	Solubility /(gSO_2/100g solvent)	Mole ratio* SO_2/Solvent	Mole fraction* x_{SO_2}
276.15	57.75	14.9	0.588	0.370
287.65		10.8	0.426	0.299
298.95		6.62	0.261	0.207
308.75		4.77	0.188	0.158
322.15		3.18	0.125	0.112
347.15		1.80	0.0710	0.0663
276.65	21.00	6.95	0.274	0.215
287.95		4.74	0.187	0.158
298.15		3.11	0.123	0.109
308.75		2.21	0.0872	0.0802
322.75		1.46	0.0576	0.0550
347.75		0.70	0.0276	0.0269
281.65	7.08	2.17	0.0856	0.0789
287.65		1.55	0.0612	0.0576
299.15		1.058	0.0417	0.0401
309.15		0.721	0.0284	0.0277
316.95		0.589	0.0232	0.0227
329.45		0.393	0.0155	0.0153
339.15		0.328	0.0129	0.0128
348.35		0.275	0.0109	0.0107
293.45	1.7	0.321	0.0127	0.0125
298.15		0.26	0.0103	0.0102
308.15		0.119	0.0047	0.0047

* Calculated by compiler.

COMPONENTS:

1. Sulfur dioxide; SO_2; [7446-09-5]
2. Pinene; $C_{10}H_{16}$; (isomer not specified, used as representing turpentine).

ORIGINAL MEASUREMENTS:

Gerrard, W. *"Solubility of Gases and Liquids,"* Plenum Press, New York, 1976, p. 239.

VARIABLES:

Temperature

PREPARED BY:

W. Gerrard.

EXPERIMENTAL VALUES:

T/K	Weight of Pinene g.	Weight of SO_2 absorbed,g	Mole ratio SO_2 /Pinene	Mole fraction x_{SO_2}
			(Total Pressure, 1 atm)	
273.15	3.4166	0.4009	0.250	0.200
293.15	3.4166	0.2201	0.124	0.110

760 mmHg = 1 atm = 101.325 kPa.

AUXILIARY INFORMATION

METHOD:

Sulfur dioxide was bubbled into a weighed amount of component (2) in a bubbler tube. The amount of gas absorbed was determined by re-weighing the tube. The total pressure was barometric, very nearly 1 atm. See ref. (1).

SOURCE AND PURITY OF MATERIALS:

1. Sulfur dioxide was the best grade contained in a small cylinder. Its high degree of purity was attested by the measurement of vapor pressure from 258-268 K.
2. The best commercial specimen was purified and attested.

ESTIMATED ERROR:

REFERENCES:

1. Gerrard, W. *J. Appl. Chem. Biotechnol.* 1972, *22*, 623

COMPONENTS:

1. Sulfur dioxide; SO_2; [7446-09-5]
2. Terpentin

ORIGINAL MEASUREMENTS:

Weissenberger, G.; Hadwiger, H.

Z. Angew. Chem. 1927, *40*, 734-736.

VARIABLES:

PREPARED BY:

W. Gerrard

EXPERIMENTAL VALUES:

T/K	Volume of component(2) /cm^3	Volume of SO_2 absorbed,/cm^3	*Mole ratio SO_2/Comp.(2) (pressure assumed to be barometric).	*Mole fraction x_{SO_2}
293	1	8.3	**0.056	0.053

* Calculated by compiler

** Based on pinene.

AUXILIARY INFORMATION

METHOD/APPARATUS/PROCEDURE:

Used an absorption vessel of 200cm^3 capacity, and a vessel with a two-way tap.

Simply stated that the absorptionmeter was like that used by Manchot, but the reference was not cited, probably ref.(1).

SOURCE AND PURITY OF MATERIALS:

No details given.

ESTIMATED ERROR:

REFERENCES:

1. Manchot, W. *Z. anorg. Chem.* 1924, *141*, 38.

COMPONENTS:

1. Sulfur dioxide; SO_2; [7446-09-5]
2. Hydroterpin.

ORIGINAL MEASUREMENTS:

Weissenberger, G.; Hadwiger, H.
Z. angew. Chem. 1927, *40*, 734-736.

VARIABLES:

PREPARED BY:

W. Gerrard

EXPERIMENTAL VALUES:

T/K	Volume of component (2) /cm^3	Volume of SO_2 absorbed, /cm^3	*Mole ratio SO_2/Comp.(2)	*Mole fraction x_{SO_2}
			(pressure assumed to be barometric)	
293.15	1	14.3 14.8	** 0.108	0.097

* Calculated by compiler

** "Hydroterpin" was stated to be obtained by the hydrogenation of "Terpentinol" In *Chem. Abs.* *21*, 3052, the compound is named hydroterpinol.

The authors state that 1 g has a volume of 1.133cm^3. The mole ratio given above is based on an assumed molar weight of 154, the assumed formula being $C_{10}H_{18}O$.

AUXILIARY INFORMATION

METHOD/APPARATUS/PROCEDURE:

Used an absorption vessel of 200cm^3 capacity, and a vessel with a two-way tap.

Simply stated that the absorptiometer was like that used by Manchot, but the reference was not cited, probably ref. (1).

SOURCE AND PURITY OF MATERIALS:

No details given.

ESTIMATED ERROR:

REFERENCES:

1. Manchot, W. *Z. anorg. Chem.* 1924, *141*, 38.

COMPONENTS:

1. Sulfur dioxide; SO_2; [7446-09-5]
2. Kerosene (petroleum, boiling range unspecified).

ORIGINAL MEASUREMENTS:

Yuferev, R.F.; Maluigin, P.V.

Z. Chim. Prom. 1930, *7*, 553-555

VARIABLES:

Temperature

PREPARED BY:

W. Gerrard

EXPERIMENTAL VALUES:

T/K	Weight of SO_2 per cm^3 of solvent g	Volume of SO_2 per cm^3 of solvent cm^3	* Mole fraction, x_{SO_2}, approximate, based on an assumed molecular weight of 200, and a density of 0.7 for the solvent.
273.15	0.0771	26.98	0.256
292.15	0.0439	15.36	0.164
Kerosene saturated with water			
273.15	0.0801	28.03	
292.15	0.04472	15.65	

*Calculated by compiler.

AUXILIARY INFORMATION

METHOD/APPARATUS/PROCEDURE:

Sulfur dioxide was passed through the kerosene (5 cm^3) until saturation was reached. The whole was then transferred to an aqueous solution of I_2 + KI. The formed sulfuric acid was determined gravimetrically. The pressure was probably barometric.

SOURCE AND PURITY OF MATERIALS:

1. Sulfur dioxide was dried by sulfuric acid.
2. Kerosene was dried over sodium, and, without distillation, was stored over sodium.

ESTIMATED ERROR:

REFERENCES:

COMPONENTS:

1. Sulfur dioxide; SO_2; [7446-09-5]

2. Soft paraffin (see note).

ORIGINAL MEASUREMENTS:

Sobolev, I.A.; Kukarin, V.A.; Dzhagatspanyan, R.V.; Kosorotov, V.I.; Zogorets, P.A.; Popov, A.I.

Khim. Prom. 1970, *46*, 668-70.

VARIABLES:

Temperature

PREPARED BY:

C.L. Young

EXPERIMENTAL VALUES:

T/K	Solubility*, S /mol l^{-1}
303.15	0.15
313.15	0.128
323.15	0.115
333.15	0.0937

* at atmospheric pressure (authors claim linear relationship between pressure and solubility up to 1 atmosphere partial pressure of SO_2)

$\ln S = 690/(T/K) - 5.98$ (where S is in units of mol $l^{-1}atm^{-1}$)

AUXILIARY INFORMATION

METHOD/APPARATUS/PROCEDURE:

Bubbler method. Samples of saturated liquid taken and added to sodium hydroxide solution and then excess alkali back titrated.

NOTE: Mixture of C_{12}-C_{22} hydrocarbons (86.7% normal paraffins, 11.3% branched paraffins, 0.4% aromatic and 1.6% naphthenic hydrocarbons).

SOURCE AND PURITY OF MATERIALS:

1. Purity at least 95 mole per cent.

ESTIMATED ERROR:

$\delta T/K = \pm 0.1$; δ(Solubility) = ±2% (estimated by compiler).

REFERENCES:

COMPONENTS:

1. Sulfur dioxide; SO_2; [7446-09-5]

2. Methanol; CH_4O; [67-56-1]

EVALUATOR:

Colin L. Young,
School of Chemistry,
University of Melbourne,
Parkville, Victoria, 3052
AUSTRALIA

June, 1981

CRITICAL EVALUATION:

The data of Lobry de Bruyn (1) and that of Gerrard (2) are in reasonable agreement and are deemed acceptable. These data are for a total pressure of 1 atmosphere. The vapor pressures of pure methanol are 29.6 mmHg (273 K) 54.7 mmHg (283 K) 96.0 mmHg (293. K) 160.0 (303 K) 260.5 mmHg (313 K). Obviously as the temperature increases there will be an increasingly important difference between the mole fraction at 1 atmosphere total pressure and 1 atmosphere partial pressure. Bekarek and Hala (3) give their data as mole fractions in gas and liquid phases for a series of total pressures, these being considerable less than 1 atmoshpere. It is possible to calculate so called Henry's constants from these data (see Table 1) by assuming the partial pressure of methanol in the vapor phase is equal to the vapor pressure of pure methanol times the mole fraction in the liquid phase. From inspection of Table 1 it is immediately apparent that in the present situation these vary and there is not a linear relationship between partial pressure and mole fraction of gas dissolved in the liquid.

Table 1. Henry's law constants calculated from the data of Bekarek and Hala (3)

T/K	p/mmHg	x_{SO_2}	y_{SO_2}	H/atm	H^{-1}/atm^{-1}
293.2	101.40	0.0043	0.0508	1.580	0.634
	136.20	0.0314	0.3100	1.769	0.565
303.2	169.40	0.0050	0.0542	2.416	0.414
	294.30	0.0619	0.4370	2.734	0.366
313.2	278.60	0.0058	0.0567	3.641	0.275
	418.70	0.0518	0.3990	4.144	0.236

Tokunaga (4) reported data for pure water, pure methanol and mixtures of these, as grams of sulfur dioxide "per 100g of solution (solute gas free) at 1 atm. partial pressure". The evaluator has calculated the mole fractions for 1 atm partial pressure as follows: 283.15K (0.397); 293.15K (0.301); 303.15 (0.223); 313.15K (0.169). There is fair agreement of these data for a partial pressure of 1 atmosphere with those of Lobry de Bruyn (1) and Gerrard (2) at 283.15 K for a total pressure of 1 atmosphere. The increasing difference between partial and total pressure causes the two lines to diverge more and more as the temperature increases. The datum of Sano and Nakamoto (5) is considerably smaller than those of Gerrard (2) and Lobry de Bruyn (1) and is classified as doubtful.

REFERENCES:

1. Lobry de Bruyn, C.A. *Rec. Trav. Chim.* 1892, *11*, 112.

2. Gerrard, W. *J. Appl. Chem. Biotechnol.* 1972, *22*, 623.

3. Bekarek, V.; Hala, E. *Coll. Czech. Chem. Comm.* 1968 *33*, 2598.

4. Tokunaga, J. *J. Chem. Eng. Data.* 1974, *19*, 162.

5. Sano, H.; Nakamoto, Y. *Nippon Kagaku Zasshi* 1968, *89*, 369.

COMPONENTS:

1. Sulfur dioxide; SO_2; [7446-09-5]

2. Methanol; CH_4O; [67-56-1]

ORIGINAL MEASUREMENTS:

Lobry de Bruyn, C.A.

(a) *Rec.Trav.Chim.* 1892, *11*, 112-157.

(b) *Z.Phys. Chem.* 1892, *10*, 782-789.

VARIABLES:

PREPARED BY:

W. Gerrard.

EXPERIMENTAL VALUES:

	T/K	Weight %, SO_2	Mole ratio, ROH/SO_2	*Mole ratio SO_2/ROH	Mole fraction* x_{SO_2}
Ref. (a)	273.15	71.1	0.81	1.23	0.552
	280.15	59.9	1.33	0.747	0.428
	285.45	52.2	1.83	0.546	0.353
	290.95	44	2.55	0.393	0.282
	299.15	31.7	4.25	0.232	0.188
Ref. (b)	273.15	247 g SO_2/100 g CH_3OH		1.235	0.553
	299.15	47 g SO_2/100 g CH_3OH		0.235	0.190

(The primary data for the expression in (b) were probably the same as in (a).)

* Calculated by compiler, from wt %.

Total pressure = atmospheric pressure.

AUXILIARY INFORMATION

METHOD/APPARATUS/PROCEDURE:

The dry gas was passed into a weighed amount of methanol in a bulb of 2-3 cm^3 capacity until equilibrium had been reached at the observed temperature and at a total pressure (barometric, 760-770 mmHg). The bulb was sealed "a la lampe", and weighed. A titration was also mentioned, but the statement is vague.

SOURCE AND PURITY OF MATERIALS:

1. Simply stated that the dry gas was used.

2. It was stated that the pure alcohol was used.

ESTIMATED ERROR:

REFERENCES:

COMPONENTS:

1. Sulfur dioxide; SO_2; [7446-09-5]
2. Methanol; CH_4O; [67-56-1]
   Ethanol; C_2H_6O; [64-17-5]

ORIGINAL MEASUREMENTS:

Sano, H.; Nakamoto, Y.
Nippon Kagaku Zasshi
1968, *89*, 369-373.

VARIABLES:

PREPARED BY:

C. L. Young

EXPERIMENTAL VALUES:

T/°C = 30 T/K = 303

Vapor pressure of methanol = 0.125 atm
Vapor pressure of ethanol = 0.069 atm

Absorption coefficient, $S^{a,b}$	Bunsen coefficient,[b] α	Bunsen coefficient,[c] α	$\ell^{c,d}$	Mole ratio	Mole fraction[e]
		Methanol			
116.2	122.1	124.5	99.9	0.241	0.194
		Ethanol			
88.1	85.9	84.0	75.9	0.227	0.185

[a] Volume of sulfur dioxide absorbed by unit volume of solvent at experimental temperature and a total pressure of 1 atmosphere.

[b] Determined by static absorption method.

[c] Determined by flow method.

[d] $\alpha = \ell(1 - p_s^\circ)$ where p_s° is the vapor pressure of solvent.

[e] Calculated by compiler.

AUXILIARY INFORMATION

METHOD/APPARATUS/PROCEDURE:

Two techniques were used: (a) static absorption method and (b) flow method. Static method consisted of equilibrating solvent and sulfur dioxide in a 50 ml flask and measuring the absorption volumetrically. The flow method consists of passing sulfur dioxide through a micro gas absorption tube and estimating sulfur dioxide by iodometric titration. The vapor pressure of the solvents were measured roughly and were assumed to be equal to the partial pressure of the solvent by the authors when calculating the Bunsen coefficient.

SOURCE AND PURITY OF MATERIALS:

1. Purity 99.0-99.8 mole per cent.

ESTIMATED ERROR:

$\delta T/K = \pm 0.5$; $\delta x_{SO_2} = \pm 5\%$

(estimated by compiler).

REFERENCES:

COMPONENTS:

1. Sulfur dioxide; SO_2; [7446-09-5]
2. Methanol; CH_4O; [67-56-1]

ORIGINAL MEASUREMENTS:

Bekarek, V.; Hala, E. *Coll. Czech. Chem. Comm.* 1968, *33*, 2598-2607.

VARIABLES:

Temperature, pressure

PREPARED BY:

C.L. Young

EXPERIMENTAL VALUES:

T/K	P/Torr[+]	P/kPa[+]	Mole fraction of sulfur dioxide in liquid, x_{SO_2}	in gas, y_{SO_2}
293.2	101.40	13.519	0.0043	0.0508
	109.10	14.545	0.0111	0.1230
	113.50	15.132	0.0144	0.1570
	136.20	18.159	0.0314	0.3100
303.2	169.40	22.585	0.0050	0.0542
	182.6	24.34	0.0118	0.1230
	183.00	24.398	0.0122	0.1260
	186.00	24.798	0.0135	0.1390
	208.6	27.81	0.0248	0.239
	234.20	31.224	0.0375	0.3310
	294.30	39.237	0.0619	0.4370
313.2	278.60	37.144	0.0058	0.0576
	298.30	39.770	0.0134	0.1280
	300.80	40.103	0.0139	0.1320
	344.50	45.930	0.0288	0.2480
	418.70	55.822	0.0518	0.3990

[+] Total pressure
760 Torr = 760 mmHg = 1 atm = 101.325 kPa

AUXILIARY INFORMATION

METHOD/APPARATUS/PROCEDURE:

A modified Gillespie still was used. Sulfur dioxide in liquid and gaseous phases estimated by iodometric titration. A small bulb containing a sample of the solution (0.5 - 1g) was broken in a solution of iodine and potassium iodide. Details in source and ref. (1). Same data given in ref. (2).

SOURCE AND PURITY OF MATERIALS:

1. Dried with sulfuric acid and phosphorus pentoxide.
2. Analytical grade, fractionally distilled; d^4_{20} = 0.7915.

ESTIMATED ERROR:

$\delta T/K = \pm 0.1$; $\delta P/kPa = \pm 0.1\%$
δx_{SO_2}, $\delta y_{SO_2} = \pm 0.5\%$.
(estimated by compiler.

REFERENCES:

1. Dvorak, K.; Boublik, T. *Coll. Czech. Chem. Comm.* 1963, *28*, 1252.
2. Bekarek, V.; Hala, E. *Coll. Czech. Chem. Comm.* 1968, *33*, 2608.

COMPONENTS:

1. Sulfur dioxide; SO_2; [7446-09-5]
2. Methanol; CH_4O; [67-56-1]
3. 2-Propanone (Acetone); C_3H_6O; [67-64-1]

ORIGINAL MEASUREMENTS:

Bekarek, V.

Coll. Czech. Chem. Comm. 1968, *33*, 2608-2619.

VARIABLES:

Temperature, pressure

PREPARED BY:

C.L. Young

EXPERIMENTAL VALUES:

T/K	p/Torr	p/kPa	Mole fractions in liquid x_{SO_2}	in liquid $x_{C_3H_6O}$	in gas y_{SO_2}	in gas $y_{C_3H_6O}$
293.2	161.80	21.572	0.0407	0.1500	0.2330	0.2380
	174.20	23.225	0.0547	0.1050	0.3450	0.1640
	198.70	26.491	0.0732	0.3600	0.2670	0.3880
	206.00	27.464	0.0797	0.2010	0.3840	0.2390
	222.30	29.638	0.0918	0.0942	0.5216	0.1090
	245.70	32.757	0.1410	0.5790	0.3960	0.4460
303.2	278.00	37.064	0.0449	0.2620	0.1880	0.3300
	264.70	35.290	0.0457	0.0975	0.2950	0.1550
	268.80	35.837	0.0462	0.1570	0.2510	0.2260
	311.60	41.543	0.0722	0.2120	0.3330	0.2510
	316.50	42.197	0.0737	0.1280	0.4100	0.1600
	335.70	44.756	0.0930	0.4190	0.3010	0.3950
	403.20	53.756	0.1620	0.5900	0.4340	0.4180
313.2	390.20	52.022	0.0348	0.1440	0.1930	0.2160
	390.20	52.022	0.0590	0.1920	0.2800	0.2430
	452.50	60.328	0.0579	0.3140	0.2200	0.3560
	533.00	71.060	0.0915	0.1420	0.4520	0.1510
	524.10	69.874	0.1020	0.4190	0.3200	0.3660
	583.50	77.794	0.1190	0.2440	0.4600	0.2110
	634.20	84.553	0.1800	0.6020	0.4690	0.4150

AUXILIARY INFORMATION

METHOD/APPARATUS/PROCEDURE:

A modified Gillespie still was used. Sulfur dioxide in liquid and gaseous phases estimated by iodometric titration. Details in source and ref. (1).

To determine the acetone concentration a bulb containing (0.5 - 1.0g) of mixture was broken in a solution containing iodine just sufficient to convert sulfur dioxide to sulfuric acid. Acetone was then determined by Messinger's method (ref.(2))

SOURCE AND PURITY OF MATERIALS:

1. Dried with sulfuric acid and phosphorus pentoxide.
2. Analytical grade, fractionally distilled; d_{20}^{4} = 0.7915.
3. Analytical grade, fractionally distilled; d_{20}^{4} = 0.7902.

ESTIMATED ERROR:

$\delta T/K = \pm 0.1$; $\delta p/kPa = \pm 0.1\%$; δx_{SO_2}, $\delta x_{C_3H_6O}$, δy_{SO_2}, $\delta y_{C_3H_6O}$ = ±1%. (estimated by compiler).

REFERENCES:

1. Dvorak, K.; Boublik, T. *Coll. Czech. Chem. Comm.* 1963, *28*, 1252.
2. Goodwin, L.F.; *J. Am. Chem. Soc.*, 1920, *42*, 39.

COMPONENTS:	ORIGINAL MEASUREMENTS:
1. Sulfur dioxide; SO_2; [7446-09-5] 2. Methanol; CH_4O; [67-56-1] 3. Acetic acid, methyl ester (Methyl acetate); $C_3H_6O_2$; [79-20-9]	Bekarek, V. *Coll. Czech. Chem. Comm.* 1968, *33*, 2608-2619
VARIABLES: Temperature, pressure	PREPARED BY: C.L. Young

EXPERIMENTAL VALUES:

T/K	p/Torr	p/kPa	Mole fractions in liquid x_{SO_2}	in liquid $x_{C_3H_6O_2}$	in gas y_{SO_2}	in gas $y_{C_3H_6O_2}$
293.2	131.00	17.465	0.0130	0.0544	0.1070	0.2000
	134.80	17.972	0.0144	0.0591	0.1130	0.2100
	153.00	20.398	0.0245	0.0939	0.1630	0.2620
	148.00	19.732	0.0249	0.0685	0.1850	0.2230
	158.80	21.172	0.0107	0.1970	0.0552	0.4300
	160.60	21.412	0.0390	0.0572	0.2870	0.1640
	181.10	24.145	0.0448	0.1570	0.2420	0.3070
	223.80	29.838	0.0853	0.2660	0.3530	0.3290
303.2	222.10	29.611	0.0162	0.0653	0.1210	0.2060
	234.10	31.211	0.0112	0.1170	0.0693	0.3100
	228.00	30.397	0.0173	0.0696	0.1230	0.2160
	246.60	32.877	0.0300	0.0747	0.2060	0.2040
	260.60	34.744	0 0306	0.1170	0.1820	0.2650
	269.50	35.930	0.0447	0.0594	0.3080	0.1540
	306.50	40.863	0.0547	0.1820	0.2650	0.3150
	368.30	49.103	0.0936	0.2140	0.3960	0.2610
313.2	358.90	47.849	0.0123	0.0960	0.0802	0.2540
	367.70	49.023	0.0207	0.0799	0.1360	0.2130
	357.00	47.596	0.0210	0.0847	0.1350	0.2200
	399.40	53.249	0.0358	0.0692	0.2350	0.1740
	428.10	57.075	0.0374	0.1410	0.2010	0.2580
	440.50	58.729	0.0521	0.0573	0.3360	0.1170
	508.30	67.768	0.0699	0.2320	0.2900	0.3100
	597.10	79.607	0.1080	0.1870	0.4450	0.2110

AUXILIARY INFORMATION

METHOD/APPARATUS/PROCEDURE:

A modified Gillespie still was used. Sulfur dioxide in liquid and gaseous phases estimated by iodometric titration. Details in source and ref. (1).

To determine the concentration of methyl acetate, a bulb containing a sample of saturated solution was broken in a known amount of sodium hydroxide soln., heated to hydrolyse the ester, sulfur dioxide oxidised to sulfuric acid and the excess alkali was back titrated

SOURCE AND PURITY OF MATERIALS:

1. Dried with sulfuric acid and phosphorus pentoxide.
2. Analytical grade, fractionally distilled, d_{20}^{4} = 0.7915.
3. Added to 5 per cent acetic anhydride and fractionally distilled.

ESTIMATED ERROR:

$\delta T/K = \pm 0.1$; $\delta p/kPa = \pm 0.1\%$. δx_{SO_2}, $\delta x_{C_3H_8O_2}$, δy_{SO_2}, $\delta y_{C_3H_8O_2} = \pm 1\%$ (estimated by compiler).

REFERENCES:

1. Dvorak, K.; Boublik, T. *Coll. Czech. Chem. Comm.* 1963, *28*, 1252.

COMPONENTS:	ORIGINAL MEASUREMENTS:
1. Sulfur dioxide; SO_2; [7446-09-5] 2. Methanol; CH_4O; [67-56-1]	Gerrard, W. *J. Appl. Chem. Biotechnol.* 1972, *22*, 623-650.
VARIABLES:	PREPARED BY:
Temperature	C. L. Young

EXPERIMENTAL VALUES:

T/K	*P*/kPa	Mole ratio SO_2/CH_4O	* Mole fraction x_{SO_2}
273.15	101.3	1.23	0.581
283.15		0.620	0.383
293.15		0.355	0.262
298.15		0.27	0.21

* Calculated by compiler.

P is the total pressure.

AUXILIARY INFORMATION

METHOD/APPARATUS/PROCEDURE:

Sulfur dioxide was bubbled into a weighed amount of component 2 in a bubbler tube as described in detail, diagram given, in the original paper. The amount of gas absorbed at equilibrium and at the observed temperature and pressure was weighed. By means of a manometer assembly, to which the bubbler tube was attached, the weight of gas absorbed at successively lower pressure was measured. Eventually the pressure was reduced to that of the component 2. The refractive index, and the infrared spectrum of the residue showed it to be essentially pure component 2.

SOURCE AND PURITY OF MATERIALS:

1. Sulfur dioxide was the best grade contained in a small cylinder. Its high purity was attested by the measurement of vapor pressure from 258 to 268 K.

2. The best obtainable specimen was carefully purified and rigorously attested.

ESTIMATED ERROR:

REFERENCES:

COMPONENTS:	EVALUATOR:
1. Sulfur dioxide; SO_2; [7446-09-5] 2. Ethanol; C_2H_6O; [64-17-5]	Colin L. Young School of Chemistry, University of Melbourne Parkville, Victoria, 3052 AUSTRALIA June, 1981

CRITICAL EVALUATION:

The mole fraction values based on the data of Carius (1), Lobry de Bruyn (2) and Gerrard (3) are for a total pressure of 1 atmosphere. It is again important to distinguish between measurements reported for a total pressure of 1 atmosphere and a partial pressure of 1 atmosphere particularly at the higher temperatures. The vapor pressure of pure ethanol are 12.2 mmHg (273K); 23.8 mmHg (283K); 44.0 mmHg (293K) 78.1 mmHg (303K) and 133.5 mmHg (313K). The data of Tokunaga (4) are for a partial pressure of 1 atm. There is fair agreement between the data of Tokunaga (4) and the above three sets of data at 273 to 283 K. The data of Barclay and Butler (5) are for low partial pressures of sulfur dioxide. These data cannot be extrapolated up to 1 atmosphere pressure because there is not a linear relationship between pressure and mole fraction i.e. Henry's law is not obeyed up to 1 atmosphere. This point is also obvious when the data of Nitta *et al* (6) are considered. These workers studied the mixture over the whole composition range at 293.15K. The data of Yuferev and Maluigin (7) are in reasonable agreement with the data from references (1), (2), (3), (4) and (6) if it is assumed the mole fraction solubility refers to 101.3kPa total pressure. The data of Nitta *et al* (6) are thought to be most reliable at 293.15K whereas those of Gerrard (3) are the most reliable at the lower temperatures. The data of de Saussure (8) are of historical interest only. The mole fraction solubility of Sano and Nakamoto (9) is slightly smaller than the value obtained by extrapolation of the data of Gerrard (3) to 303 K.

REFERENCES:

1. Carius, L. *Annalen*, 1855, *94*, 129.
2. Lobry de Bruyn, C.A. *Rec. Trav. Chim.* 1892, *11*, 112 and *Z. Phys. Chem.* 1892, *10*, 782.
3. Gerrard, W. *J. Appl. Chem. Biotechnol.* 1972, *22*, 623.
4. Tokunaga, J. *J. Chem. Eng. Data.* 1974, *19*, 162.
5. Barclay, I.M.; Butler, J.A.V.; *Trans Faraday Soc.* 1938. *34*, 1445
6. Nitta, T.; Itami, J.; Katayama, T.; *J. Chem. Engng. Japan.* 1973, *6*, 303.
7. Yuferev, R.F.; Maluigan, P.V. *Z. Chim. Prom.* 1930, *7*, 553.
8. de Saussure, T. *Gilbert's Annalen der Physik.* 1814, *47*, 113. *Annals of Phil*, 1815, *6*, 241.
9. Sano, H.; Nakamoto, Y. *Nippon Kagaku Zasshi* 1968, *89*, 369.

COMPONENTS:

1. Sulfur dioxide; SO_2; [7446-09-5]
2. Ethanol: C_2H_6O; [64-17-5]

ORIGINAL MEASUREMENTS:

Carius, L. *Annalen*, 1855, *94*, 129-166

VARIABLES:

PREPARED BY:

W. Gerrard.

EXPERIMENTAL VALUES: 760 mmHg = 1 atm = 101.325 kPa

Absorption coefficient, Ab.Coef., is the volume of sulfur dioxide absorbed by 1 volume of ethanol or solution at a total pressure of 1 atm. The Ab. coef. refers to the volume of gas after adjustment to 273.15 K. To convert the Ab.coef. into the mole ratio, M.R., the molar volume of sulfur dioxide was taken to be 21887 cm^3 at 273.15 K and 1 atm, based on the published density of the gas at 273.15 K and 1 atm. The density of ethanol at the stated temperature was that cited by Timmermans (1). Carius gave a table showing the Ab.Coef. for a saturated solution; the data are combined in the table given here. Carius also gave a table showing density of the solution at a stated temperature, shown below.

T/K	Ab.Coef. m^3/m^3 (soln)	Ab.Coef. m^3/m^3 (ethanol)	* M.R. SO_2/EtOH	*Mole fraction x_{SO_2}
276.35	188.34	276.76	0.720	0.419
278.95	170.67	239.66	0.625	0.385
284.15	135.48	176.93	0.464	0.317
287.15	120.00	149.98	0.395	0.283
290.15	106.03	129.33	0.342	0.255
293.15	96.34	114.50	0.304	0.233
297.55	84.03	97.43	0.260	0.206

T/K	Density of saturated solution g/cm^3
277.15	1.0622
284.75	0.9880
289.15	0.9527
293.25	0.9402
296.65	0.9282

*Calculated by compiler.

Carius plotted the Ab.coef.(volume of saturated solution) vs temperature, and read off values for each degree from 273 to 298 K. These were tabulated and attached to a smoothing equation. From the densities of the saturated solutions, a table showing the Ab.coef. for the volume of alcohol itself was constructed. Again, from a plot of Ab.coef.(ethanol) vs temperature, a table showing the value at each degree was given, and accompanied by a smoothing equation which is not correct.

AUXILIARY INFORMATION

METHOD/APPARATUS/PROCEDURE:

The amount of sulfur dioxide absorbed was determined by iodometry. The Bunsen absorptiometer was mentioned.

SOURCE AND PURITY OF MATERIALS:

1. Sulfur dioxide was stated to be dry.
2. Ethanol was water free, density 0.792 g/cm^3 at 293 K.

ESTIMATED ERROR:

REFERENCES:

1. Timmermans, J. "*Physico-chemical Constants of Pure Organic Compounds*," Elsevier, New York, 1950.

COMPONENTS:	ORIGINAL MEASUREMENTS:
1. Sulfur dioxide; SO_2; [7446-09-5] 2. Ethanol; C_2H_6O: [64-17-5]	Lobry de Bruyn, C.A. (a) *Rec. Trav. Chim.* 1892, *11*, 112-157. (b) *Z. Phys. Chem.* 1892, *10*, 782-789.
VARIABLES:	PREPARED BY: W. Gerrard.

EXPERIMENTAL VALUES:

	T/K	Weight % SO_2	Mole ratio ROH / SO_2	*Mole ratio SO_2/ROH	Mole fraction* x_{SO_2}
Ref. (a)	273.15	53.3	1.21	0.827	0.453
	280.15	45	1.70	0.588	0.370
	285.45	39.9	2.09	0.477	0.323
	291.35	32.8	2.85	0.351	0.260
	299.15	24.4	4.31	0.232	0.188
Ref. (b)	273.15	115 g SO_2 / 100 g ROH		0.827	0.453
	299.15	32.3g SO_2/ 100 g ROH		0.232	0.188

(The primary data for expression (b) were probably the same as in (a).)

* Calculated by compiler.

Total pressure = atmospheric pressure.

AUXILIARY INFORMATION

METHOD APPARATUS/PROCEDURE:	SOURCE AND PURITY OF MATERIALS:
The dry gas was passed into a weighed amount of ethanol in a bulb of 2-3 cm^3 capacity until equilibrium had been reached at the observed temperature, and at a total pressure, barometric, 760-770 mmHg. The bulb was sealed "a la lampe," and weighed. A titration was mentioned, but the statement is vague.	1. Simply stated that the dry gas was used. 2. Probably purified.
	ESTIMATED ERROR:
	REFERENCES:

COMPONENTS:

1. Sulfur dioxide; SO_2; [7446-09-5]
2. Ethanol; C_2H_6O; [64-1705]

ORIGINAL MEASUREMENTS:

Yuferev, R.F.; Maluigin, P.V.

Z. Chim. Prom. 1930, *7*, 553-555.

VARIABLES:

Temperature

PREPARED BY:

W. Gerrard

EXPERIMENTAL VALUES:

T/K	Weight of SO_2 per cm^3 of solvent g.	Volume of SO_2 per cm^3 of solvent cm^3	*Mole ratio SO_2/ethanol	Mole fraction x_{SO_2}
273.15	0.9528	333.48	0.851	0.461
292.65	0.3036	106.26	0.276	0.216
"Wet" ethanol				
273.15	0.9758	341.53		
292.65	0.3066	107.31		

* Calculated by compiler.

AUXILIARY INFORMATION

METHOD/APPARATUS/PROCEDURE:

Sulfur dioxide was passed into 5 cm^3 of liquid (component 2) until saturated. The whole solution was then transferred to a stoppered flask containing an aqueous solution of I_2 + KI. The sulfuric acid formed was determined gravimetrically. The total pressure was presumably 1 atm; but this was not stated.

SOURCE AND PURITY OF MATERIALS:

1. Sulfur dioxide dried by sulfuric acid.
2. Heated over calcium oxide, distilled, and stored over pulverized calcium.

   ("Wet" ethanol was ethanol dried over calcium oxide, but not distilled.)

ESTIMATED ERROR:

REFERENCES:

COMPONENTS:

1. Sulfur dioxide; SO_2; [7446-09-5]
2. Ethanol; C_2H_6O; [64-17-5]

ORIGINAL MEASUREMENTS:

Barclay, I.M.; Butler, J.A.V.;

Trans. Faraday. Soc. 1938, *34*, 1445-1454.

VARIABLES:

PREPARED BY:

W. Gerrard

EXPERIMENTAL VALUES:

T/K	Pressure, p_{SO_2}, mmHg.	Mole fraction x_{SO_2}	p_{SO_2}/x_{SO_2} mmHg*
298.15	4.731	0.002223	2129
	4.699	0.002206	2130
308.15	6.85	0.002259	3035
	6.74	0.002224	3029

760 mmHg = 1 atm = 101.325 kPa.

AUXILIARY INFORMATION

METHOD: /APPARATUS/PROCEDURE:

Dry nitrogen was passed through a bubbler containing the solution at a stated T/K. The saturated vapor emerging was condensed in a tube cooled by solid carbon dioxide and ether. The amount of sulfur dioxide in the condensate was determined by an iodemetric titration.

SOURCE AND PURITY OF MATERIALS:

Stated that the materials were carefully purified and dried.

Ethanol had b.p. 78.47^{0}C at 1 atm.

ESTIMATED ERROR:

REFERENCES:

COMPONENTS:	ORIGINAL MEASUREMENTS:
1. Sulfur dioxide; SO_2; [7446-09-5] 2. Ethanol; C_2H_6O; [64-17-5]	Gerrard, W. *J. Appl. Chem. Biotechnol.* 1972, *22*, 623-650.
VARIABLES:	PREPARED BY:
Temperature	W. Gerrard.

EXPERIMENTAL VALUES:

T/K	P/kPa	Mole ratio SO_2 / ROH	*Mole fraction x_{SO_2}
273.15	102.5	1.10	0.524
283.15		0.564	0.361
293.15		0.357	0.263

* Calculated by compiler.

P is the total pressure.

AUXILIARY INFORMATION

METHOD/APPARATUS/PROCEDURE:

Sulfur dioxide was bubbled into a weighed amount of component (2) in a bubbler tube as described in detail, diagram given, in the original paper. The amount of gas absorbed at equilibrium and at the observed temperature and pressure was weighed. By means of a manometer assembly, to which the bubbler tube was attached, the weight of gas absorbed at successively lower pressure was measured. Eventually the pressure was reduced to that of the component (2). The refractive index, and the infrared spectrum of the residue showed it to be essentially pure component (2).

SOURCE AND PURITY OF MATERIALS:

1. Sulfur dioxide was the best grade contained in a small cylinder. Its high purity was attested by the measurement of vapor pressure from 258 to 268 K.
2. The best obtainable specimen was carefully purified and rigorously attested.

ESTIMATED ERROR:

REFERENCES:

COMPONENTS:	ORIGINAL MEASUREMENTS:
1. Sulfur dioxide; SO_2; [7446-09-5] 2. Ethanol; C_2H_6O; [64-17-5]	Nitta, T.; Itami, J.; Katayama, T., *J. Chem. Engng Japan*, 1973, *6*, 303-9.
VARIABLES:	PREPARED BY:
Pressure	C. L. Young

EXPERIMENTAL VALUES:

T/K	p/mmHg[+]	p/kPa[+]	Mole fraction of sulfur dioxide in liquid, x_{SO_2}
293.15	414.8	55.30	0.1415
	796.6	106.2	0.2516
	1223	163.1	0.3866
	1379	183.9	0.4468
	1438	191.7	0.4705
	1526	203.4	0.5042
	1694	225.8	0.5847
	1739	231.8	0.5881
	1846	246.1	0.6507
	2052	273.6	0.7580
	2236	298.1	0.8617
	2452	326.9	1.000

[+] Total pressure

AUXILIARY INFORMATION

METHOD/APPARATUS/PROCEDURE:

Static equilibrium cell with magnetic stirrer. After equilibrium established, pressure measured with Bourdon gauge and approximately 2 cm^3 of liquid analysed by iodometric titration.

SOURCE AND PURITY OF MATERIALS:

1. Seitetsu Chemical Co. sample purity better than 99.9 mole per cent.

2. Nakarai Chemicals spectrograde sample.

ESTIMATED ERROR:

$\delta T/K = \pm 0.05$; $\delta p/kPa = \pm 0.5$;

$\delta x_{SO_2} = \pm 1\%$ (estimated by compiler)

REFERENCES:

COMPONENTS:

1. Sulfur dioxide; SO_2; [7446-09-5]
2. Ethanol; C_2H_6O; [64-17-5] (containing water, *d*, 0.84)

ORIGINAL MEASUREMENTS:

de Saussure, T.

Gilbert's Annalen der Physik 1814, *47*, 113-183.

Annals of Phil. 1815, *6*, No. 4, 241.

VARIABLES:

PREPARED BY:

W. Gerrard

EXPERIMENTAL VALUES:

291.15 K — 100 volumes of ethanol absorb 11577 volumes of sulfur dioxide at, presumably, the barometric pressure. If these were the data for pure ethanol, the mole ratio would be 0.272, and the mole fraction, x_{SO_2}, would be 0.214 (calculated by compiler).

NOTE: The data for water are: 100 volumes of water absorb 4378 volumes of sulfur dioxide at barometric pressure. The mole ratio is 0.338, and the mole fraction, x_{SO_2}, is 0.0327 (calculated by compiler).

AUXILIARY INFORMATION

METHOD/APPARATUS/PROCEDURE:

The ethanol was freed as much as possible from air by boiling. The gas was held over mercury in a tube of 4 cm internal diameter. The alcohol was added, and the tube was shaken. The final volume was read after several days.

SOURCE AND PURITY OF MATERIALS:

No specific information.

ESTIMATED ERROR:

REFERENCES:

COMPONENTS:	EVALUATOR:
1. Sulfur dioxide; SO_2; [7446-09-5] 2. Organic compounds containing oxygen	C. L. Young, School of Chemistry, University of Melbourne, Parkville, Victoria 3052, Australia. November 1982

CRITICAL EVALUATION:

In general the measurements on these systems are such that a detailed critical evaluation of each system is not possible.

1-Octanol; $C_8H_{18}O$; [111-87-5]

Gerrard's (1) values of the mole fraction solubility extrapolated to 303 K give a value which is significantly larger than that found by Sano and Nakamoto (2) but it is difficult to assess which is the more accurate. A comparison of the mole fraction solubilities for the alkanes determined by Makranczy *et al.* (3) with data for decane (Gerrard) and hexane (Sano) suggests Gerrard's value (5) to be high and Sano's (4) to be low.

1,2-Ethanediol; $C_2H_6O_2$; [107-21-1]

This system has been investigated by four groups of workers (1,2, 6,7). The results of Foote and Fleisher (6) and Gerrard (1) are consistent but the degree of overlap in experimental conditions is small. The value of Sano and Nakamoto (2) is again significantly smaller than the value obtained by extrapolation of Gerrard's data.

2-Propanone; C_3H_6O; [67-64-1]

This system has been investigated by five groups (4, 8-11) but detailed comparisons between sets of data are difficult to make in view of the different range of temperatures employed. The data of Bekarek and Hala (10) are in general agreement with those of Horiuti (9). The isolated value of Sano (4) appears to be consistent with those of Bekarek and Hola although extrapolation over such a large range of pressure could obscure any discrepancy. The data of Lorimer *et al.* (11) cannot be meaningfully compared in detail with other data on this system because of the differences in temperature.

Acetic acid, methyl ester; $C_3H_6O_3$; [79-20-9]

The data of Horiuti (9) at 298.15 K and Bekarek and Hala (10) at 293.2, 303.2 and 313.2 K are consistent and are both classified as tentative. The isolated value of Sano (4) appears to be slightly low compared to a value obtained by extrapolation of the data of Bekarek and Hala (10).

Acetic anhydride; $C_4H_6O_3$; [108-24-7]

The mole fraction solubilities of Lloyd (12) for this system are considerably lower than the values at comparable temperatures and pressures obtained by Gerrard (1) and Sano (4) and are almost certainly in error. The isolated value of Sano (4) appears to be somewhat low when compared with Gerrard's (1) values. The data of Nitta, Kido and

(cont.)

COMPONENTS:	EVALUATOR:
1. Sulfur dioxide; SO_2; [7446-09-5] 2. Organic compounds containing oxygen	Colin L. Young, School of Chemistry, University of Melbourne, Parkville, Victoria 3052, Australia. November 1982

CRITICAL EVALUATION:

Katayama (13) are consistent with those of Gerrard (1).

4-Methyl-1,3-dioxolan-2-one (*Propylene carbonate*); $C_4H_6O_3$; [108-32-7]

The data of Lenoir *et al.* (7) and Benoit and Milanova (14) on this system were both determined at low partial pressures of sulfur dioxide. There is a discrepancy of about 25 per cent between the values of the Henry's law constant obtained by these workers at 298 K.

1,1'-Oxybisethane (*Diethyl ether*); $C_4H_{10}O$; [60-29-7]

There is reasonable agreement between the data of Nitta *et al.* (13) at 293.15 K and the value of Yuferev and Maluigin (15) at 291.65 K. The value of Sano (4) at 303 K appears to be too high.

Cyclohexanone; $C_6H_{10}O$; [108-94-1]

The value of Weissenberger and Hadwiger (16) at 293.15 K is much too low to be consistent with the data of Sano (4) at 303 K and that of Gerrard (5) at 283.15 K and 293.15 K and is almost certainly in error. Sano's value is somewhat lower than would be expected from any reasonable extrapolation of Gerrard's values but it is not possible to state which of the values are in error.

1,7,7-Trimethylbicyclo[2.2.1]heptan-2-one (*Camphor*); $C_{10}H_{16}O$; [464-49-3]

The data of Bineau (17) and the isolated values of Schulze (8) and Sano (5) are consistent within ±8 per cent.

References

1. Gerrard, W. *J. Appl. Chem. Biotechnol.* 1972, *22*, 623.
2. Sano, H.; Nakamoto, Y. *Nippon Kagaku Zasshi* 1968, *89*, 369.
3. Makranczy, J.; Megyery-Balog, K.; Rusz, L.; Patyi, L. *Hung. J. Ind. Chem.* 1976, *4*, 269.
4. Sano, H. *Nippon Kagaku Zasshi* 1968, *89*, 362.
5. Gerrard, W. *Solubility of Gases and Liquids*, Plenum Press, New York, 1976, p.239.
6. Foote, H. W.; Fleisher, J. *J. Am. Chem. Soc.* 1934, *56*, 870.
7. Lenoir, J.-Y.; Renault, P.; Renon, H. *J. Chem. Engng. Data* 1971, *16*, 340.
8. Schulze, H. *J. Prakt. Chem.* 1881, *24*, 168.
9. Horiuti, J. *Sci. Pap. Inst. Phys. Chem. Res. (Jpn.)* 1931/32, *17*, 125.
10. Bekarek, V.; Hala, E. *Coll. Czech. Chem. Comm.* 1968, *33*, 2598.
11. Lorimer, J. W.; Smith, B. C.; Smith, G. H. *J. C. S. Faraday I* 1975, *71*, 2232.

(cont.)

COMPONENTS:

1. Sulfur dioxide; SO_2; [7446-09-5]

2. Organic compounds containing oxygen

EVALUATOR:

Colin L. Young,
School of Chemistry,
University of Melbourne,
Parkville, Victoria 3052,
Australia.

November 1982

CRITICAL EVALUATION:

References (cont.)

12. Lloyd, S. J. *J. Phys. Chem.* 1918, *22*, 200.

13. Nitta, T.; Kido, O.; Katayama, T. *J. Chem. Engng. Japan* 1976, *9*, 317.

14. Benoit, R. L.; Milanova, E. *Can. J. Chem.* 1979, *57*, 1319.

15. Yuferev, R. F.; Maluigin, P. V. *Z. Chim. Prom.* 1930, *7*, 553.

16. Weissenberger, G.; Hadwiger, H. *Z. angew. Chem.* 1927, *40*, 734.

17. Bineau, A. *Ann. Chim. Phys.* 1848 (3), *24*, 326.

COMPONENTS:

1. Sulfur dioxide; CO_2; [7446-09-5]
2. 2-Propanol; C_3H_8O; [67-63-0]
   2-Methyl-2-propanol (*tert*-butanol); $C_4H_{10}O$; [75-65-0]

ORIGINAL MEASUREMENTS:

Sano, H.; Nakamoto, Y.
Nippon Kagaku Zasshi
1968, *89*, 369-373.

VARIABLES:

PREPARED BY:

C. L. Young

EXPERIMENTAL VALUES:

T/°C = 30 T/K = 303

Vapor pressure of 2-Propanol = 0.035 atm
Vapor pressure of 2-Methyl-2-propanol = 0.041 atm
Vapor pressure of 1-Butanol = 0.008 atm

Absorption coefficient, $S^{a,b}$	Bunsen coefficient,[b] α	Bunsen coefficient,[c] α	$\ell^{c,d}$	Mole ratio	Mole fraction[e]
		2-Propanol			
53.9	50.7	52.6	48.6	0.184	0.155
		2-Methyl-2-propanol			
47.9	45.7	45.4	42.2	0.191	0.160
		1-Butanol			
53.8	45.5	49.2	48.4	0.203	0.169

[a] Volume of sulfur dioxide absorbed by unit volume of solvent at experimental temperature and a total pressure of 1 atmosphere.
[b] Determined by static absorption method.
[c] Determined by flow method.
[d] $\alpha = \ell(1 - p_s^\circ)$ where p_s° is the vapor pressure of solvent.
[e] Calculated by compiler.

AUXILIARY INFORMATION

METHOD/APPARATUS/PROCEDURE:

Two techniques were used: (a) static absorption method and (b) flow method. Static method consisted of equilibrating solvent and sulfur dioxide in a 50 ml flask and measuring the absorption volumetrcally. The flow method consists of passing sulfur dioxide through a micro gas absorption tube and estimating sulfur dioxide by iodometric titration. The vapor pressure of the solvents were measured roughly and were assumed to be equal to the partial pressure of the solvent by the authors when calculating the Bunsen coefficient.

SOURCE AND PURITY OF MATERIALS:

1. Purity 99.0-99.8 mole per cent.

ESTIMATED ERROR:

$\delta T/K = \pm 0.5$; $\delta x_{SO_2} = \pm 5\%$ (estimated by compiler).

REFERENCES:

COMPONENTS:

1. Sulfur dioxide; SO_2; [7446-09-5]
2. 3-Methyl-1-butanol (iso-amyl alcohol); $C_5H_{12}O$; [123-51-3]
   1-Hexanol; $C_6H_{14}O$; [111-27-3]
   1-Octanol; $C_8H_{18}O$; [111-87-5]

ORIGINAL MEASUREMENTS:

Sano, H.; Nakamoto, Y.

Nippon Kagaku Zasshi

1968, *89*, 369-373.

VARIABLES:

PREPARED BY:

C. L. Young

EXPERIMENTAL VALUES:

$T/°C = 30$ $T/K = 303$

Vapor pressure of 3-Methyl-1-butanol = 0.006 atm
Vapor pressure of 1-Hexanol = 0.004 atm

Absorption coefficient, $S^{a,b}$	Bunsen coefficient,[b] α	Bunsen coefficient,[c,f] α	$\ell^{c,d}$	Mole ratio	Mole fraction[e]
		3-Methyl-1-butanol			
40.9	37.4	37.5	37.4	0.183	0.155
		1-Hexanol			
41.5	37.9	37.7	37.6	0.212	0.175
		1-Octanol			
28.7	26.1	25.5	25.5	0.183	0.155

[a] Volume of sulfur dioxide absorbed by unit volume of solvent at experimental temperature and a total pressure of 1 atmosphere.

[b] Determined by static absorption method.

[c] Determined by flow method.

[d] $\alpha = \ell(1 - p_s^\circ)$ where p_s° is the vapor pressure of solvent.

[e] Calculated by compiler.

[f] Assuming the vapor pressure of 1-octanol is negligible.

AUXILIARY INFORMATION

METHOD/APPARATUS/PROCEDURE:

Two techniques were used: (a) static absorption method and (b) flow method. Static method consisted of equilibrating solvent and sulfur dioxide in a 50 ml flask and measuring the absorption volumetrically. The flow method consists of passing sulfur dioxide through a micro gas absorption tube and estimating sulfur dioxide by iodometric titration. The vapor pressure of the solvents were measured roughly and were assumed to be equal to the partial pressure of the solvent by the authors when calculating the Bunsen coefficient.

SOURCE AND PURITY OF MATERIALS:

1. Purity 99.0-99.8 mole per cent.

ESTIMATED ERROR:

$\delta T/K = \pm 0.5$; $\delta x_{SO_2} = \pm 5\%$

(estimated by compiler).

REFERENCES:

COMPONENTS:	ORIGINAL MEASUREMENTS:
1. Sulfur dioxide, SO_2; [7446-09-5] 2. 1-Heptanol; $C_7H_{16}O$; [111-70-6]	Albright, L.F.; Shannon, P.T.; Yu, S.-N.; Chueh, P.L. *Chem. Engng. Prog. Symp. Ser.* 1963, *59*, (44), 66-74.
VARIABLES:	PREPARED BY:
Temperature, pressure	C.L. Young

EXPERIMENTAL VALUES:

T/K	P/bar	Mole fraction of sulfur dioxide in liquid, x_{SO_2}
298.15	2.04	0.377
	3.01	0.626
	3.48	0.831
310.93	2.76	0.363
	2.83	0.373
	3.85	0.482
	4.39	0.617
	5.21	0.831
	5.55	0.874
338.71	5.36	0.363
	8.43	0.589
	10.96	0.825
366.48	8.54	0.343
	14.17	0.586
	20.13	0.824

AUXILIARY INFORMATION

METHOD/APPARATUS/PROCEDURE:

Static equilibrium cell fitted with Bourdon pressure gauge. Temperature measured with mercury in glass thermometer. The equipment used in this investigation was stated to be similar to that used in previous studies, (1), (2).

SOURCE AND PURITY OF MATERIALS:

1. and 2. Commercial grade samples no other details given.

ESTIMATED ERROR:

$\delta T/K = \pm 0.2$; $\delta P/\text{bar} = \pm 0.3$; $\delta x_{SO_2} = \pm 1\%$.

REFERENCES:

1. Albright, L. F.; Shannon, P. T.; Terrier, F.; Chueh, P. L. *Am. Inst. Chem. Engrs. J.* 1962, *8*, 668.
2. Albright, L. F.; Buclez, P. C.; Pluche, C. R.; Doody, T. C. *ASHRAE Trans.* 1960, *66*, 423.

COMPONENTS:

1. Sulfur dioxide; SO_2; [7446-09-5]

2. 1-Octanol; $C_8H_{18}O$; [111-87-5]

ORIGINAL MEASUREMENTS:

Gerrard, W.

J. Appl. Chem. Biotechnol.

1972, *22*, 623-650.

VARIABLES:

Temperature

PREPARED BY:

W. Gerrard.

EXPERIMENTAL VALUES:

760 mmHg = 1 atm = 101.325 kPa

T/K	P/kPa	Mole ratio SO_2 /ROH	*Mole fraction x_{SO_2}
273.15	101.3	0.710	0.415
283.15		0.412	0.292
293.15		0.292	0.226
298.15		0.250	0.200

* Calculated by compiler.

P is the total pressure.

AUXILIARY INFORMATION

METHOD/APPARATUS/PROCEDURE:

Sulfur dioxide was bubbled into a weighed amount of component (2) in a bubbler tube as described in detail, diagram given, in the original paper. The amount of gas absorbed at equilibrium and at the observed temperature and pressure was weighed. By means of a manometer assembly, to which the bubbler tube was attached, the weight of gas absorbed at successively lower pressure was measured. Eventually the pressure was reduced to that of the component (2). The refractive index, and the infrared spectrum of the residue showed it to be essentially pure component (2).

SOURCE AND PURITY OF MATERIALS:

1. Sulfur dioxide was the best grade contained in a small cylinder. Its high purity was attested by the measurement of vapor pressure from 258 to 268 K.

2. The best obtainable specimen was carefully purified and rigorously attested.

ESTIMATED ERROR:

REFERENCES:

COMPONENTS:

1. Sulfur dioxide; SO_2; [7446-09-5]
2. Cyclohexanol; $C_6H_{12}O$; [108-93-0]
   Benzenemethanol (Benzyl alcohol); C_7H_8O; [100-51-6]

ORIGINAL MEASUREMENTS:

Sano, H.; Nakamoto, Y.
Nippon Kagaku Zasshi
1968, *89*, 369-373.

VARIABLES:

PREPARED BY:

C. L. Young

EXPERIMENTAL VALUES:

T/°C = 30   T/K = 303

Vapor pressure of cyclohexanol = 0.001 atm

Absorption coefficient, $S^{a,b}$	Bunsen coefficient,[b] α	Bunsen coefficient,[c,f] α	$\ell^{c,d}$	Mole ratio	Mole fraction[e]
		Cyclohexanol			
38.0	34.6	34.3	34.3	0.160	0.139
		Benzenemethanol			
66.3	60.3	59.5	59.5	0.280	0.219

[a] Volume of sulfur dioxide absorbed by unit volume of solvent at experimental temperature and a total pressure of 1 atmosphere.
[b] Determined by static absorption method.
[c] Determined by flow method.
[d] $\alpha = \ell(1 - p_s^\circ)$ where p_s° is the vapor pressure of solvent.
[e] Calculated by compiler.
[f] Assuming the vapor pressure of benzenemethanol is negligible.

AUXILIARY INFORMATION

METHOD/APPARATUS/PROCEDURE:

Two techniques were used: (a) static absorption method and (b) flow method. Static method consisted of equilibrating solvent and sulfur dioxide in a 50 ml flask and measuring the absorption volumetrically. The flow method consists of passing sulfur dioxide through a micro gas absorption tube and estimating sulfur dioxide by iodometric titration. The vapor pressure of the solvents were measured roughly and were assumed to be equal to the partial pressure of the solvent by the authors when calculating the Bunsen coefficient.

SOURCE AND PURITY OF MATERIALS:

1. Purity 99.0-99.8 mole per cent.

ESTIMATED ERROR:

$\delta T/K = \pm 0.5$; $\delta x_{SO_2} = \pm 5\%$

(estimated by compiler).

REFERENCES:

COMPONENTS:

1. Sulfur dioxide; SO_2; [7446-09-5]
2. Cyclohexanol; $C_6H_{12}O$; [108-93-0]

ORIGINAL MEASUREMENTS:

Gerrard, W. "*Solubility of Gases and Liquids,*" Plenum Press, New York, 1976, p.239

VARIABLES:

PREPARED BY:

W. Gerrard.

EXPERIMENTAL VALUES:

T/K	Weight of Component (2) g	Weight of SO_2 absorbed, g	Mole ratio SO_2 /Comp. (2)	Mole fraction x_{SO_2}
			(Total pressure = 1 atm)	
293.15	2.5570	0.4489	0.275	0.216

760 mmHg = 1 atm = 101.325 kPa.

AUXILIARY INFORMATION

METHOD:

Sulfur dioxide was bubbled into a weighed amount of component (2) in a bubbler tube. The amount of gas absorbed was determined by re-weighing the tube. The total pressure was barometric, very nearly 1 atm. See ref. (1).

SOURCE AND PURITY OF MATERIALS:

1. Sulfur dioxide was the best grade contained in a small cylinder. Its high degree of purity was attested by the measurement of vapor pressure from 258-268 K.
2. The best commercial specimen was purified and attested.

ESTIMATED ERROR:

REFERENCES:

1. Gerrard, W. *J. Appl. Chem. Biotechnol.* 1972, *22*, 623.

COMPONENTS:	ORIGINAL MEASUREMENTS:
1. Sulfur dioxide; SO_2; [7446-09-5] 2. Formic acid; CH_2O_2; [64-18-6] or Acetic acid, $C_2H_4O_2$; [64-19-7]	Schulze, H.; *J. Prakt. Chem.* 1881, *24*, 168-183.
VARIABLES:	PREPARED BY: W. Gerrard.

EXPERIMENTAL VALUES:

760 mmHg = 101.325 kPa.

T/K	Weight of SO_2/ g of comp.(2) g	*Mole ratio SO_2/Comp.(2)	*Mole fraction x_{SO_2}
	Formic Acid		
273.15	0.821	0.590	0.371
	Acetic Acid		
273.15	0.961	0.901	0.474

Pressure of SO_2 stated to be 725 mmHg.

1 volume of formic acid absorbed 351 volumes of SO_2

1 volume of acetic acid absorbed 318 volumes of SO_2

* Calculated by compiler.

AUXILIARY INFORMATION

METHOD/APPARATUS/PROCEDURE:

The gas was passed into a known weight of formic acid in a long thin walled test-tube fitted with a rubber stopper carrying an inlet and an outlet tube. The tube was cooled by melting snow. The amount of gas absorbed was determined by weighing.

SOURCE AND PURITY OF MATERIALS:

No details given.

ESTIMATED ERROR:

REFERENCES:

COMPONENTS:

1. Sulfur dioxide; SO_2 [7446-09-5]
2. Water; H_2O; [7732-18-5]
3. Acetic acid; $C_2H_4O_2$; [64-19-7]

ORIGINAL MEASUREMENTS:

Cupr, V.

Rec. Trav. Chim.

1928, *47*, 55-72.

VARIABLES:

PREPARED BY:

W. Gerrard

EXPERIMENTAL VALUES:

1 atm = 760 mmHg = 101.325 kPa

The absorption coefficient, designated by Cupr as L, appears to be the Bunsen coefficient; but there is some doubt about the use of the term "solution", which could mean the aqueous solution of acetic acid, i.e., taken as the original liquid; the initial water itself would then be simply the original liquid for that data, and the final acetic acid itself would be the original liquid for its data. However, Seidell and Linke (1) define this L as "cc. SO_2 per 1 cc sat. sol." when reporting this work.

T/K = 300.15 Total pressure = 1 atm

Concentration of acetic acid in aqueous solution % *	SO_2 in g absorbed by 100 g of acetic acid solution (liquid indicated in col. 1)	"Absorption coefficient" L
0.00	7.98	25.7
	8.01	25.8
	8.19	26.4
	8.12	26.2
	8.18	26.4
	8.11	26.1
16.72	8.86	31.5
16.74	8.80	31.3
45.58	12.04	44.1
44.73	11.92	43.7
63.51	16.35	61.6
83.59	22.83	84.8
91.89	26.49	97.8
100.00	31.15	112.8

* It is not clear if this is a weight or mole per cent.

AUXILIARY INFORMATION

METHOD/APPARATUS/PROCEDURE:

Sulfur dioxide was passed into the original liquid (the aqueous solution in the acetic acid + water mixtures) as for a total pressure of 1 atm. The contents of the final solution were determined by chemical titrations, acid-base, and iodine-thiosulfate.
The "absorption coefficient, L" was determined by calculation from the weight data, 22400 being apparently taken as the molar volume in cm^3 at 1 atm and 273.15 K. Based on the real molar volume, these L values are 2% too large.

SOURCE AND PURITY OF MATERIALS:

1. Sulfur dioxide was obtained from Merck, quality sodium sulfite and concentrated sulfuric acid, by which it was dried.
2. Water may be taken as of satisfactory purity.
3. Acetic acid was of Merck standard, and gave no residue when ignited on a platinum capsule.

ESTIMATED ERROR:

REFERENCES:

1. Seidell, A.; Linke, W. F.
Solubilities of Inorganic and Organic Compounds,
1958, 1965.

COMPONENTS:

1. Sulfur dioxide; SO_2; [7446-09-5]
2. Acetic acid; $C_2H_4O_2$; [64-19-7]
   Formic acid; CH_2O_2; [64-18-6]

ORIGINAL MEASUREMENTS:

Sano, H.; Nakamoto, Y.
Nippon Kagaku Zasshi
1968, *89*, 369-373.

VARIABLES:

PREPARED BY:

C. L. Young.

EXPERIMENTAL VALUES:

$T/°C = 30$ $T/K = 303$

Vapor pressure of acetic acid = 0.010 atm
Vapor pressure of formic acid = 0.040 atm

Absorption coefficient, $S^{a,b}$	Bunsen coefficient,[b] α	Bunsen coefficient,[c] α	$\ell^{c,d}$	Mole ratio	Mole fraction[e]
		Acetic acid			
106.0	97.4	96.2	95.2	0.251	0.201
		Formic acid			
61.2	57.8	60.0	55.5	0.103	0.094

[a] Volume of sulfur dioxide absorbed by unit volume of solvent at experimental temperature and a total pressure of 1 atmosphere.
[b] Determined by static absorption method.
[c] Determined by flow method.
[d] $\alpha = \ell(1 - p_s^\circ)$ where p_s° is the vapor pressure of solvent.
[e] Calculated by compiler.

AUXILIARY INFORMATION

METHOD/APPARATUS/PROCEDURE:

Two techniques were used: (a) static absorption method and (b) flow method. Static method consisted of equilibrating solvent and sulfur dioxide in a 50 ml flask and measuring the absorption volumetrically. The flow method consists of passing sulfur dioxide through a micro gas absorption tube and estimating sulfur dioxide by iodometric titration. The vapor pressure of the solvents were measured roughly and were assumed to be equal to the partial pressure of the solvent by the authors when calculating the Bunsen coefficient.

SOURCE AND PURITY OF MATERIALS:

1. Purity 99.0-99.8 mole per cent.

ESTIMATED ERROR:

$\delta T/K = \pm 0.5$; $\delta x_{SO_2} = \pm 5\%$
(estimated by compiler).

REFERENCES:

COMPONENTS:

1. Sulfur dioxide, SO_2; [7446-09-5]
2. Acetic acid; $C_2H_4O_2$; [64-19-7]

ORIGINAL MEASUREMENTS:

Nitta, T.; Kido , O; Katayama, T.;

J. Chem. Engng. Japan. 1976, *9*, 317-8

VARIABLES:

Pressure

PREPARED BY:

C.L. Young

EXPERIMENTAL VALUES:

T/K	P/bar	Mole fraction of sulfur dioxide in liquid, x_{SO_2}
293.15	0.0800	0.0207
	0.2758	0.0796
	0.7079	0.1971
	0.9758	0.2651
	1.176	0.3173
	1.537	0.4065
	1.901	0.4981
	2.338	0.6171
	2.613	0.7013
	2.936	0.8238

AUXILIARY INFORMATION

METHOD:

Static method liquid phase composition estimated by iodometry. Few details given in source. Method similar to that used in ref. (1).

SOURCE AND PURITY OF MATERIALS:

1. Seitetsu Chemical Co. sample, purity better than 99.9 mole per cent.
2. Merck analytical grade sample.

ESTIMATED ERROR:

$\delta T/K = \pm 0.1$; $\delta P/\text{bar} = \pm 0.001$ (below 1.3 bar) ± 0.005 (above 1.3 bar); $\delta x_{SO_2} = \pm 1.5\%$

REFERENCES:

Quitzich, K.; Ulbrecht, H.; Geiseler, G.

Z. Physik Chem. 1967, *234*, 33.

COMPONENTS:	ORIGINAL MEASUREMENTS:
1. Sulfur dioxide; SO_2; [7446-09-5] 2. Esters of formic acid	Sano, H. *Nippon Kagaku Zasshi* 1968, *89*, 362-368.
VARIABLES:	PREPARED BY: C. L. Young

EXPERIMENTAL VALUES:

T/°C = 30 T/K = 303

Bunsen coefficient, α	Mole ratio	Mole fraction[a]
Methyl formate; $C_2H_4O_2$; [107-31-3]		
331	0.920	0.479
Ethyl formate; $C_3H_6O_2$; [109-94-4]		
217	0.781	0.439

[a] Calculated by compiler.

AUXILIARY INFORMATION

METHOD/APPARATUS/PROCEDURE:

Static absorption method consisted of equilibrating solvent and sulfur dioxide in a 50 ml flask and measuring the absorption volumetrically. The vapor pressure of the solvents were measured roughly and were assumed to be equal to the partial pressure of the solvent by the authors when calculating the Bunsen coefficient.

SOURCE AND PURITY OF MATERIALS:

1. Purity 99.0-99.8 mole per cent.

ESTIMATED ERROR:

$\delta T/K = \pm 0.5$; $\delta x_{SO_2} = \pm 5\%$ (estimated by compiler).

REFERENCES:

COMPONENTS:

1. Sulfur dioxide; SO_2; [7446-09-5]
2. 1,2-Ethanediol (dihydroxyethane, ethylene glycol, glycol); $C_2H_6O_2$; [107-21-1]

ORIGINAL MEASUREMENTS:

Foote, H.W.; Fleisher, J.

J. Am. Chem. Soc. 1934, *56*, 870-873

VARIABLES:

PREPARED BY:

W. Gerrard

EXPERIMENTAL VALUES:

T/K	p_{SO_2}/mmHg	Mole fraction x_{SO_2}
273.15	1055	0.707
	1028	0.673
	951	0.575
	857	0.499
	758	0.436
	571	0.340
	427	0.272
	291	0.206
	168	0.138
	99.1	0.093
	66.7	0.068

NOTE: A mole fraction of 0.707 is equivalent to a mole ratio of 2.413 moles of SO_2 per mole of 1,2-ethanediol.

760 mmHg = 1 atm = 101.325 kPa. Observed pressure was taken to be p_{SO_2}.

AUXILIARY INFORMATION

METHOD:/APPARATUS/PROCEDURE:

The apparatus and method had been previously described (1). A glass bulb of 29.3 cm^3 capacity was connected via a tap to an open manometer, a vacuum flask of 2 dm^3 capacity, and a source of sulfur dioxide. Component (2) was weighed in the bulb. Sulfur dioxide was condensed in it at 253 K. Equilibrium was attained at 273.15K and at the observed pressure p_{Total}, assumed to be equal to p_{SO_2}.

SOURCE AND PURITY OF MATERIALS:

(1) Sulfur dioxide was dried by sulfuric acid.

(2) The glycol was of Eastman C.P. grade, or purified by customary method.

ESTIMATED ERROR:

REFERENCES:

1. Foote, H.W.; Fleischer, J.

J. Am. Chem. Soc. 1931, *53*, 1752.

COMPONENTS:

1. Sulfur dioxide; SO_2; [7446-09-5]
2. Diols and glycerol

ORIGINAL MEASUREMENTS:

Sano, H.; Nakamoto, Y.
Nippon Kagaku Zasshi
1968, *89*, 369-373.

VARIABLES:

PREPARED BY:

C. L. Young

EXPERIMENTAL VALUES: T/°C = 30 T/K = 303

Absorption coefficient, S[a,b]	Bunsen coefficient,[b] α	Bunsen coefficient,[c,f] α	ℓ[c,d]	Mole ratio	Mole fraction[e]
1,2-Ethanediol (*ethylene glycol*); $C_2H_6O_2$; [107-21-1]					
59.2	53.9	51.8	51.8	0.135	0.119
2,2'-Oxybisethanol (*diethylene glycol*); $C_4H_{10}O_3$; [111-46-6]					
130	119	119	119	0.500	0.333
1,2-Propanediol (*propylene glycol*); $C_3H_8O_2$; [57-55-6]					
68.0	61.9	61.7	61.7	0.204	0.169
1,2,3-Propanetriol (*glycerol*); $C_3H_8O_3$; [56-81-5]					
-	57.5	-	-	0.188	0.158

[a] Volume of sulfur dioxide absorbed by unit volume of solvent at experimental temperature and a total pressure of 1 atmosphere.
[b] Determined by static absorption method.
[c] Determined by flow method.
[d] $\alpha = \ell(1 - p_s^\circ)$ where p_s° is the vapor pressure of solvent.
[e] Calculated by compiler.
[f] Assuming vapor pressure of solvent to be negligible.

AUXILIARY INFORMATION

METHOD/APPARATUS/PROCEDURE:

Two techniques were used: (a) static absorption method and (b) flow method. Static method consisted of equilibrating solvent and sulfur dioxide in a 50 ml flask and measuring the absorption volumetrically. The flow method consists of passing sulfur dioxide through a micro gas absorption tube and estimating sulfur dioxide by iodometric titration. The vapor pressure of the solvents were measured roughly and were assumed to be equal to the partial pressure of the solvent by the authors when calculating the Bunsen coefficient.

SOURCE AND PURITY OF MATERIALS:

1. Purity 99.0-99.8 mole per cent.

ESTIMATED ERROR:

$\delta T/K = \pm 0.5$; $\delta x_{SO_2} = \pm 5\%$

(estimated by compiler).

REFERENCES:

COMPONENTS:	ORIGINAL MEASUREMENTS:
1. Sulfur dioxide; SO_2 [7446-09-5] 2. 1,2-Ethanediol (Ethylene glycol) $C_2H_6O_2$; [107-21-1]	Lenoir, J-Y.; Renault, P.; Renon, H. *J. Chem. Eng. Data* 1971, *16*, 340-342.
VARIABLES:	PREPARED BY: W. Gerrard.

EXPERIMENTAL VALUES:

760 mmHg = 1 atm = 101.325 kPa.

T/K	Henry's constant H_{SO_2}/atm	*Mole fraction at 1 atm.
298	3.41	0.293

* Calculated by compiler assuming a linear function of P_{SO_2} vs x_{SO_2}, i.e. x_{SO_2} (1 atm = $1/H_{SO_2}$.

The authors gave "Henry's constant" as

$$H_{SO_2} = \lim_{x_{SO_2} \to 0} \frac{f_{SO_2}(P,T)}{x_{SO_2}}$$

x_{SO_2} is the mole fraction. f_{SO_2} is the fugacity of SO_2 "in the equilibrium condition"of pressure and temperature. H_{SO_2} is related to experimental chromatographic parameters based on the number of moles of solvent in in the column, and the absolute retention time of the solute. Helium was the carrier gas.

AUXILIARY INFORMATION

METHOD/APPARATUS/PROCEDURE:	SOURCE AND PURITY OF MATERIALS:
A conventional gas-liquid chromatographic unit was used. The chromatograph was an IGC 12 M Intersmat model, equipped with a thermal conductivity detector. The temperature control was stated to be within ± 0.1°C. The pressure drop in the column was measured by a mercury manometer. Helium was the carrier gas. There is serious uncertainty in the estimation of accuracy.	1. Sulfur dioxide was provided by l'Air Liquide, and had a stated purity of 99.9%. It was used as provided. 2. Provided by Touzart & Matignon and Serlabo, and stated to have a minimum purity of 99%. Distillation was carried out when necessary, and the refractive index was measured.
	ESTIMATED ERROR:
	REFERENCES:

COMPONENTS:	ORIGINAL MEASUREMENTS:
1. Sulfur dioxide; SO_2; [7446-09-5] 2. 1,2-Ethanediol, (ethylene glycol); $C_2H_6O_2$; [107-21-1]	Gerrard, W. *J. Appl. Chem. Biotechnol.* 1972, *22*, 623-650.
VARIABLES: Temperature	PREPARED BY: W. Gerrard

EXPERIMENTAL VALUES:

T/K	*P*/kPa	Mole ratio SO_2 / ROH	*Mole fraction x_{SO_2}
273.15	101.5	0.83	0.448
283.15		0.420	0.296
293.15		0.310	0.237
298.15		0.255	0.203

* Calculated by compiler.

P is the total pressure.

AUXILIARY INFORMATION

METHOD /APPARATUS/PROCEDURE:

Sulfur dioxide was bubbled into a weighed amount of component (2) in a bubbler tube as described in detail, diagram given, in the original paper. The amount of gas absorbed at equilibrium and at the observed temperature and pressure was weighed. By means of a manometer assembly, to which the bubbler tube was attached, the weight of gas absorbed at successively lower pressure was measured. Eventually the pressure was reduced to that of the component (2). The refractive index, and the infrared spectrum of the residue showed it to be essentially pure component (2).

SOURCE AND PURITY OF MATERIALS:

1. Sulfur dioxide was the best grade contained in a small cylinder. Its high purity was attested by the measurement of vapor pressure from 258 to 268 K.
2. The best obtainable specimen was carefully purified and rigorously attested.

ESTIMATED ERROR:

REFERENCES:

COMPONENTS:

1. Sulfur dioxide; SO_2; [7446-09-5]
2. 2-Propanone (acetone, dimethyl ketone); C_3H_6O; [67-64-1]

ORIGINAL MEASUREMENTS:

Schulze, H.

J. Prakt. Chem. 1881, *24*, 168-183.

VARIABLES:

PREPARED BY:

W. Gerrard.

EXPERIMENTAL VALUES:

760 mmHg = 1 atm = 101.325 kPa.

T/K	Weight of SO_2 /1g of Comp. (2).	*Mole ratio SO_2/Comp. (2)	*Mole fraction x_{SO_2}
273.15	2.07	1.88	0.653

* Calculated by compiler.

Pressure of SO_2 stated to be 725 mmHg.

1 Volume of acetone absorbed 589 volumes of SO_2.

AUXILIARY INFORMATION

METHOD/APPARATUS/PROCEDURE:

The gas was passed into a weighed amount of acetone in a long thin-walled test-tube carrying a rubber stopper and an inlet tube and an outlet tube. The test-tube was cooled by melting snow. The amount of gas absorbed was determined by weighing.

SOURCE AND PURITY OF MATERIALS:

No details given.

ESTIMATED ERROR:

REFERENCES:

COMPONENTS:

1. Sulfur dioxide; SO_2; [7446-09-5]

2. 2-Propanone (acetone); C_3H_6O; [67-64-1]

ORIGINAL MEASUREMENTS:

Horiuti, J.

Sci. Pap. Inst. Phys. Chem. Res. (Jpn.), 1931/32, *17*, 125-256.

VARIABLES:

Temperature, pressure

PREPARED BY:

C. L. Young

EXPERIMENTAL VALUES:

T/K	$P^{\S}$/bar	Mole fraction of sulfur dioxide in liquid, x_{SO_2}
298.15	0.3053	0.000
	0.4053	0.142
	0.4142	0.160
	0.5086	0.237
	0.6016	0.294
	0.6307	0.308
	0.7647	0.368
	0.7682	0.372
	0.9415	0.430
	0.9860	0.445
	1.012	0.470
	1.141	0.483
	1.232	0.506
	1.325	0.519

T/K	$P^{\dagger}$/bar	Mole fraction of sulfur dioxide in liquid, x_{SO_2}	Ostwald coefficient, L
283.15	0.9531	0.615	276.4
298.15	0.8440	0.448	216.4
313.15	0.6008	0.267	171.3

§ total pressure.

† partial pressure, total pressure = 1 atmosphere = 1.01325×10^5 Pa.

Ostwald coefficient (as $x \rightarrow 0$, T/K = 298.15) = 398.

Henry's law constant, dp/dx (as $x \rightarrow 0$, T/K = 298.15) = 631 mmHg = 0.841 bar.

AUXILIARY INFORMATION

METHOD/APPARATUS/PROCEDURE:

Apparatus consisted of a gas buret, a solvent reservoir and an absorption pipet. Dry gas introduced into degassed solvent. System mixed using a magnetic stirrer until saturation occurred. Care was taken to prevent solvent vapor from mixing with gas in the gas buret. Volume of gas determined from gas buret reading; volume of liquid determined from height of meniscus in absorption pipet.

SOURCE AND PURITY OF MATERIALS:

1. Sample obtained from Nippon Sulphuric Acid Manufacturing Co. Dried and fractionated several times.
2. Extra pure sample, recrystallized with sodium sulfite. Dried and fractionated. B. pt. 56.09 °C.

ESTIMATED ERROR:

$\delta T/K = \pm 0.1$; $\delta P/\text{bar} = \pm 0.001$;

$\delta x_{SO_2} = \pm 0.001$ (estimated by compiler).

REFERENCES:

COMPONENTS:	ORIGINAL MEASUREMENTS:
1. Sulfur dioxide; SO_2; [7446-09-5] 2. 2-Propanone, (Acetone); C_3H_6O; [67-64-1]	Bekarek, V.; Hala, E. *Coll. Czech. Chem. Comm.* 1968 *33*, 2598-2607.
VARIABLES:	PREPARED BY:
Temperature, pressure.	C.L. Young.

EXPERIMENTAL VALUES:

T/K	P/Torr	P/kPa	Mole fraction of sulfur dioxide in liquid, x_{SO_2}	in gas, x_{SO_2}
293.2	192.20	25.625	0.0374	0.0914
	200.20	26.691	0.0600	0.1450
	205.30	27.371	0.0741	0.1790
	227.60	30.344	0.1230	0.3000
	237.00	31.597	0.1390	0.3430
	253.90	33.851	0.1660	0.4100
	280.00	37.330	0.2020	0.4980
	397.30	52.969	0.3130	0.7200
303.2	294.70	39.290	0.0343	0.0814
	306.60	40.877	0.0562	0.1340
	322.40	42.983	0.0829	0.2030
	350.10	46.676	0.1217	0.3010
	397.30	52.969	0.1759	0.4310
	444.40	59.248	0.2150	0.5180
	558.70	74.487	0.2902	0.6780
313.2	440.40	58.715	0.0325	0.0776
	454.80	60.635	0.0513	0.1240
	470.00	62.662	0.0693	0.1670
	527.00	70.261	0.1240	0.3030
	548.80	73.167	0.1410	0.3510
	572.50	76.327	0.1600	0.3940
	592.70	79.020	0.2340	0.5570

N.B. Same data given in ref.(2)

AUXILIARY INFORMATION

METHOD/APPARATUS/PROCEDURE:

A modified Gillespie still was used. Sulfur dioxide in liquid and gaseous phases estimated by iodometric titration. Details in source and ref. (1).

SOURCE AND PURITY OF MATERIALS:

1. Dried with sulfuric acid and phosphorus pentoxide.
2. Analytical grade, fractionally distilled, d_{20}^{4} = 0.7902.

ESTIMATED ERROR:

$\delta T/K = \pm 0.1$; $\delta P/kPa = \pm 0.1\%$; δx_{SO_2}, $\delta y_{SO_2} = \pm 0.5\%$.

(estimated by compiler).

REFERENCES:

1. Dvorak, K.; Boublik, T. *Coll. Czech. Chem. Comm.* 1963, *28*, 1252.
2. Bekarek, V.; Hala, E. *Coll. Czech. Chem. Comm.* 1968, *33*, 2608.

COMPONENTS:	ORIGINAL MEASUREMENTS:
1. Sulfur dioxide; SO_2; [7446-09-5] 2. 2-Propanone (*acetone*); C_3H_6O; [67-64-1] 2-Butanone (*methyl ethyl ketone*); C_4H_8O; [78-93-3] 4-Methyl-2-pentanone (*methyl isobutyl ketone*); $C_6H_{12}O$; [108-10-1]	Sano, H. *Nippon Kagaku Zasshi* 1968, *89*, 362-368.
VARIABLES:	PREPARED BY: C. L. Young

EXPERIMENTAL VALUES:

T/°C = 30 T/K = 303

Absorption coefficient, s[a]	Bunsen coefficient, α	Mole ratio	Mole fraction[b]
	2-Propanone		
182.1	201.2	0.665	0.400
	2-Butanone		
161.9	163.2	0.656	0.396
	4-Methyl-2-pentanone		
109.3	100.5	0.563	0.360

[a] Volume of sulfur dioxide absorbed by unit volume of solvent at experimental temperature and a total pressure of 1 atmosphere.

[b] Calculated by compiler.

AUXILIARY INFORMATION

METHOD/APPARATUS/PROCEDURE:

Static absorption method consisted of equilibrating solvent and sulfur dioxide in a 50 ml flask and measuring the absorption volumetrically. The vapor pressure of the solvents were measured roughly and were assumed to be equal to the partial pressure of the solvent by the authors when calculating the Bunsen coefficient.

SOURCE AND PURITY OF MATERIALS:

1. Purity 99.0-99.8 mole per cent.

ESTIMATED ERROR:

$\delta T/K = \pm 0.5$; $\delta x_{SO_2} = \pm 5\%$ (estimated by compiler).

REFERENCES:

COMPONENTS:

1. Sulfur dioxide; SO_2; [7446-09-5]

2. 2-Propanone, (Acetone); C_3H_6O; [67-64-1]

ORIGINAL MEASUREMENTS:

Lorimer, J.W.; Smith, B.C.; Smith, G.H.

J.C.S. Faraday I, 1975, *71*, 2232-50.

VARIABLES:

Temperature, pressure

PREPARED BY:

C.L. Young

EXPERIMENTAL VALUES:

T/K	p/mmHg	p/kPa	Mole fraction of sulfur dioxide in liquid, x_{SO_2}	in vapor, y_{SO_2}
227.6	3.76	0.501	0.0000	0.000
	3.99	0.532	0.1088	0.187
	5.32	0.709	0.2123	0.518
	8.77	1.169	0.3132	0.792
	10.79	1.439	0.3899	0.880
	24.11	3.214	0.4926	0.969
	39.87	5.316	0.5889	0.990
	46.95	6.259	0.6241	0.994
	72.24	9.631	0.7419	0.998
	96.61	12.880	0.8623	1.000
	116.72	15.561	1.0000	1.000
237.4	7.53	1.004	0.0000	0.000
	8.66	1.155	0.1090	0.238
	11.93	1.591	0.2109	0.544
	18.42	2.456	0.3120	0.778
	26.00	3.466	0.3861	0.881
	45.93	6.123	0.4884	0.959
	73.19	9.758	0.5848	0.985
	84.40	11.252	0.6166	0.990
	128.49	17.131	0.7364	0.997
	172.24	22.963	0.8602	0.999
	208.81	27.839	1.0000	1.000

AUXILIARY INFORMATION

METHOD/APPARATUS/PROCEDURE:

Static vapor-liquid equilibrium apparatus. Total pressure measured using mercury manometers and cathetometer. Temperature measured using a sulfur dioxide vapor pressure thermometer. Some details in source.

SOURCE AND PURITY OF MATERIALS:

No details given.

ESTIMATED ERROR:

$\delta T/K = \pm 0.1$; $\delta p/kPa = \pm 0.01$
δx_{SO_2}, $\delta y_{SO_2} = \pm 1\%$.

REFERENCES:

COMPONENTS:	ORIGINAL MEASUREMENTS:
1. Sulfur dioxide; SO_2; [7446-09-5] 2. 2-Propanone, (Acetone); C_3H_6O; [67-64-1]	Lorimer, J.W.; Smith, B.C.; Smith, G.H. *J.C.S. Faraday I*, 1975, *71*, 2232-50.

EXPERIMENTAL VALUES:

T/K	p/mmHg	p/kPa	Mole fraction of sulfur dioxide in liquid, x_{SO_2}	in vapor, y_{SO_2}
250.0	17.92	2.389	0.0000	0.000
	21.16	2.821	0.1076	0.257
	28.59	3.812	0.2092	0.543
	41.91	5.588	0.3094	0.761
	57.83	7.710	0.3792	0.864
	94.94	12.658	0.4797	0.946
	145.57	19.408	0.5750	0.979
	164.17	21.888	0.6016	0.984
	247.81	33.039	0.7248	0.995
	335.03	44.667	0.8554	0.999
	409.65	54.616	1.0000	1.000

COMPONENTS:

1. Sulfur dioxide; SO_2; [7446-09-5]
2. Propanoic acid (propionic acid); $C_3H_6O_2$; [79-09-4]

   Hexanoic acid (caproic acid); $C_6H_{12}O_2$; [142-62-1]

ORIGINAL MEASUREMENTS:

Sano, H.; Nakamoto, Y.
Nippon Kagaku Zasshi
1968, *89*, 369-373.

VARIABLES:

PREPARED BY:

C. L. Young

EXPERIMENTAL VALUES:

$T/°C = 30$ $T/K = 303$

Vapor pressure of propanoic acid = 0.004 atm

Absorption coefficient, S[a,b]	Bunsen coefficient,[b] α	Bunsen coefficient,[c,f] α	ℓ[c,d]	Mole ratio	Mole fraction[e]
		Propanoic acid			
75.7	69.0	-	-	0.232	0.188
		Hexanoic acid			
37.0	33.5	33.0	33.0	0.187	0.158

[a] Volume of sulfur dioxide absorbed by unit volume of solvent at experimental temperature and a total pressure of 1 atmosphere.

[b] Determined by static absorption method.

[c] Determined by flow method.

[d] $\alpha = \ell(1 - p_s^\circ)$ where p_s° is the vapor pressure of solvent.

[e] Calculated by compiler.

[f] Assuming the vapor pressure of hexanoic acid is negligible.

AUXILIARY INFORMATION

METHOD/APPARATUS/PROCEDURE:

Two techniques were used: (a) static absorption method and (b) flow method. Static method consisted of equilibrating solvent and sulfur dioxide in a 50 ml flask and measuring the absorption volumetrically. The flow method consists of passing sulfur dioxide through a micro gas absorption tube and estimating sulfur dioxide by iodometric titration. The vapor pressure of the solvents were measured roughly and were assumed to be equal to the partial pressure of the solvent by the authors when calculating the Bunsen coefficient.

SOURCE AND PURITY OF MATERIALS:

1. Purity 99.0-99.8 mole per cent.

ESTIMATED ERROR:

$\delta T/K = \pm 0.5$; $\delta x_{SO_2} = \pm 5\%$ (estimated by compiler).

REFERENCES:

COMPONENTS:

1. Sulfur dioxide; SO_2; [7446-09-5]

2. Acetic acid, methyl ester (Methyl acetate); $C_3H_6O_2$; [79-20-9]

ORIGINAL MEASUREMENTS:

Horiuti, J.

Sci. Pap. Inst. Phys. Chem. Res. (Jpn.), 1931/32, *17*, 125-256.

VARIABLES:

Temperature, pressure

PREPARED BY:

C. L. Young

EXPERIMENTAL VALUES:

T/K	$P^{\S}$/bar	Mole fraction of sulfur dioxide in liquid, x_{SO_2}
298.15	0.2843	0.000
	0.4237	0.137
	0.4322	0.143
	0.5887	0.243
	0.6084	0.254
	0.7659	0.322
	0.8027	0.336
	0.9471	0.387
	1.005	0.406
	1.123	0.438
	1.191	0.456
	1.295	0.478
	1.384	0.497

T/K	$P^{\dagger}$/bar	Mole fraction of sulfur dioxide in liquid, x_{SO_2}	Ostwald coefficient, L
283.15	0.9568	0.602	254.9
298.15	0.8442	0.408	182.1
313.15	0.6000	0.227	133.8

§ total pressure.

† partial pressure, total pressure = 1 atmosphere = 1.01325×10^5 Pa.

Ostwald coefficient (as $x \to 0$, T/K = 298.15) = 270.

Henry's law constant, dp/dx (as $x \to 0$, T/K = 298.15) = 861 mmHg = 1.148 bar.

AUXILIARY INFORMATION

METHOD/APPARATUS/PROCEDURE:

Apparatus consisted of a gas buret, a solvent reservoir and an absorption pipet. Dry gas introduced into degassed solvent. System mixed using a magnetic stirrer until saturation occurred. Care was taken to prevent solvent vapor from mixing with gas in the gas buret. Volume of gas determined from gas buret reading; volume of liquid determined from height of meniscus in absorption pipet.

SOURCE AND PURITY OF MATERIALS:

1. Sample obtained from Nippon Sulphufic Acid Manufacturing Co. Dried and fractionated several times.

2. Merck, extra pure sample, dried and distilled several times. B. pt. 57.12 °C.

ESTIMATED ERROR:

$\delta T/K = \pm 0.1$; $\delta P/\text{bar} = \pm 0.001$;

$\delta x_{SO_2} = \pm 0.001$ (estimated by compiler).

REFERENCES:

COMPONENTS:

1. Sulfur dioxide; SO_2; [7446-09-5]

2. Acetic acid, methyl ester, (Methyl acetate); $C_3H_6O_2$; [79-20-9]

ORIGINAL MEASUREMENTS:

Bekarek, V.; Hala, E.

Coll. Czech. Chem. Comm. 1968, *33*, 2598-2607.

VARIABLES:

Temperature, pressure

PREPARED BY:

C.L. Young

EXPERIMENTAL VALUES:

T/K	P/Torr	P/kPa	Mole fraction of sulfur dioxide in liquid, x_{SO_2}	in gas, y_{SO_2}
293.2	174.50	23.265	0.0110	0.0387
	200.90	26.784	0.0636	0.2110
	273.90	36.517	0.1660	0.4920
	360.60	48.076	0.2510	0.6650
	403.50	53.796	0.2840	0.7160
	410.00	54.662	0.2860	0.7200
	553.00	73.727	0.3720	0.8290
303.2	274.60	36.610	0.0169	0.0580
	329.60	43.943	0.0841	0.2660
	362.40	48.316	0.1150	0.3600
	381.00	50.796	0.1340	0.4010
	397.30	52.969	0.1480	0.4370
	476.00	63.461	0.2070	0.5700
	485.60	64.741	0.2120	0.5830
	646.20	86.153	0.3030	0.7260
313.2	413.20	55.089	0.0120	0.0401
	466.00	62.128	0.0587	0.1860
	483.60	64.475	0.0737	0.2330
	554.40	73.914	0.1240	0.3700
	556.00	74.127	0.1260	0.3710
	632.90	84.380	0.1710	0.4710

N.B. Same data given in ref.(2)

AUXILIARY INFORMATION

METHOD/APPARATUS/PROCEDURE:

A modified Gillespie still was used. Sulfur dioxide in liquid and gaseous phases estimated by iodometric titration. Details in source and ref. (1).

SOURCE AND PURITY OF MATERIALS:

1. Dried with sulfuric acid and phosphorus pentoxide.
2. Added to 5 per cent acetic anhydride and fractionally distilled.

ESTIMATED ERROR:

$\delta T/K = \pm 0.1$; $\delta P/kPa = \pm 0.1\%$
δx_{SO_2}, δy_{SO_2} = $\pm 0.5\%$. (estimated by compiler).

REFERENCES:

1. Dvorak, K.; Boublik, T.
   Coll. Czech. Chem. Comm. 1963, *28*, 1252.
2. Bekarek, V.; Hala, E.
   Coll. Czech. Chem. Comm. 1968, *33*, 2608.

COMPONENTS:

1. Sulfur dioxide; SO_2; [7446-09-5]
2. 2-Propanone, C_3H_6O; [67-64-1]
3. Acetic acid, methyl ester, (Methyl acetate); $C_3H_6O_2$; [79-20-9]

ORIGINAL MEASUREMENTS:

Bekarek, V.; Hala, E.

Coll. Czech. Chem. Comm. 1968, *33*, 2598-2607.

VARIABLES:

Pressure

PREPARED BY:

C.L. Young

EXPERIMENTAL VALUES:

T/K	*p*/Torr	*p*/kPa	Mole fractions in liquid x_{SO_2}	Mole fractions in liquid $x_{C_3H_8O}$	Mole fractions in gas y_{SO_2}	Mole fractions in gas $y_{C_3H_8O}$
313.2	450.3	60.035	0.0426	0.829	0.107	0.769
	477.1	63.608	0.0609	0.383	0.169	0.352
	477.1	63.608	0.0736	0.847	0.183	0.738
	494.3	65.901	0.0801	0.411	0.214	0.353
	501.5	66.861	0.0864	0.407	0.231	0.342
	535.0	71.327	0.114	0.459	0.302	0.361
	506.7	67.554	0.107	0.834	0.262	0.683
	686.5	91.526	0.222	0.591	0.529	0.360

AUXILIARY INFORMATION

METHOD/APPARATUS/PROCEDURE:

A modified Gillespie still was used. Sulfur dioxide in liquid and gaseous phases estimated by iodometric titration. Details in source and ref. (1).

To determine the acetone concentration a bulb containing (0.5 - 1.0g) of mixture was broken in a solution containing iodine just sufficient to convert sulfur dioxide to sulfuric acid. Acetone was then determined by Messinger's method (ref.(2)).

SOURCE AND PURITY OF MATERIALS:

1. Dried with sulfuric acid and phosphorus pentoxide
2. Analytical grade, fractionally distilled, d_{20}^{4} = 0.7902.
3. Added to 5 per cent acetic anhydride and fractionally distilled.

ESTIMATED ERROR:

$\delta T/K = \pm 0.1$; $\delta p/kPa = \pm 0.1\%$. δx_{SO_2}, $\delta x_{C_3H_8O}$, δy_{SO_2}, $\delta y_{C_3H_8O}$ $= \pm 2\%$.
(estimated by compiler).

REFERENCES:

1. Dvorak, K,; Boublik, T. *Coll. Czech. Chem. Comm.* 1963, *28*, 1252.
2. Goodwin, L.F.; *J. Am. Chem. Soc.*, 1920, *42*, 39.

COMPONENTS:	ORIGINAL MEASUREMENTS:
1. Sulfur dioxide; SO_2; [7446-09-5] 2. Cellosolves	Sano, H.; Nakamoto, Y. *Nippon Kagaku Zasshi* 1968, *89*, 369-373.
VARIABLES:	PREPARED BY: C. L. Young

EXPERIMENTAL VALUES:

T/°C = 30 T/K = 303

Vapor pressure of 2-Methoxyethanol = 0.008 atm
Vapor pressure of 2-Butoxyethanol = 0.001 atm

Absorption coefficient, S[a,b]	Bunsen coefficient,[b] α	Bunsen coefficient,[c] α	ℓ[c,d]	Mole ratio	Mole fraction[e]
2-Methoxyethanol (*methyl cellosolve*); $C_3H_8O_2$; [109-86-4]					
201	184.2	185.9	185.1	0.652	0.395
2-Butoxyethanol (*butyl cellosolve*); $C_6H_{14}O_2$; [111-76-2]					
113.5	103.2	103.7	103.6	0.605	0.377
2-Phenoxyethanol (*phenyl cellosolve*); $C_8H_{10}O_2$; [122-99-6]					
96.1	87.5	-	-	0.491	0.329

[a] Volume of sulfur dioxide absorbed by unit volume of solvent at experimental temperature and a total pressure of 1 atmosphere.
[b] Determined by static absorption method.
[c] Determined by flow method.
[d] $\alpha = \ell(1 - p_s^\circ)$ where p_s° is the vapor pressure of solvent.
[e] Calculated by compiler.

AUXILIARY INFORMATION

METHOD/APPARATUS/PROCEDURE:

Two techniques were used: (a) static absorption method and (b) flow method. Static method consisted of equilibrating solvent and sulfur dioxide in a 50 ml flask and measuring the absorption volumetrically. The flow method consists of passing sulfur dioxide through a micro gas absorption tube and estimating sulfur dioxide by iodometric titration. The vapor pressure of the solvents were measured roughly and were assumed to be equal to the partial pressure of the solvent by the authors when calculating the Bunsen coefficient.

SOURCE AND PURITY OF MATERIALS:

1. Purity 99.0-99.8 mole per cent.

ESTIMATED ERROR:

$\delta T/K = \pm 0.5$; $\delta x_{SO_2} = \pm 5\%$ (estimated by compiler).

REFERENCES:

COMPONENTS:

1. Sulfur dioxide; SO_2; [7446-09-5]
2. Esters of acetic acid

ORIGINAL MEASUREMENTS:

Sano, H.

Nippon Kagaku Zasshi

1968, *89*, 362-368.

VARIABLES:

PREPARED BY:

C. L. Young

EXPERIMENTAL VALUES:

T/°C = 30 T/K = 303

Vapor pressure of ethyl acetate = 0.097 atm
Vapor pressure of butyl acetate = 0.023 atm
Vapor pressure of *iso*-amyl acetate = 0.007 atm

Absorption coefficient, $S^{a,b}$	Bunsen coefficient,[b] α	Bunsen coefficient,[c] α	$\ell^{c,d}$	Mole ratio	Mole fraction[e]
	Methyl acetate; $C_3H_6O_2$; [79-20-9]				
-	142	-	-	0.509	0.337
	Ethyl acetate; $C_4H_8O_2$; [141-78-6]				
118.9	119.7	110.5	100.5	0.531	0.347
	Butyl acetate; $C_6H_{12}O_2$; [123-86-4]				
92.0	85.0	84.0	82.0	0.500	0.333
3-Methyl-1-butanol acetate (*iso-amyl acetate*); $C_7H_{14}O_2$; [123-92-2]					
83.0	75.1	74.8	75.8	0.500	0.333

[a] Volume of sulfur dioxide absorbed by unit volume of solvent at experimental temperature and a total pressure of 1 atmosphere.

[b] Determined by static absorption method.

[c] Determined by flow method.

[d] $\alpha = \ell(1 - p_s^\circ)$ where p_s° is the vapor pressure of solvent.

[e] Calculated by compiler.

AUXILIARY INFORMATION

METHOD/APPARATUS/PROCEDURE:

Two techniques were used: (a) static absorption method and (b) flow method. Static method consisted of equilibrating solvent and sulfur dioxide in a 50 ml flask and measuring the absorption volumetrically. The flow method consists of passing sulfur dioxide through a micro gas absorption tube and estimating sulfur dioxide by iodometric titration. The vapor pressure of the solvents were measured roughly and were assumed to be equal to the partial pressure of the solvent by the authors when calculating the Bunsen coefficient.

SOURCE AND PURITY OF MATERIALS:

1. Purity 99.0-99.8 mole per cent.

ESTIMATED ERROR:

$\delta T/K = \pm 0.5$; $\delta x_{SO_2} = \pm 5\%$

(estimated by compiler).

REFERENCES:

COMPONENTS:

1. Sulfur dioxide; SO_2; [7446-09-5]
2. 1,2,3-Propanetriol (glycerol); $C_3H_8O_3$; [56-81-5]

ORIGINAL MEASUREMENTS:

Gerrard, W.

J. Appl. Chem. Biotechnol. 1972, *22*, 623-650.

VARIABLES:

Temperature

PREPARED BY:

W. Gerrard.

EXPERIMENTAL VALUES:

T/K	P/kPa	Mole ratio SO_2/ ROH	*Mole fraction x_{SO_2}
273.15	101.3	0.39	0.281
283.15		0.20	0.167
293.15		0.117	0.105

* Calculated by compiler

P is the total pressure.

AUXILIARY INFORMATION

METHOD/APPARATUS/PROCEDURE:

Sulfur dioxide was bubbled into a weighed amount of component (2) in a bubbler tube as described in detail, diagram given, in the original paper. The amount of gas absorbed at equilibrium and at the observed temperature and pressure was weighed. By means of a manometer assembly, to which the bubbler tube was attached, the weight of gas absorbed at successively lower pressure was measured. Eventually the pressure was reduced to that of the component (2). The refractive index, and the infrared spectrum of the residue showed it to be essentially pure component (2).

SOURCE AND PURITY OF MATERIALS:

1. Sulfur dioxide was the best grade contained in a small cylinder. Its high purity was attested by the measurement of vapor pressure from 258 to 268 K.
2. The best obtainable specimen was carefully purified and rigorously attested.

ESTIMATED ERROR:

REFERENCES:

COMPONENTS:

1. Sulfur dioxide; SO_2; [7446-09-5]
2. Acetic acid, anhydride; $C_4H_6O_3$; [108-24-7]

ORIGINAL MEASUREMENTS:

Lloyd, S.J.

J. Phys. Chem. 1918, *22*, 300-302.

VARIABLES:

PREPARED BY:

W. Gerrard.

EXPERIMENTAL VALUES:

T/K	Weight of SO_2 per dm^3 of solution/g.	*Mole ratio, SO_2/ Comp.(2)	*Mole fraction (as for 1 atm) x_{SO_2}
268.15	196	0.342	0.255
273.15	148	0.247	0.198
		(0.220)	(0.181)**
278.15	136	0.225	0.184
283.15	122	0.200	0.167
288.15	114	0.187	0.157
293.15	106	0.173	0.148
298.15	99	0.162	0.139
303.15	90	0.147	0.128

*Speculative values

760 mmHg = 1 atm = 101.325 kPa

* Calculated by compiler by assuming that the volume of solution is the same as the volume of the component (2) before absorption of gas.

** Based on the original author's statement that the specific gravity of the "saturated acetic anhydride solution" is 1.22 at 273.15 K.

AUXILIARY INFORMATION

METHOD:/APPARATUS/PROCEDURE:

Gas was bubbled through component (2) in a long graduated tube of small diameter. After equilibrium had been attained at a total pressure of about 1 atm (barometric pressure, 756-760 mmHg), the volume of the solution (about 5 cm^3) was measured, and the weight of absorbed sulfur dioxide was determined by an iodometric titration.

SOURCE AND PURITY OF MATERIALS:

(1) Prepared from concentrated sulfuric acid and a saturated aqueous solution of sodium bisulfite. Dried by sulfuric acid and phosphorus pentoxide.

(2) Distilled, dried by calcium chloride, and redistilled before use.

ESTIMATED ERROR:

REFERENCES:

COMPONENTS:

1. Sulfur dioxide; SO_2; [7446-09-5]
2. Acetic acid anhydride; $C_4H_6O_3$; [108-24-7]

ORIGINAL MEASUREMENTS:

Sano, H.
Nippon Kagaku Zasshi
1968, *89*, 362-368.

VARIABLES:

PREPARED BY:

C. L. Young

EXPERIMENTAL VALUES:

$T/°C = 30$ $T/K = 303$

Vapor pressure of acetic anhydride = 0.006 atm

Absorption coefficient, $S^{a,b}$	Bunsen coefficient,[b] α	Bunsen coefficient,[c] α	$\ell^{c,d}$	Mole ratio	Mole fraction[e]
136.8	125.2	120.7	120.1	0.607	0.378

[a] Volume of sulfur dioxide absorbed by unit volume of solvent at experimental temperature and a total pressure of 1 atmosphere.

[b] Determined by static absorption method.

[c] Determined by flow method.

[d] $\alpha = \ell(1 - p_s^o)$ where p_s^o is the vapor pressure of solvent.

[e] Calculated by compiler.

AUXILIARY INFORMATION

METHOD/APPARATUS/PROCEDURE:

Two techniques were used: (a) static absorption method and (b) flow method. Static method consisted of equilibrating solvent and sulfur dioxide in a 50 ml flask and measuring the absorption volumetrically. The flow method consists of passing sulfur dioxide through a micro gas absorption tube and estimating sulfur dioxide by iodometric titration. The vapor pressure of the solvents were measured roughly and were assumed to be equal to the partial pressure of the solvent by the authors when calculating the Bunsen coefficient.

SOURCE AND PURITY OF MATERIALS:

1. Purity 99.0-99.8 mole per cent.

ESTIMATED ERROR:

$\delta T/K = \pm 0.5$; $\delta x_{SO_2} = \pm 5\%$ (estimated by compiler).

REFERENCES:

COMPONENTS:	ORIGINAL MEASUREMENTS:
1. Sulfur dioxide; SO_2; [7446-09-5] 2. Acetic acid, anhydride; $C_4H_6O_3$; [108-24-7]	Gerrard, W. *J. Appl. Chem. Biotechnol.* 1972, *22*, 623-650.
VARIABLES:	PREPARED BY:
Temperature, pressure	W. Gerrard

EXPERIMENTAL VALUES:

T/K	Total pressure /mmHg	Mole ratio SO_2/Component (2)	*Mole fraction x_{SO_2}
273.15	760	3.08	0.753
	747	2.89	0.743
	649	2.33	0.700
	526	1.69	0.628
	448	1.35	0.574
	357	1.03	0.507
	253	0.703	0.413
	149	0.405	0.288
	70	0.183	0.155
283.15	747	1.57	0.611
293.15	747	0.95	0.487
298.15	747	0.76	0.432

NOTE: The plot of pressure vs solubility given in the original paper is based on the above data.

760 mmHg= 1 atm = 101.325 kPa.

* Calculated by compiler.

AUXILIARY INFORMATION

METHOD/APPARATUS/PROCEDURE:

Sulfur dioxide was bubbled into a weighed amount of component (2) in a bubbler tube as described in detail, diagram given, in the original paper. The amount of gas absorbed at equilibrium and at the observed temperature and pressure was weighed. By means of a manometer assembly, to which the bubbler tube was attached, the weight of gas absorbed at successively lower pressure was measured. Eventually the pressure was reduced to that of the component (2). The refractive index, and the infrared spectrum of the residue showed it to be essentially pure component (2).

SOURCE AND PURITY OF MATERIALS:

1. Sulfur dioxide was the best grade contained in a small cylinder. Its high purity was attested by the measurement of vapor pressure from 258 to 268 K.
2. The best obtainable specimen was carefully purified and rigorously attested.

ESTIMATED ERROR:

REFERENCES:

COMPONENTS:	ORIGINAL MEASUREMENTS:
1. Sulfur dioxide; SO_2; [7446-09-5] 2. Acetic acid anhydride; $C_4H_6O_3$; [108-24-7]	Nitta, T.; Kido, O.; Katayama, T., *J. Chem. Engng. Japan*, 1976, *9*, 317-8.
VARIABLES:	PREPARED BY:
Pressure	C.L. Young

EXPERIMENTAL VALUES:

T/K	P/bar	Mole fraction of sulfur dioxide in liquid, x_{SO_2}
293.15	0.0547	0.0411
	0.2004	0.1430
	0.7366	0.3940
	1.213	0.5378
	1.628	0.6388
	2.029	0.7288
	2.572	0.8350
	2.908	0.9061

AUXILIARY INFORMATION

METHOD:

Static method liquid phase composition estimated by iodometry. Few details given in source. Method similar to that used in ref. (1).

SOURCE AND PURITY OF MATERIALS:

1. Seitetsu Chemical Co. sample, purity better than 99.9 mole per cent.
2. Merck analytical grade sample.

ESTIMATED ERROR:

$\delta T/K = \pm 0.1$; δP/bar ± 0.001 (below 1.3 bar), ± 0.005 (above 1.3 bar); $\delta x_{SO_2} = \pm 1.5\%$

REFERENCES:

Quitzich, K.; Ulbrecht, H.; Geiseler, G.
Z. Physik Chem. 1967, *234*, 33.

COMPONENTS:

1. Sulfur dioxide, SO_2; [7446-09-5]
2. 4-Methyl-1,3-dioxolan-2-one, (Propylene carbonate); $C_4H_6O_3$; [108-32-7]

ORIGINAL MEASUREMENTS:

Lenoir, J-Y.; Renault, P.; Renon, H.

J. Chem. Eng. Data 1971, *16*, 340-342.

VARIABLES:

PREPARED BY:

W. Gerrard

EXPERIMENTAL VALUES:

760 mmHg = 1 atm = 101.325 kPa.

T/K	Henry's Constant H_{SO_2}/atm	*Mole fraction at 1 atm
298	1.20	0.833
343	3.17	0.315

* Calculated by compiler assuming a linear function of p_{SO_2} vs x_{SO_2}, i.e. x_{SO_2}(1 atm) = $1/H_{SO_2}$.

The authors gave "Henry's constant" as $$H_{SO_2} = \lim_{x_{SO_2}\to 0} \frac{f_{SO_2}(P,T)}{x_{SO_2}}.$$

x_{SO_2} is the mole fraction. f_{SO_2} is the fugacity of SO_2 "in the equilibrium condition" of pressure and temperature. H_{SO_2} is related to experimental chromatographic parameters based on the number of moles of solvent in the column, and the absolute retention time of the solute. Helium was the carrier gas.

AUXILIARY INFORMATION

METHOD/APPARATUS/PROCEDURE:

A conventional gas-liquid chromatographic unit was used. The chromatograph was an IGC 12 M Intersmat model, equipped with a thermal conductivity detector. The temperature control was stated to be within ±0.1 °C. The pressure drop in the column was measured by a mercury manometer. Helium was the carrier gas.

There is serious uncertainty in the estimation of accuracy.

SOURCE AND PURITY OF MATERIALS:

1. Sulfur dioxide was provided by l'Air Liquide, and had a stated purity of 99.9%. It was used as provided.
2. Provided by Touzart & Matignon and Serlabo, and stated to have a minimum purity of 99%. Distillation was carried out when necessary, and the refractive index was measured.

ESTIMATED ERROR:

REFERENCES:

COMPONENTS:	ORIGINAL MEASUREMENTS:
1. Sulfur dioxide; SO_2; [7446-09-5] 2. 4-Methyl-1,3-dioxolan-2-one, (Propylene carbonate); $C_4H_6O_3$; [108-32-7]	Benoit, R.L.; Milanova, E. *Can. J. Chem.*, 1979, *57*, 1319-1323
VARIABLES:	PREPARED BY: C.L. Young

EXPERIMENTAL VALUES:

T/K	Henry's law constant[+] H/atm
298.15	1.51

[+] Defined in the original as

$$H = \frac{\text{partial pressure of sulfur dioxide}}{\text{mole fraction of sulfur dioxide}}$$

AUXILIARY INFORMATION

METHOD/APPARATUS/PROCEDURE:

Few details given. Pressure measured with quartz spiral gauge. Partial pressure calculated assuming Raoult's law. Concentration of sulfur dioxide determined by iodometric titration. Concentration of sulfur dioxide ranged from 0.2 to 1.0 mol l^{-1}

SOURCE AND PURITY OF MATERIALS:

1. Anhydrous sample from Canadian Liquid Air, purity 99.98 mole per cent.

2. Dried over 4A molecular sieve

ESTIMATED ERROR:

$\delta T/K = \pm 0.1$; $\delta H/\text{atm} = \pm 4\%$ (estimated by compiler)

REFERENCES:

COMPONENTS:

1. Sulfur dioxide; SO_2; [7446-09-5]
2. Cyclic ethers

ORIGINAL MEASUREMENTS:

Sano, H.

Nippon Kagaku Zasshi

1968, *89*, 362-368.

VARIABLES:

PREPARED BY:

C. L. Young

EXPERIMENTAL VALUES:

T/°C = 30 T/K = 303

Vapor pressure of 1,4-Dioxane = 0.032 atm

Absorption coefficient, $S^{a,b}$	Bunsen coefficient,[b] α	Bunsen coefficient,[c] α	$\ell^{c,d}$	Mole ratio	Mole fraction[e]
	Tetrahydrofuran; $C_4H_8O_2$; [109-99-9]				
-	256.5	-	-	0.550	0.355
	1,4-Dioxane; $C_4H_8O_2$; [123-91-1]				
264.6	248.6	257	238.2	0.95	0.487

[a] Volume of sulfur dioxide absorbed by unit volume of solvent at experimental temperature and a total pressure of 1 atmosphere.

[b] Determined by static absorption method.

[c] Determined by flow method.

[d] $\alpha = \ell(1 - p_s^\circ)$ where p_s° is the vapor pressure of solvent.

[e] Calculated by compiler.

AUXILIARY INFORMATION

METHOD/APPARATUS/PROCEDURE:

Two techniques were used: (a) static absorption method and (b) flow method. Static method consisted of equilibrating solvent and sulfur dioxide in a 50 ml flask and measuring the absorption volumetrically. The flow method consists of passing sulfur dioxide through a micro gas absorption tube and estimating sulfur dioxide by iodometric titration. The vapor pressure of the solvents were measured roughly and were assumed to be equal to the partial pressure of the solvent by the authors when calculating the Bunsen coefficient.

SOURCE AND PURITY OF MATERIALS:

1. Purity 99.0-99.8 mole per cent.

ESTIMATED ERROR:

$\delta T/K = \pm 0.5$; $\delta x_{SO_2} = \pm 5\%$

(estimated by compiler).

REFERENCES:

COMPONENTS:	ORIGINAL MEASUREMENTS:
1. Sulfur dioxide; SO_2; [7446-09-5] 2. Tetrahydrofuran; C_4H_8O; [109-99-9]	Benoit, R.L.; Milanova, E. *Can. J. Chem.*, 1979, *57*, 1319-1323
VARIABLES:	PREPARED BY: C.L. Young

EXPERIMENTAL VALUES:

T/K	Henry's law constant[+] H/atm
298.15	0.392

[+] Defined in the original as

$$H = \frac{\text{partial pressure of sulfur dioxide}}{\text{mole fraction of sulfur dioxide}}$$

AUXILIARY INFORMATION

METHOD/APPARATUS/PROCEDURE:	SOURCE AND PURITY OF MATERIALS:
Few details given. Pressure measured with quartz spiral gauge. Partial pressure calculated assuming Raoult's law. Concentration of sulfur dioxide determined by iodometric titration. Concentration of sulfur dioxide ranged from 0.1 to 0.9 mol l^{-1}	1. Anhydrous sample from Canadian Liquid Air, purity 99.98 mole per cent. 2. Dried over 4A molecular sieve
	ESTIMATED ERROR: $\delta T/K = \pm 0.1$; $\delta H/\text{atm} = \pm 4\%$ (estimated by compiler)
	REFERENCES:

COMPONENTS:

1. Sulfur dioxide; SO_2; [7446-09-5]
2. 1,4-Dioxane; $C_4H_8O_2$; [123-91-1]

ORIGINAL MEASUREMENTS:

Gerrard, W.

J. Appl. Chem. Biotechnol. 1972, *22*, 623-650.

VARIABLES:

PREPARED BY:

W. Gerrard.

EXPERIMENTAL VALUES:

T/K	Pressure, total mmHg	Mole ratio SO_2/dioxane	*Mole fraction x_{SO_2}
288.15	760	1.75	0.636

760 mmHg = 1 atm = 101.325 kPa.

* Calculated by compiler.

AUXILIARY INFORMATION

METHOD/APPARATUS/PROCEDURE:

Sulfur dioxide was bubbled into a weighed amount of component (2) in a bubbler tube as described in detail, diagram given, in the original paper. The amount of gas absorbed at equilibrium and at the observed temperature and pressure was weighed. By means of a manometer assembly, to which the bubbler tube was attached, the weight of gas absorbed at successively lower pressure was measured. Eventually the pressure was reduced to that of the component (2). The refractive index, and the infrared spectrum of the residue showed it to be essentially pure component (2).

SOURCE AND PURITY OF MATERIALS:

1. Sulfur dioxide was the best grade contained in a small cylinder. Its high purity was attested by the measurement of vapor pressure from 258 to 268 K.
2. The best obtainable specimen was carefully purified and rigorously attested.

ESTIMATED ERROR:

REFERENCES:

COMPONENTS:

1. Sulfur dioxide, SO_2; [7446-09-5]

2. 1,4-Dioxane; $C_4H_8O_2$; [123-91-1]

ORIGINAL MEASUREMENTS:

Nitta, T.; Kido, O.; Katayama, T.;

J. Chem. Engng. Japan. 1976, *9*, 317-8.

VARIABLES:

Pressure

PREPARED BY:

C.L. Young

EXPERIMENTAL VALUES:

T/K	P/bar	Mole fraction of sulfur dioxide in liquid, x_{SO_2}
293.15	0.0720	0.0920
	0.1108	0.1645
	0.1477	0.2162
	0.2329	0.2996
	0.3976	0.3988
	0.6514	0.4960
	1.139	0.6102
	1.624	0.6928
	2.220	0.7829
	2.824	0.8789

AUXILIARY INFORMATION

METHOD

Static method liquid phase composition estimated by iodometry. Few details given in source method similar to that used in ref. (1).

SOURCE AND PURITY OF MATERIALS:

1. Seitetsu Chemical Co. sample, purity better than 99.9 mole per cent.

2. Nakarai Chemicals Ltd. sample, spectral grade.

ESTIMATED ERROR:

$\delta T/K = \pm 0.1$; $\delta P/\text{bar} = \pm 0.001$ (up to 1.3 bar), ± 0.005 (above 1.3 bar); $\delta x_{SO_2} = \pm 1.5\%$

REFERENCES:

Quitzich K.; Ulbrecht, H.; Geiseler, G.
Z. Physik Chem. 1967, *234*, 33.

COMPONENTS:

1. Sulfur dioxide; SO_2; [7446-09-5]

2. Acetic acid, ethyl ester (ethyl acetate); $C_4H_8O_2$; [141-78-6]

ORIGINAL MEASUREMENTS:

Benoit, R.L.; Milanova, E.

Can. J. Chem., 1979, *57*, 1319-1323

VARIABLES:

PREPARED BY:

C.L. Young

EXPERIMENTAL VALUES:

T/K	Henry's law constant+ H/atm
298.15	1.09

+ Defined in the original as

$$H = \frac{\text{partial pressure of sulfur dioxide}}{\text{mole fraction of sulfur dioxide}}$$

AUXILIARY INFORMATION

METHOD/APPARATUS/PROCEDURE:

Few details given. Pressure measured with quartz spiral gauge. Partial pressure calculated assuming Raoult's law. Concentration of sulfur dioxide determined by iodometric titration. Concentration of sulfur dioxide ranged from 0.1 to 0.5 mol l^{-1}

SOURCE AND PURITY OF MATERIALS:

1. Anhydrous sample from Canadian Liquid Air, purity 99.98 mole per cent.

2. Dried over 4A molecular sieve

ESTIMATED ERROR:

$\delta T/K = \pm 0.1$; $\delta H/atm = \pm 4\%$ (estimated by compiler)

REFERENCES:

COMPONENTS:

1. Sulfur dioxide, SO_2; [7446-09-5]
2. 1,1'-Oxybisethane, (Diethyl-ether); $C_4H_{10}O$; [60-29-7]

ORIGINAL MEASUREMENTS:

Yuferev, R.F.; Maluigin, P.V.

Z. Chim. Prom. 1930, *7*, 553-555

VARIABLES:

Temperature

PREPARED BY:

W. Gerrard

EXPERIMENTAL VALUES:

T/K	Weight of SO_2 per cm^3 of solvent g	Volume of SO_2 per cm^3 of solvent cm^3	*Mole ratio SO_2/diethyl ether.	*Mole fraction x_{SO_2}
273.15	1.090	381.51	1.712	0.631
291.65	0.2828	98.98	0.457	0.314
Ether saturated with water;				
273.15	0.8238	288.33		
291.65	0.3252	113.82		

* Calculated by compiler.

AUXILIARY INFORMATION

METHOD/APPARATUS/PROCEDURE:

Sulfur dioxide was passed into 5 cm^3 of ether until saturated. The whole solution was then transferred to a stoppered flask containing an aqueous solution of I_2 + KI. The sulfuric acid formed was determined gravimetrically. The total pressure was presumably 1 atm; but this was not stated.

SOURCE AND PURITY OF MATERIALS:

1. Sulfur dioxide was dried by sulfuric acid.

2. The ether was dried over "calcined potash," and then over sodium, distilled, and stored over sodium.

ESTIMATED ERROR:

REFERENCES:

COMPONENTS:

1. Sulfur dioxide; SO_2; [7446-09-5]
2. 1,1'-Oxybisethane, (Diethyl ether); $C_4H_{10}O$; [60-29-7]

ORIGINAL MEASUREMENTS:

Nitta, T.; Itami, J.; Katayama, T.,

J. Chem. Engng. Japan, 1973, *6*, 303-9.

VARIABLES:

Pressure

PREPARED BY:

C. L. Young

EXPERIMENTAL VALUES:

T/K	p/mmHg[+]	p/kPa[+]	Mole fraction of sulfur dioxide in liquid, x_{SO_2}
293.15	635.9	84.78	0.2209
	761.3	101.5	0.3058
	887.0	118.3	0.3696
	1049	139.9	0.4542
	1269	169.2	0.5496
	1551	206.8	0.6490
	1804	240.5	0.7342
	2209	294.5	0.8960

[+] Total pressure

AUXILIARY INFORMATION

METHOD/APPARATUS/PROCEDURE:

Static equilibrium cell with magnetic stirrer. After equilibrium established, pressure measured with Bourdon gauge and approximately 2 cm^3 of liquid analysed by iodometric titration.

SOURCE AND PURITY OF MATERIALS:

1. Seitetsu Chemical Co. sample purity better than 99.9 mole per cent.
2. Merck Uvasol spectral grade.

ESTIMATED ERROR:

$\delta T/K = \pm 0.05$; $\delta p/kPa = \pm 0.5$;

$\delta x_{SO_2} = \pm 1\%$ (estimated by compiler)

REFERENCES:

COMPONENTS:

1. Sulfur dioxide; SO_2; [7446-09-5]
2. 2-Furanmethanol (*furfuryl alcohol*); $C_5H_6O_2$; [98-00-0]

   Tetrahydro-2-furanmethanol (*tetrahydrofurfuryl alcohol*); $C_5H_{10}O_2$; [97-99-4]

ORIGINAL MEASUREMENTS:

Sano, H.; Nakamoto, Y.

Nippon Kagaku Zasshi

1968, *89*, 369-373.

VARIABLES:

PREPARED BY:

C. L. Young

EXPERIMENTAL VALUES:

T/°C = 30 T/K = 303

Vapor pressure of 2-Furanmethanol = 0.001 atm

Vapor pressure of Tetrahydro-2-furanmethanol = 0.001 atm

Absorption coefficient, $S^{a,b}$	Bunsen coefficient,[b] α	Bunsen coefficient,[c] α	$\ell^{c,d}$	Mole ratio	Mole fraction[e]
		2-Furanmethanol			
96	86			0.335	0.251
		Tetrahydro-2-furanmethanol			
184	167	160.0	165.8	0.724	0.420

[a] Volume of sulfur dioxide absorbed by unit volume of solvent at experimental temperature and a total pressure of 1 atmosphere.

[b] Determined by static absorption method.

[c] Determined by flow method.

[d] $\alpha = \ell(1 - p_s^\circ)$ where p_s° is the vapor pressure of solvent.

[e] Calculated by compiler.

AUXILIARY INFORMATION

METHOD/APPARATUS/PROCEDURE:

Two techniques were used: (a) static absorption method and (b) flow method. Static method consisted of equilibrating solvent and sulfur dioxide in a 50 ml flask and measuring the absorption volumetrically. The flow method consists of passing sulfur dioxide through a micro gas absorption tube and estimating sulfur dioxide by iodometric titration. The vapor pressure of the solvents were measured roughly and were assumed to be equal to the partial pressure of the solvent by the authors when calculating the Bunsen coefficient.

SOURCE AND PURITY OF MATERIALS:

1. Purity 99.0-99.8 mole per cent.

ESTIMATED ERROR:

$\delta T/K = \pm 0.5$; $\delta x_{SO_2} = \pm 5\%$ (estimated by compiler).

REFERENCES:

COMPONENTS:

1. Sulfur dioxide; SO_2; [7446-09-5]
2. Ethers

ORIGINAL MEASUREMENTS:

Sano, H.

Nippon Kagaku Zasshi

1968, *89*, 362-368.

VARIABLES:

PREPARED BY:

C. L. Young

EXPERIMENTAL VALUES:

T/°C = 30 T/K = 303

Vapor pressure of diisopropyl ether = 0.202 atm
Vapor pressure of 1,2-dimethoxyethane = 0.056 atm

Absorption coefficient, S[a,b]	Bunsen coefficient,[b] α	Bunsen coefficient,[c] α	ℓ[c,d]	Mole ratio	Mole fraction[e]
1,1'-Oxybisethane (*diethyl ether*); $C_4H_{10}O$; [60-29-7]					
-	148	-	-	0.687	0.407
2,2'-Oxybispropane (*isopropyl ether*); $C_6H_{14}O$; [108-20-3]					
32.85	37.4	38.9	31.0	0.235	0.190
Dimethoxymethane (*methylal*); $C_3H_8O_2$; [109-87-5]					
-	304	-	-	1.22	0.550
1,2-Dimethoxyethane (*ethylene glycol dimethyl ether*); $C_4H_{10}O_2$; [110-71-4]					
230.4	221.5	226.0	215.0	1.03	0.507

[a] Volume of sulfur dioxide absorbed by unit volume of solvent at experimental temperature and a total pressure of 1 atmosphere.

[b] Determined by static absorption method.

[c] Determined by flow method.

[d] $\alpha = \ell(1 - p_s^\circ)$ where p_s° is the vapor pressure of solvent.

[e] Calculated by compiler.

AUXILIARY INFORMATION

METHOD/APPARATUS/PROCEDURE:

Two techniques were used: (a) static absorption method and (b) flow method. Static method consisted of equilibrating solvent and sulfur dioxide in a 50 ml flask and measuring the absorption volumetrically. The flow method consists of passing sulfur dioxide through a micro gas absorption tube and estimating sulfur dioxide by iodometric titration. The vapor pressure of the solvents were measured roughly and were assumed to be equal to the partial pressure of the solvent by the authors when calculating the Bunsen coefficient.

SOURCE AND PURITY OF MATERIALS:

1. Purity 99.0-99.8 mole per cent.

ESTIMATED ERROR:

$\delta T/K = \pm 0.5$; $\delta x_{SO_2} = \pm 5\%$

(estimated by compiler).

REFERENCES:

COMPONENTS:

1. Sulfur dioxide; SO_2; [7746-09-5]
2. Phenol; C_6H_6O; [108-95-2]

ORIGINAL MEASUREMENTS:

Lenoir, J-Y.; Renault, P.; Renon, H.

J. Chem. Eng. Data. 1971, *16*, 340-342

VARIABLES:

PREPARED BY:

W. Gerrard.

EXPERIMENTAL VALUES:

T/K	Henry's constant H_{SO_2}/atm	Mole fraction at 1 atm *
323	7.07	0.141

* Calculated by compiler assuming a linear function of p_{SO_2} vs x_{SO_2}, i.e. x_{SO_2} (1 atm) = $1/H_{SO_2}$.

The authors gave "Henry's constant" as $H_{SO_2} = \lim_{x_{SO_2}\to 0} \dfrac{f_{SO_2}(P,T)}{x_{SO_2}}$

x_{SO_2} is the mole fraction. f_{SO_2} is the fugacity of SO_2 "in the equilibrium condition" of pressure and temperature. H_{SO_2} is related to experimental chromatographic parameters based on the number of moles of solvent in the column, and the absolute retention time of the solute. Helium was the carrier gas.

AUXILIARY INFORMATION

METHOD/APPARATUS/PROCEDURE:

A conventional gas-liquid chromatographic unit was used. The chromatograph was an IGC 12 M Intersmat model, equipped with a thermal conductivity detector. The temperature control was stated to be within ±0.1° C. The pressure drop in the column was measured by a mercury manometer. Helium was the carrier gas.

There is serious uncertainty in the estimation of accuracy.

SOURCE AND PURITY OF MATERIALS:

1. Sulfur dioxide was provided by l'Air Liquide, and had a stated purity of 99.9%. It was used as provided.
2. Provided by Touzart & Matignon and Serlabo, and stated to have a minimum purity of 99%. Distillation was carried out when necessary, and the refractive index was measured.

ESTIMATED ERROR:

REFERENCES:

COMPONENTS:

1. Sulfur dioxide; SO_2; [7446-09-5]
2. Phenol; C_6H_6O; [108-95-2]
   1,2-Benzenediol (*1,2-Cresol*); $C_6H_6O_2$; [120-80-9]
   2-Methoxyphenol (*Guaiacol*); $C_7H_8O_2$; [90-05-1]

ORIGINAL MEASUREMENTS:

Sano, H.; Nakamoto, Y.
Nippon Kagaku Zasshi
1968, *89*, 369-373.

VARIABLES:

PREPARED BY:

C. L. Young

EXPERIMENTAL VALUES:

T/°C = 30 T/K = 303

Absorption coefficient, $S^{a,b}$	Bunsen coefficient,[b] α	Bunsen coefficient,[c,f] α	$\ell^{c,d}$	Mole ratio	Mole fraction[e]
		Phenol			
75.1	68.6	68.1	68.1	0.273	0.214
		1,2-Benzenediol			
75.0	68.3	68.0	68.0	0.315	0.240
		2-Methoxyphenol			
108	98.4	-	-	0.489	0.328

[a] Volume of sulfur dioxide absorbed by unit volume of solvent at experimental temperature and a total pressure of 1 atmosphere.

[b] Determined by static absorption method.

[c] Determined by flow method.

[d] $\alpha = \ell(1 - p_s^\circ)$ where p_s° is the vapor pressure of solvent.

[e] Calculated by compiler.

[f] Assuming the vapor pressure of phenol and 1,2-benzenediol to be negligible.

AUXILIARY INFORMATION

METHOD/APPARATUS/PROCEDURE:

Two techniques were used: (a) static absorption method and (b) flow method. Static method consisted of equilibrating solvent and sulfur dioxide in a 50 ml flask and measuring the absorption volumetrically. The flow method consists of passing sulfur dioxide through a micro gas absorption tube and estimating sulfur dioxide by iodometric titration. The vapor pressure of the solvents were measured roughly and were assumed to be equal to the partial pressure of the solvent by the authors when calculating the Bunsen coefficient.

SOURCE AND PURITY OF MATERIALS:

1. Purity 99.0-99.8 mole per cent.

ESTIMATED ERROR:

$\delta T/K = \pm 0.5$; $\delta x_{SO_2} = \pm 5\%$ (estimated by compiler).

REFERENCES:

COMPONENTS:

1. Sulfur dioxide; SO_2; [7446-09-5]
2. Esters

ORIGINAL MEASUREMENTS:

Sano, H.

Nippon Kagaku Zasshi

1968, *89*, 362-368.

VARIABLES:

PREPARED BY:

C. L. Young

EXPERIMENTAL VALUES:

T/°C = 30 T/K = 303

Absorption coefficient, $S^{a,b}$	Bunsen coefficient,[b] α	Bunsen coefficient,[c,f] α	$\ell^{c,d}$	Mole ratio	Mole fraction[e]
Dimethyl maleate (z)-2-Butenedioic acid, dimethyl ester; $C_6H_8O_4$; [624-48-6]					
99.1	90.1	-	-	0.508	0.337
Dimethyl phthalate 1,2-Benzenecarboxylic acid, dimethyl ester; $C_{10}H_{10}O_4$; [131-11-3]					
96.5	87.9	88.0	88.0	0.643	0.391

[a] Volume of sulfur dioxide absorbed by unit volume of solvent at experimental temperature and a total pressure of 1 atmosphere.

[b] Determined by static absorption method.

[c] Determined by flow method.

[d] $\alpha = \ell(1 - p_s^\circ)$ where p_s° is the vapor pressure of solvent.

[e] Calculated by compiler.

[f] Assuming vapor pressure of dimethyl phthalate to be negligible.

AUXILIARY INFORMATION

METHOD/APPARATUS/PROCEDURE:

Two techniques were used; (a) static absorption method and (b) flow method. Static method consisted of equilibrating solvent and sulfur dioxide in a 50 ml flask and measuring the absorption volumetrically. The flow method consists of passing sulfur dioxide through a micro gas absorption tube and estimating sulfur dioxide by iodometric titration. The vapor pressure of the solvents were measured roughly and were assumed to be equal to the partial pressure of the solvent by the authors when calculating the Bunsen coefficient.

SOURCE AND PURITY OF MATERIALS:

1. Purity 99.0-99.8 mole per cent.

ESTIMATED ERROR:

$\delta T/K = \pm 0.5$; $\delta x_{SO_2} = \pm 5\%$

(estimated by compiler).

REFERENCES:

COMPONENTS:

1. Sulfur dioxide; SO_2; [7446-09-5]
2. Cyclohexanone; $C_6H_{10}O$; [108-94-1] or 3-Methylcyclohexanone; $C_7H_{14}O$; [591-23-1]

ORIGINAL MEASUREMENTS:

Weissenberger, G.; Hadwiger, H.

Z. angew. Chem. 1927, *40*, 734-736.

VARIABLES:

PREPARED BY:

W. Gerrard

EXPERIMENTAL VALUES:

T/K	Volume of Comp.(2)/cm^3	Volume of SO_2 absorbed/cm^3	*Mole ratio SO_2/Comp.(2)	*Mole fraction x_{SO_2}
			(pressure assumed to be barometric)	
		cyclohexanone		
293.15	103.24	2353.87	0.10	0.0911
		3-methylcyclohexanone		
293.15	124.6	2082.82	0.089	0.0817

* Calculated by compiler

AUXILIARY INFORMATION

METHOD/APPARATUS/PROCEDURE:

Used an absorption vessel of 200 cm^3 capacity, and a vessel with a two-way tap. Simply stated that the absorptiometer was like that used by Manchot, but the reference was not cited; probably ref. (1).

SOURCE AND PURITY OF MATERIALS:

No details given.

ESTIMATED ERROR:

REFERENCES:

1. Manchot, W. *Z. anorg. Chem.* 1924, *141*, 38.

COMPONENTS:

1. Sulfur dioxide; SO_2; [7446-09-5]
2. Cyclohexanone; $C_6H_{10}O$; [108-94-1]

   1-Phenylethanone (*Acetophenone*); C_8H_8O; [98-86-2]

ORIGINAL MEASUREMENTS:

Sano, H.

Nippon Kagaku Zasshi

1968, *89*, 362-368.

VARIABLES:

PREPARED BY:

C. L. Young

EXPERIMENTAL VALUES:

$T/°C = 30$ $T/K = 303$

Vapor pressure of cyclohexanone = 0.005 atm

Vapor pressure of acetophenone = 0.001 atm

Absorption coefficient, S^a	Bunsen coefficient, α	Mole ratio	Mole fraction[b]
	Cyclohexanone		
133.4	122.1	0.567	0.362
	1-Phenylethanone		
96.4	87.9	0.460	0.315

[a] Volume of sulfur dioxide absorbed by unit volume of solvent at experimental temperature and a total pressure of 1 atmosphere.

[b] Calculated by compiler.

AUXILIARY INFORMATION

METHOD/APPARATUS/PROCEDURE:

Static absorption method consisted of equilibrating solvent and sulfur dioxide in a 50 ml flask and measuring the absorption volumetrically. The vapor pressure of the solvents were measured roughly and were assumed to be equal to the partial pressure of the solvent by the authors when calculating the Bunsen coefficient.

SOURCE AND PURITY OF MATERIALS:

1. Purity 99.0-99.8 mole per cent.

ESTIMATED ERROR:

$\delta T/K = \pm 0.5$; $\delta x_{SO_2} = \pm 5\%$

(estimated by compiler).

REFERENCES:

COMPONENTS:

1. Sulfur dioxide; SO_2 [7446-09-5]

2. Cyclohexanone; $C_6H_{10}O$; [108-94-1] or 2-Methylcyclohexanone ; $C_7H_{12}O$; [583-60-8]

ORIGINAL MEASUREMENTS:

Gerrard, W.; "*Solubility of Gases and Liquids*,: Plenum Press, New York, 1976, p. 239.

VARIABLES:

Temperature

PREPARED BY:

W. Gerrard.

EXPERIMENTAL VALUES:

T/K	Weight of component (2) g	Weight of SO_2 absorbed, g	Mole ratio SO_2/Comp. (2)	Mole fraction x_{SO_2}
		Cyclohexanone		
			(Total pressure = 1 atm)	
283.15	2.0601	2.1639	1.611	0.617
293.15	2.0601	1.4730	1.100	0.521
		Methyl cyclohexanone		
			(Total pressure = 1 atm)	
283.15	2.5852	1.9010	1.290	0.563
293.15	2.5852	1.3808	0.937	0.484

760 mmHg = 1 atm = 101.325 kPa.

AUXILIARY INFORMATION

METHOD:

Sulfur dioxide was bubbled into a weighed amount of component (2) in a bubbler tube. The amount of gas absorbed was determined by re-weighing the tube. The total pressure was barometric, very nearly 1 atm. See ref. (1).

SOURCE AND PURITY OF MATERIALS:

1. Sulfur dioxide was the best grade contained in a small cylinder. Its high degree of purity was attested by the measurement of vapor pressure from 258-268 K.

2. The best commercial specimen was purified and attested.

ESTIMATED ERROR:

REFERENCES:

1. Gerrard, W. *J. Appl. Chem. Biotechnol.* 1972, *22*, 623.

COMPONENTS:

1. Sulfur dioxide; SO_2; [7446-09-5]
2. 1,2-Ethanediol diacetate (*Ethylene glycol diacetate*); $C_6H_{10}O_2$; [111-55-7]
   2-Ethoxyethanol acetate (*Ethylene glycol monoethyl monoacetate*); $C_6H_{12}O_3$; [111-15-9]

ORIGINAL MEASUREMENTS:

Sano, H.
Nippon Kagaku Zasshi
1968, *89*, 362-368.

VARIABLES:

PREPARED BY:

C. L. Young

EXPERIMENTAL VALUES:

T/°C = 30 T/K = 303

Vapor pressure of ethylene glycol diacetate = 0.001 atm
Vapor pressure of ethylene glycol monoethyl monoacetate = 0.002 atm

Absorption coefficient, S[a]	Bunsen coefficient, α	Mole ratio	Mole fraction[b]
	Ethylene glycol diacetate		
150.4	137.0	0.808	0.447
	Ethylene glycol monoethyl monoacetate		
141.7	129.4	0.786	0.440

[a] Volume of sulfur dioxide absorbed by unit volume of solvent at experimental temperature and a total pressure of 1 atmosphere.

[b] Calculated by compiler.

AUXILIARY INFORMATION

METHOD/APPARATUS/PROCEDURE:

Static absorption method consisted of equilibrating solvent and sulfur dioxide in a 50 ml flask and measuring the absorption volumetrically. The vapor pressure of the solvents were measured roughly and were assumed to be equal to the partial pressure of the solvent by the authors when calculating the Bunsen coefficient.

SOURCE AND PURITY OF MATERIALS:

1. Purity 99.0-99.8 mole per cent.

ESTIMATED ERROR:

$\delta T/K = \pm 0.5$; $\delta x_{SO_2} = \pm 5\%$

(estimated by compiler).

REFERENCES:

COMPONENTS:

1. Sulfur dioxide; SO_2; [7446-09-5]
2. 3-Oxobutanoic acid, ethyl ester (*Ethyl acetoacetate*); $C_6H_{10}O_3$; [141-97-9]

ORIGINAL MEASUREMENTS:

Sano, H.
Nippon Kagaku Zasshi
1968, *89*, 362-368.

VARIABLES:

PREPARED BY:

C. L. Young

EXPERIMENTAL VALUES:

T/°C = 30 T/K = 303

Absorption coefficient, S[a]	Bunsen coefficient, α	Mole ratio	Mole fraction[b]
133.3	120.7	0.687	0.407

[a] Volume of sulfur dioxide absorbed by unit volume of solvent at experimental temperature and a total pressure of 1 atmosphere.

[b] Calculated by compiler.

AUXILIARY INFORMATION

METHOD/APPARATUS/PROCEDURE:

Static absorption method consisted of equilibrating solvent and sulfur dioxide in a 50 ml flask and measuring the absorption volumetrically. The vapor pressure of the solvents were measured roughly and were assumed to be equal to the partial pressure of the solvent by the authors when calculating the Bunsen coefficient.

SOURCE AND PURITY OF MATERIALS:

1. Purity 99.0-99.8 mole per cent.

ESTIMATED ERROR:

$\delta T/K = \pm 0.5$; $\delta x_{SO_2} = \pm 5\%$ (estimated by compiler).

REFERENCES:

COMPONENTS:

1. Sulfur dioxide; SO_2; [7446-09-5]

2. Hexanoic acid; $C_6H_{12}O_2$; [142-62-1]

ORIGINAL MEASUREMENTS:

Gerrard, W.

J. Appl. Chem. Biotechnol. 1972, *22*, 623-650

VARIABLES:

Temperature, pressure

PREPARED BY:

W. Gerrard

EXPERIMENTAL VALUES:

T/K	Pressure, total /mmHg	Mole ratio SO_2/Comp.(2)	*Mole fraction x_{SO_2}
273.15	755	0.86	0.462
	** 712	0.824	0.452
	660	0.690	0.408
	566	0.521	0.343
	206	0.143	0.125
	124	0.076	0.0706
	45	0.028	0.0272
283.15	755	0.440	0.306
	***756	0.442	0.307
	700	0.417	0.294
	620	0.353	0.261
	510	0.277	0.217
	370	0.182	0.154
	244	0.113	0.102
	158	0.070	0.065
	42	0.017	0.017
293.15	755	0.276	0.216
298.15	755	0.25	0.200

760 mmHg = 1 atm = 101.325 kPa.

* Calculated by compiler

** Plot of pressure vs solubility given in Ref. (1) is based on these data.

*** Plot of pressure vs solubility given in original paper is based on these data.

AUXILIARY INFORMATION

METHOD/APPARATUS/PROCEDURE:

Sulfur dioxide was bubbled into a weighed amount of component (2) in a bubbler tube as described in detail, diagram given, in the original paper. The amount of gas absorbed at equilibrium and at the observed temperature and pressure was weighed. By means of a manometer assembly, to which the bubbler tube was attached, the weight of gas absorbed at successively lower pressure was measured. Eventually the pressure was reduced to that of the component (2). The refractive index, and the infrared spectrum of the residue showed it to be essentially pure component (2).

SOURCE AND PURITY OF MATERIALS:

1. Sulfur dioxide was the best grade contained in a small cylinder. Its high purity was attested by the measurement of vapor pressure from 258 to 268K.

2. The best obtainable specimen was carefully purified and rigorously attested.

ESTIMATED ERROR:

REFERENCES:

1. Gerrard, W. *Solubility of Gases and Liquids*, Plenum Press, New York, 1976.

COMPONENTS:	ORIGINAL MEASUREMENTS:
1. Sulfur dioxide, SO_2; [7446-09-5] 2. Benzenemethanol (Benzyl alcohol); C_7H_8O; [100-51-6]	Lenoir, J-Y.; Renault, P.; Renon, H. *J. Chem. Eng. Data* 1971, *16*, 340-342.
VARIABLES:	PREPARED BY: W. Gerrard

EXPERIMENTAL VALUES: 760 mmHg = 1 atm = 101.325 kPa.

T/K	Henry's Constant H_{SO_2}/atm	*Mole fraction at 1 atm
298	2.08	0.481

* Calculated by compiler assuming a linear function of p_{SO_2} vs x_{SO_2}, i.e. x_{SO_2}(1 atm) = $1/H_{SO_2}$.

The authors gave "Henry's constant" as

$$H_{SO_2} = \lim_{x_{SO_2} \to 0} \frac{f_{SO_2}(P,T)}{x_{SO_2}}$$

H_{SO_2} is the mole fraction. f_{SO_2} is the fugacity of SO_2 "in the equilibrium condition" of pressure and temperature. H_{SO_2} is related to experimental chromatographic parameters based on the number of moles of solvent in the column, and the absolute retention time of the solute. Helium was the carrier gas.

AUXILIARY INFORMATION

METHOD/APPARATUS/PROCEDURE:

A conventional gas-liquid chromatographic unit was used. The chromatograph was an IGC 12 M Intersmat model, equipped with a thermal conductivity detector. The temperature control was stated to be within ±0.1°C. The pressure drop in the column was measured by a mercury manometer. Helium was the carrier gas. There is serious uncertainty in the estimation of accuracy..

SOURCE AND PURITY OF MATERIALS:

1. Sulfur dioxide was provided by l'Air Liquide, and had a stated purity of 99.9%. It was used as provided.
2. Provided by Touzart & Matignon and Serlabo, and stated to have a minimum purity of 99%. Distillation was carried out when necessary and the refractive index was measured.

ESTIMATED ERROR:

REFERENCES:

COMPONENTS:

1. Sulfur dioxide; SO_2; [7446-09-5]
2. Benzoic acid, methyl ester (*methyl benzoate*); $C_8H_8O_2$; [93-58-3]

ORIGINAL MEASUREMENTS:

Sano, H.

Nippon Kagaku Zasshi

1968, *89*, 362-368.

VARIABLES:

PREPARED BY:

C. L. Young

EXPERIMENTAL VALUES:

T/°C = 30 T/K = 303

Vapor pressure of methyl benzoate = 0.001 atm

Absorption coefficient, S^a	Bunsen coefficient, α	Mole ratio	Mole fraction[b]
72.1	65.5	0.365	0.267

[a] Volume of sulfur dioxide absorbed by unit volume of solvent at experimental temperature and a total pressure of 1 atmosphere.

[b] Calculated by compiler.

AUXILIARY INFORMATION

METHOD/APPARATUS/PROCEDURE:

Static absorption method consisted of equilibrating solvent and sulfur dioxide in a 50 ml flask and measuring the absorption volumetrically. The vapor pressure of the solvents were measured roughly and were assumed to be equal to the partial pressure of the solvent by the authors when calculating the Bunsen coefficient.

SOURCE AND PURITY OF MATERIALS:

1. Purity 99.0-99.8 mole per cent.

ESTIMATED ERROR:

$\delta T/K = \pm 0.5$; $\delta x_{SO_2} = \pm 5\%$

(estimated by compiler).

REFERENCES:

COMPONENTS:

1. Sulfur dioxide; SO_2; [7446-09-5]
2. 2-Hydroxybenzoic acid, methyl ester (*methyl salicylate*); $C_8H_8O_3$; [119-36-8]

ORIGINAL MEASUREMENTS:

Sano, H.; Nakamoto, Y.

Nippon Kagaku Zasshi

1968, *89*, 369-373.

VARIABLES:

PREPARED BY:

C. L. Young

EXPERIMENTAL VALUES:

T/°C = 30 T/K = 303

Absorption coefficient, S[a]	Bunsen coefficient, α	Mole ratio	Mole fraction[b]
84.3	76.7	0.441	0.306

[a] Volume of sulfur dioxide absorbed by unit volume of solvent at experimental temperature and a total pressure of 1 atmosphere.

[b] Calculated by compiler.

AUXILIARY INFORMATION

METHOD/APPARATUS/PROCEDURE:

Static absorption method consisted of equilibrating solvent and sulfur dioxide in a 50 ml flask and measuring the absorption volumetrically. The vapor pressure of the solvents were measured roughly and were assumed to be equal to the partial pressure of the solvent by the authors when calculating the Bunsen coefficient.

SOURCE AND PURITY OF MATERIALS:

1. Purity 99.0-99.8 mole per cent.

ESTIMATED ERROR:

$\delta T/K = \pm 0.5$; $\delta x_{SO_2} = \pm 5\%$ (estimated by compiler).

REFERENCES:

COMPONENTS:

1. Sulfur dioxide; SO_2; [7446-09-5]
2. 2-Hydroxybenzoic acid, methyl ester; (methyl salicylate) $C_8H_8O_3$; [119-36-8]

ORIGINAL MEASUREMENTS:

Gerrard, W. *J. Appl. Chem. Biotechnol.* 1972, *22*, 623-650.

VARIABLES:

Temperature, pressure

PREPARED BY:

W. Gerrard

EXPERIMENTAL VALUES:

T/K	Pressure, total /mmHg	Mole ratio SO_2/Component (2)	*Mole fraction x_{SO_2}
273.15	761	2.21	0.688
	** 616	1.43	0.588
	477	0.949	0.487
	289	0.477	0.323
	146	0.222	0.182
283.15	761	1.03	0.507

760 mmHg = 1 atm = 101.325 kPa.

* Calculated by compiler

** The plot of pressure and solubility given in the original paper, and that in Ref. (1) are based on these data.

AUXILIARY INFORMATION

METHOD/APPARATUS/PROCEDURE:

Sulfur dioxide was bubbled into a weighed amount of component (2) in a bubbler tube as described in detail, diagram given, in the original paper. The amount of gas absorbed at equilibrium and at the observed temperature and pressure was weighed. By means of a manometer assembly, to which the bubbler tube was attached, the weight of gas absorbed at successively lower pressure was measured. Eventually the pressure was reduced to that of the component (2). The refractive index, and the infrared spectrum of the residue showed it to be essentially pure component (2).

SOURCE AND PURITY OF MATERIALS:

1. Sulfur dioxide was the best grade contained in a small cylinder. Its high purity was attested by the measurement of vapor pressure from 258 to 268 K.
2. The best obtainable specimen was carefully purified and rigorously attested.

ESTIMATED ERROR:

REFERENCES:

1. Gerrard, W. *Solubility of Gases and Liquids*, Plenum Press, New York, 1976.

COMPONENTS:

1. Sulfur dioxide; SO_2; [7446-09-5]
2. Aromatic ethers

ORIGINAL MEASUREMENTS:

Sano, H.

Nippon Kagaku Zasshi

1968, *89*, 362-368.

VARIABLES:

PREPARED BY:

C. L. Young

EXPERIMENTAL VALUES:

T/°C = 30 T/K = 303

Vapor pressure of methoxybenzene = 0.004 atm

Absorption coefficient, $S^{a,b}$	Bunsen coefficient,[b] α	Bunsen coefficient,[c,f] α	$\ell^{c,d}$	Mole ratio	Mole fraction[e]
1,1'-(Methoxymethylene)bisbenzene (*benzyl ether*); $C_{14}H_{14}O$; [1016-09-7]					
68.8	62.5	63.3	63.3	0.549	0.354
Methoxybenzene (*anisole*); C_7H_8O; [100-66-3]					
94.9	86.8	87.8	87.5	0.424	0.298

[a] Volume of sulfur dioxide absorbed by unit volume of solvent at experimental temperature and a total pressure of 1 atmosphere.

[b] Determined by static absorption method.

[c] Determined by flow method.

[d] $\alpha = \ell(1 - p_s^\circ)$ where p_s° is the vapor pressure of solvent.

[e] Calculated by compiler.

[f] Assuming vapor pressure of benzyl ether is negligible.

AUXILIARY INFORMATION

METHOD/APPARATUS/PROCEDURE:

Two techniques were used: (a) static absorption method and (b) flow method. Static method consisted of equilibrating solvent and sulfur dioxide in a 50 ml flask and measuring the absorption volumetrically. The flow method consists of passing sulfur dioxide through a micro gas absorption tube and estimating sulfur dioxide by iodometric titration. The vapor pressure of the solvents were measured roughly and were assumed to be equal to the partial pressure of the solvent by the authors when calculating the Bunsen coefficient.

SOURCE AND PURITY OF MATERIALS:

1. Purity 99.0-99.8 mole per cent.

ESTIMATED ERROR:

$\delta T/K = \pm 0.5$; $\delta x_{SO_2} = \pm 5\%$ (estimated by compiler).

REFERENCES:

COMPONENTS:

1. Sulfur dioxide; SO_2; [7446-09-5]
2. Ethoxybenzene , (phenetole); $C_8H_{10}O$; [103-73-1]

ORIGINAL MEASUREMENTS:

Gerrard, W.

J. Appl. Chem. Biotechnol. 1972, *22*, 623-650.

VARIABLES:

Temperature, pressure

PREPARED BY:

W. Gerrard

EXPERIMENTAL VALUES:

T/K		Pressure, total /mmHg	Mole ratio SO_2/Comp.(2)	*Mole fraction x_{SO_2}
273.15		760	2.10	0.677
	**	715	1.81	0.644
		665	1.57	0.611
		600	1.27	0.559
		480	0.874	0.466
		390	0.664	0.399
		252	0.375	0.273
		148	0.213	0.176
283.15		760	0.93	0.482

760 mmHg = 1 atm = 101.325 kPa.

* Calculated by compiler

** Data on which the graph in the original paper is based.

AUXILIARY INFORMATION

METHOD/APPARATUS/PROCEDURE:

Sulfur dioxide was bubbled into a weighed amount of component (2) in a bubbler tube as described in detail, diagram given, in the original paper. The amount of gas absorbed at equilibrium and at the observed temperature and pressure was weighed. By means of a manometer assembly, to which the bubbler tube was attached, the weight of gas absorbed at successively lower pressure was measured. Eventually the pressure was reduced to that of the component (2). The refractive index, and the infrared spectrum of the residue showed it to be essentially pure component (2).

SOURCE AND PURITY OF MATERIALS:

1. Sulfur dioxide was the best grade contained in a small cylinder. Its high purity was attested by the measurement of vapor pressure from 258 to 268 K.
2. The best obtainable specimen was carefully purified and rigorously attested.

ESTIMATED ERROR:

REFERENCES:

COMPONENTS:

1. Sulfur dioxide; SO_2; [7446-09-5]
2. 2-Octanone; $C_8H_{16}O$; [111-13-7]

ORIGINAL MEASUREMENTS:

Albright, L.F.; Shannon, P.T.; Yu, S.-N.; Chueh, P.L.

Chem. Engng. Prog. Symp. Ser. 1963. *59*, (44), 66-74.

VARIABLES:

Temperature, pressure

PREPARED BY:

C.L. Young.

EXPERIMENTAL VALUES:

T/K	P/bar	Mole fraction of sulfur dioxide in liquid, x_{SO_2}
298.14	1.14	0.402
	2.03	0.609
	3.45	0.854
310.93	1.63	0.400
	2.98	0.604
	5.01	0.854
338.71	3.41	0.389
	6.04	0.587
	6.10	0.594
	10.67	0.845
366.48	6.19	0.376
	10.70	0.567
	19.44	0.833

AUXILIARY INFORMATION

METHOD/APPARATUS/PROCEDURE:

Static equilibrium cell fitted with Bourdon pressure gauge. Temperature measured with mercury in glass thermometer. The equipment used in this investigation was stated to be similar to that used in previous studies, (1), (2).

SOURCE AND PURITY OF MATERIALS:

1. and 2. Commercial grade samples, no other details given.

ESTIMATED ERROR:

$\delta T/K = \pm 0.2$; $\delta P/bar = \pm 0.3$; $\delta x_{SO_2} = \pm 1\%$.

REFERENCES:

1. Albright, L. F.; Shannon, P. T.; Terrier, F.; Chueh, P. L. *Am. Inst. Chem. Engrs. J.* 1962, *8*, 668.
2. Albright, L. F.; Buclez, P. C.; Pluche, C. R.; Doody, T.C. *ASHRAE Trans.* 1960, *66*, 423.

COMPONENTS:

1. Sulfur dioxide; SO_2; [7446-09-5]
2. 1,1´-Oxybisbutane, (dibutyl ether); $C_8H_{18}O$; [142-96-1]

ORIGINAL MEASUREMENTS:

Hill, A.E.; Fitzgerald, T.B.

J. Am. Chem. Soc. 1935, *57*, 250-254.

VARIABLES:

PREPARED BY:

W. Gerrard

EXPERIMENTAL VALUES:

T/K	Weight of SO_2 absorbed by 5 cm^3 of ether, g	Total pressure, P, mmHg	"Const.of Henry" K (Note)	*Mole Ratio SO_2/ ether	Mole fraction x_{SO_2}
293.15	0.0	11.0			
	0.0764	87.5	1.00	0.040	0.0389
	0.1541	170.5	1.04	0.082	0.0755
	0.2291	250.1	1.05	0.121	0.108
	0.3128	332.4	1.03	0.166	0.142
	0.3960	413.5	1.02	0.210	0.173
	0.4762	483.5	0.99	0.252	0.201

760 mmHg = 1 atm = 101.325 kPa

* Calculated by compiler.

NOTE: p_{SO_2} is taken to be p_{total} - 11.0 mmHg throughout. This cannot be strictly correct.

NOTE: "Constant of Henry's Law", $K = p_{SO_2}/W$

W = Milligrams SO_2 absorbed by 5 cm^3 of the ether.

AUXILIARY INFORMATION

METHOD:/APPARATUS/PROCEDURE:

The apparatus had been previously described (1).

A weighed amount of liquid was treated with a measured volume of sulfur dioxide at a constant temperature and at a measured total pressure. A gas buret and a mercury manometer were used.

SOURCE AND PURITY OF MATERIALS:

(1) Not stated.

(2) Purified by distillation.

ESTIMATED ERROR:

REFERENCES:

1. Hill, A.E. *J. Am. Chem. Soc.* 1931, *53*, 2598.

COMPONENTS:

1. Sulfur dioxide; SO_2; [7446-09-5]
2. 1,4-Dimethoxybenzene (*hydroquinone dimethyl ether*); $C_8H_{10}O_2$; [150-78-7]
   1,3-Dimethoxybenzene (*recorcinol dimethyl ether*); $C_8H_{10}O_2$; [151-10-0]
   1,1'-Oxybis(2-methoxyethane) (*diglyme*); $C_6H_{14}O_3$; [111-96-6]

VARIABLES:

ORIGINAL MEASUREMENTS:

Sano, H.

Nippon Kagaku Zasshi

1968, *89*, 362-368.

PREPARED BY:

C. L. Young

EXPERIMENTAL VALUES:

T/°C = 30 T/K = 303

Vapor pressure of 1,1'-Oxybis(2-methoxyethane) = 0.002 atm

Absorption coefficient, S^a	Bunsen coefficient, α	Mole ratio	Mole fraction[b]
	1,4-Dimethoxybenzene		
116.2	105.5	0.625	0.385
	1,3-Dimethoxybenzene		
102.6	96.3	1.02	0.505
	1,1'-Oxybis(2-methoxyethane)		
177.6	96.3	0.345	0.257

[a] Volume of sulfur dioxide absorbed by unit volume of solvent at experimental temperature and a total pressure of 1 atmosphere.

[b] Calculated by compiler.

AUXILIARY INFORMATION

METHOD/APPARATUS/PROCEDURE:

Static absorption method consisted of equilibrating solvent and sulfur dioxide in a 50 ml flask and measuring the absorption volumetrically. The vapor pressure of the solvents were measured roughly and were assumed to be equal to the partial pressure of the solvent by the authors when calculating the Bunsen coefficient.

SOURCE AND PURITY OF MATERIALS:

1. Purity 99.0-99.8 mole per cent.

ESTIMATED ERROR:

$\delta T/K = \pm 0.5$; $\delta x_{SO_2} = \pm 5\%$

(estimated by compiler).

REFERENCES:

COMPONENTS:

1. Sulfur dioxide; SO_2; [7446-09-5]
2. Acetic acid, phenylmethyl ester, (benzyl acetate); $C_9H_{10}O_2$; [140-11-4] or Benzoic acid, phenylmethyl ester, (benzyl benzoate); $C_{14}H_{12}O_2$; [120-51-4]

ORIGINAL MEASUREMENTS:

Gerrard, W.; *"Solubility of Gases and Liquids,"* Plenum Press, New York, 1976, p. 239.

VARIABLES:

PREPARED BY:

W. Gerrard

EXPERIMENTAL VALUES:

T/K	Weight of Component(2) g	Weight of SO_2 absorbed, g	Mole ratio SO_2 / Comp(2)	Mole fraction, x_{SO_2} (Total Pressure = 1 atm)
		Acetic acid, phenylmethyl ester		
293.15	6.1630	2.4370	0.928	0.481
		Benzoic acid, phenylmethyl ester		
293.15	7.2841	1.5875	0.723	0.420

760 mmHg = 1 atm = 101.325 kPa

AUXILIARY INFORMATION

METHOD:

Sulfur dioxide was bubbled into a weighed amount of component (2) in a bubbler tube. The amount of gas absorbed was determined by re-weighing the tube. The total pressure was barometric, very nearly 1 atm. See ref. (1).

SOURCE AND PURITY OF MATERIALS:

1. Sulfur dioxide was the best grade contained in a small cylinder. Its high degree of purity was attested by the measurement of vapor pressure from 258-268 K.

2. The best commercial specimen was purified and attested.

ESTIMATED ERROR:

REFERENCES:

1. Gerrard, W. *J. Appl. Chem. Biotechnol.* 1972, *22*, 623.

COMPONENTS:

1. Sulfur dioxide; SO_2; [7446-09-5]
2. 1,7,7-Trimethylbicyclo [2.2.1] heptan-2-one, (Camphor); $C_{10}H_{16}O$; [464-49-3]

ORIGINAL MEASUREMENTS:

Bineau, A.

Ann. Chim. Phys. 1848,(3),*24*,326-337.

Information repeated in :

J. Prakt. Chem. 1849, *46*,296-301.

VARIABLES:

PREPARED BY:

W. Gerrard.

EXPERIMENTAL VALUES: 760 mmHg = 1 atm = 101.325 kPa.

T/K	Pressure, total /mmHg.	Weight of SO_2/ 100 g camphor	*Mole fraction x_{SO_2}	* x_{SO_2} (700 mmHg)	* x_{SO_2} (760mmHg)
297	524	25.5	0.377		
	650	30.8	0.423		
	745	35.4	0.457	(0.440)	(0.46)
294	670	34.7	0.452		
293	730	39.7	0.485	(0.472)	
288.6	355	28.0	0.399		
	744	47.6	0.531	(0.513)	(0.536)
287	611	40.4	0.490		
	738	48.6	0.536	(0.526)	(0.543)
285.6	529	37.3	0.470		
	703	49.1	0.538		
	727	50.5	0.545	(0.537)	(0.555)
283	320	31.7	0.429		
	560	42.6	0.503		
	720	55.8	0.570	(0.562)	(0.582)
281	304	33.0	0.439		
	503	42.0	0.499		
	682	57.4	0.577	(0.582)	(0.605)
277	490	46.0	0.522		
	720	73.6	0.636	(0.626)	(0.656)
275	469	48.4	0.535		
	650	72.0	0.631	-	(0.690)

* Calculated by compiler. The column x_{SO_2} (700 mmHg) is based on the weight value of Bineau estimated for 700 mmHg. The corresponding column for x_{SO_2} (760 mmHg) shows a value read from the compiler's graph.

AUXILIARY INFORMATION

METHOD/APPARATUS/PROCEDURE:

No information

SOURCE AND PURITY OF MATERIALS:

No information.

ESTIMATED ERROR:

REFERENCES:

COMPONENTS:

1. Sulfur dioxide; SO_2; [7446-09-5]
2. 1,7,7-Trimethylbicyclo [2.2.1] heptan-2-one, (Camphor); $C_{10}H_{16}O$; [464-49-3]

ORIGINAL MEASUREMENTS:

Schulze, H.

J. Prakt. Chem. 1881, *24*, 168-183.

VARIABLES:

PREPARED BY:

W. Gerrard.

EXPERIMENTAL VALUES:

760 mmHg = 1 atm = 101.325 kPa.

T/K	Weight of SO_2 / 1 g of Comp. (2)	*Mole ratio SO_2/Comp.(2)	*Mole fraction x_{SO_2}
273.15	0.880	2.090	0.676

* Calculated by compiler.

Pressure of SO_2 stated to be 725 mmHg. 1 Volume of camphor absorbed 308 volumes of SO_2.

NOTE: Schulze quoted three values by Bineau; but one reference was wrong, and the other did not appear to apply. Bineau's data have been compiled. Markham and Kobe (1) cited only the Bineau group; but attributed it to Schulze. Bineau's name was not mentioned.

AUXILIARY INFORMATION

METHOD/APPARATUS/PROCEDURE:

The gas was passed into a weighed amount of liquid camphor in a thin-walled test-tube fitted with a rubber stopper carrying an inlet tube and an out-let tube. The test-tube was cooled in melting snow. The amount of gas absorbed was determined by weighing.

SOURCE AND PURITY OF MATERIALS:

No details given.

ESTIMATED ERROR:

REFERENCES:

1. Markham, A.E.; Kobe, K.A. *Chem. Rev.* 1941, *28*, 519.

COMPONENTS:

1. Sulfur dioxide; SO_2; [7446-09-5]
2. 1,7,7-Trimethylbicyclo[2.2.1]-heptan-2-one (*camphor*); $C_{10}H_{16}O$; [76-22-2] or [464-49-3]

ORIGINAL MEASUREMENTS:

Sano, H.

Nippon Kagaku Zasshi

1968, *89*, 362-368.

VARIABLES:

PREPARED BY:

C. L. Young

EXPERIMENTAL VALUES:

T/°C = 30 T/K = 303

Bunsen coefficient, α	Mole ratio	Mole fraction[a]
99.0	0.670	0.401

[a] Calculated by compiler.

AUXILIARY INFORMATION

METHOD/APPARATUS/PROCEDURE:

Static absorption method consisted of equilibrating solvent and sulfur dioxide in a 50 ml flask and measuring the absorption volumetrically. The vapor pressure of the solvents were measured roughly and were assumed to be equal to the partial pressure of the solvent by the authors when calculating the Bunsen coefficient.

SOURCE AND PURITY OF MATERIALS:

1. Purity 99.0-99.8 mole per cent.

ESTIMATED ERROR:

$\delta T/K = \pm 0.5$; $\delta x_{SO_2} = \pm 5\%$

(estimated by compiler).

REFERENCES:

COMPONENTS:

1. Sulfur dioxide; SO_2; [7446-09-5]
2. 1,7,7-Trimethylbicyclo [2.2.1] heptan-2-one, (Camphor); $C_{10}H_{16}O$ (in solution); [469-49-3]

ORIGINAL MEASUREMENTS:

Weissenberger, G.; Hadwiger, H.

Z. angew. Chem. 1927, *40*, 734-736.

VARIABLES:

PREPARED BY:

W. Gerrard

EXPERIMENTAL VALUES:

T/K = 293.15

"Percent solution of camphor"	Solvent	Vol. of 1 mole of mixture /cm^3	Vol. SO_2 absorbed /cm^3 mol^{-1} (soln)
10	*Tetrahydro-naphthalene $C_{10}H_{12}$; [119-64-2]	141.1	1001.81
10	Cyclohexanone $C_6H_{10}O$; [108-94-1]	111.67	1697.38
20	Cyclohexanone	120.09	1765
10	1,3-Methylcyclo-hexanol $C_8H_{16}O$;	130.9	2264.57
20	1,3-Methylcyclo-hexanol	137.18	1508.98

* Tetrahydronaphthalene itself was stated to absorb no measurable volume of gas.

AUXILIARY INFORMATION

METHOD/APPARATUS/PROCEDURE:

Used an absorption vessel of 200 cm^3 capacity, and a vessel with a two-way tap. Simply stated that the absorptiometer was like that used by Manchot, but the reference was not cited; probably ref. (1).

SOURCE AND PURITY OF MATERIALS:

No details given.

ESTIMATED ERROR:

REFERENCES:

1. Manchot, W.
Z. anorg. Chem. 1924, *141*, 38.

COMPONENTS:

1. Sulfur dioxide; SO_2; [7446-09-5]
2. 2,5,8,11,14-Pentaoxapentadecane, (Tetraethylene glycol dimethyl ether); $C_{10}H_{22}O_5$; [143-24-8]

ORIGINAL MEASUREMENTS:

Albright, L.F.; Shannon, P.T., Yu, S.-N.; Chueh, P.L.

Chem. Engng. Prog. Symp. Ser. 1963, *59*, (44), 66-74.

VARIABLES:

Temperature, pressure

PREPARED BY:

C.L. Young

EXPERIMENTAL VALUES:

T/K	P/bar	Mole fraction of sulfur dioxide in liquid, x_{SO_2}	T/K	P/bar	Mole fraction of sulfur dioxide in liquid, x_{SO_2}
298.15	0.042	0.157	310.93	5.03	0.925
	0.086	0.253	338.71	0.245	0.157
	0.159	0.361		0.448	0.252
	0.283	0.484		0.830	0.358
	0.834	0.671		1.52	0.484
	2.36	0.841		3.65	0.672
	2.45	0.844		3.76	0.676
	3.31	0.926		7.91	0.833
310.93	0.085	0.157		10.9	0.921
	0.155	0.253	366.48	0.597	0.157
	0.283	0.359		1.091	0.251
	0.508	0.483		1.98	0.355
	1.52	0.683		2.35	0.476
	1.55	0.687		7.36	0.658
	3.59	0.838		14.55	0.830
	3.69	0.841		20.31	0.915

AUXILIARY INFORMATION

METHOD/APPARATUS/PROCEDURE:

Static equilibrium cell fitted with Bourdon pressure gauge. Temperature measured with mercury in glass thermometer. The equipment used in this investigation was stated to be similar to that used in previous studies, (1), (2).

SOURCE AND PURITY OF MATERIALS:

1. and 2. Commercial grade samples, no other details given.

ESTIMATED ERROR:

$\delta T/K = \pm 0.2$; $\delta P/\text{bar} = \pm 0.3$; $\delta x_{SO_2} = \pm 1\%$.

REFERENCES:

1. Albright, L. F.; Shannon, P. T.; Terrier, F.; Chueh, P. L. *Am. Inst. Chem. Engrs. J.* 1962, *8*, 668.
2. Albright, L. F.; Buclez, P. C.; Pluche, C. R.; Doody, T. C. *ASHRAE Trans.* 1960, *66*, 423.

COMPONENTS:

1. Sulfur dioxide, SO_2; [7446-09-5]
2. Dodecanoic acid, ethyl ester (Ethyl laurate); $C_{14}H_{28}O_2$; [106-33-2]

ORIGINAL MEASUREMENTS:

Albright, L.F.; Shannon, P.T.;
Yu, S.-N.;Chueh, P.L.

Chem. Engng. Prog. Symp. Ser. 1963, *59*, (44) 66-74.

VARIABLES:

Temperature, pressure

PREPARED BY:

C.L. Young.

EXPERIMENTAL VALUES:

T/K	P/bar	Mole fraction of sulfur dioxide in liquid x_{SO_2}
298.15	1.45	0.453
	2.48	0.638
	3.45	0.839
	3.51	0.846
310.93	2.14	0.448
	3.69	0.626
	5.07	0.839
	5.21	0.844
338.71	4.30	0.432
	7.10	0.615
	10.78	0.828
	10.87	0.841
366.48	7.31	0.412
	11.85	0.591
	19.55	0.824

AUXILIARY INFORMATION

METHOD/APPARATUS/PROCEDURE:

Static equilibrium cell fitted with Bourdon pressure gauge. Temperature measured with mercury in glass thermometer. The equipment used in this investigation was stated to be similar to that used in previous studies, (1), (2).

SOURCE AND PURITY OF MATERIALS:

1 and 2. Commercial grade samples, no other details given.

ESTIMATED ERROR:

$\delta T/K = \pm 0.2$; $\delta P/\text{bar} = \pm 0.3$; $\delta x_{SO_2} = \pm 1\%$.

REFERENCES:

1. Albright, L. F.; Shannon, P. T.; Terrier, F.; Chueh, P. L. *Am. Inst. Chem. Engrs. J.* 1962, *8*, 668.
2. Albright, L.F.; Buclez, P. C.; Pluche, C. R.; Doody, T. C. *ASHRAE Trans.* 1960, *66*, 423.

COMPONENTS:

1. Sulfur dioxide; SO_2; [7446-09-5]
2. Polyethylene glycol (*α-hydro-ω-hydroxy-poly(oxy-1,2-ethanediyl)*); $(C_2H_4O)_nH_2O$; [25322-68-3]

ORIGINAL MEASUREMENTS:

Sano, H.

Nippon Kagaku Zasshi

1968, *89*, 362-368.

VARIABLES:

PREPARED BY:

C. L. Young

EXPERIMENTAL VALUES:

T/°C = 30 T/K = 303

Bunsen coefficient, α	Mole ratio	Mole fraction[a]
140.8	0.295	0.228

[a] Calculated by compiler.

AUXILIARY INFORMATION

METHOD/APPARATUS/PROCEDURE:

Static absorption method consisted of equilibrating solvent and sulfur dioxide in a 50 ml flask and measuring the absorption volumetrically. The vapor pressure of the solvents were measured roughly and were assumed to be equal to the partial pressure of the solvent by the authors when calculating the Bunsen coefficient.

SOURCE AND PURITY OF MATERIALS:

1. Purity 99.0-99.8 mole per cent.
2. Molecular weight 400.

ESTIMATED ERROR:

$\delta T/K = \pm 0.5$; $\delta x_{SO_2} = \pm 5\%$ (estimated by compiler).

REFERENCES:

COMPONENTS:	EVALUATOR:
1. Sulfur dioxide; SO_2; [7446-09-5] 2. Compounds containing nitrogen	Colin L. Young, School of Chemistry, University of Melbourne, Parkville, Victoria 3052, Australia. December 1982

CRITICAL EVALUATION:

N,N-Dimethylformamide; C_2H_7NO; [68-12-2]

Measurements on this system have been reported by five groups of workers. The data of Gerrard (1) and Albright, Shannon, Yu and Church (2) are consistent with each other and broadly consistent with the isolated value from DuPont (3). Values of mole fraction of sulfur dioxide for smooth values of temperatures estimated from the data of refs. (1) and (2) are as follows:

T/K	273	283	293	298	303	
x_{SO_2} (total pressure = 1 atm)	0.787	0.706	0.636	0.600	0.564	
T/K	313	323	333	343	353	363
x_{SO_2} (total pressure = 1 atm)	0.494	0.425	0.363	0.307	0.259	0.218

The data of Pfeifer (4) deviate considerably from these values at low temperatures but agree at 333 K. The isolated data of Sano (5) agrees with that of Pfeifer at 303 K.

The isolated value of Benoit and Milanova (6) was determined at a low partial pressure and is not considered further as extrapolation on the basis of Henry's law to 1 atmosphere total pressure would not be valid.

Pyridine; C_5H_5N; [110-86-1]

This system has been investigated by three groups of workers. The isolated value of Benoit and Milanova (6) was determined at a low partial pressure and is not considered further as extrapolation on the basis of Henry's law to 1 atmosphere pressure would not be valid. Gerrard's (1) data extrapolated to a partial pressure of 1 atmosphere and 303 K is broadly consistent with the value found by Sano and Nakamoto (7).

Nitrobenzene; $C_6H_5NO_2$; [98-95-3]

This system has been investigated by six groups of workers (1,2,5, 6,8,9). The data of Lenoir *et al.* (8) was determined using gas chromatography and corresponds to a very low partial pressure of sulfur dioxide. It is considerably larger than the data of Lloyd (9), Albright *et al.* (2), and Gerrard (1) and is classified as doubtful. Similarly the data of Benoit and Milanova correspond to a low partial pressure of sulfur dioxide.

There is moderate agreement between the data of Gerrard (1), Albright *et al.* (2), and Lloyd (9). Because of the different ranges of pressure and temperature involved and the fact that considerable approximation was incurred in calculating mole fraction solubility from Lloyd's data (9), precise comparison is not possible and the data are classified as tentative. The value of Sano (5) is considerably larger than the value of Lloyd at the same temperature.

N,N-Dimethylbenzenamine; $C_8H_{11}N$; [121-69-7]

Hill and Fitzgerald (10) gave data for this compound as mole ratio SO_2/Dimethylaniline for several pressures up to above 75 kPa at 298.15 K. Balej and Regner (11) used mixtures of nitrogen and sulfur dioxide to achieve different partial pressures of sulfur dioxide below about 35 kPa for temperatures 288 K, 293 K, 298 K and 313 K. There is qualitative agreement between the two sets of results where they overlap and both are classified as tentative. The isolated value of Sano and Nakamoto (7) is broadly consistent with the two sets of data but no precise comparison is possible.

(cont.)

COMPONENTS:	EVALUATOR:
1. Sulfur dioxide; SO_2; [7446-09-5] 2. Compounds containing nitrogen	Colin L. Young, School of Chemistry, University of Meloburne, Parkville, Victoria 3052, Australia. December 1982

CRITICAL EVALUATION:

References

1. Gerrard, W. *J. Appl. Chem. Biotechnol.* 1972, *22*, 623.
2. Albright, L. F.; Shannon, P. T.; Yu, S.-N.; Cheuh, P. L. *Chem. Engng. Prog. Symp. Ser.* 1963, *59*, 66.
3. DuPont de Nemours and Co. *Chem. Eng. New* 1955, *33*, 2366.
4. Pfeifer, G. *Magy. Kem. Folyoirat.* 1963, *69*, No. 3, 138.
5. Sano, H. *Nippon Kagaku Zasshi* 1968, *89*, 362.
6. Benoit, R. L.; Milanova, E. *Can. J. Chem.* 1979, *57*, 1319.
7. Sano, H.; Nakamoto, Y. *Nippon Kagaku Zasshi* 1968, *89*, 369.
8. Lenoir, J.-Y.; Renault, P.; Renon, H. *J. Chem. Engng. Data* 1971, *16*, 340.
9. Lloyd, S. J. *J. Phys. Chem.* 1918, *22*, 300.
10. Hill, A. E.; Fitzgerald, T. B. *J. Am. Chem. Soc.* 1935, *57*, 250.
11. Balej, J.; Regner, A. *Chem. Listy* 1956, *50*, 1374.

COMPONENTS:

1. Sulfur dioxide; SO_2; [7446-09-5]
2. Formamide; CH_3NO; [75-12-7]

ORIGINAL MEASUREMENTS:

Sano, H.; Nakamoto, Y.
Nippon Kagaku Zasshi
1968, *89*, 369-373.

VARIABLES:

PREPARED BY:

C. L. Young

EXPERIMENTAL VALUES:

$T/°C = 30$ $T/K = 303$

Vapor pressure of formamide = 0.001 atm

Absorption coefficient, $S^{a,b}$	Bunsen coefficient,[b] α	Bunsen coefficient,[c] α	$\ell^{c,d}$	Mole ratio	Mole fraction[e]
175.2	160.0	159.2	158.0	0.286	0.222

[a] Volume of sulfur dioxide absorbed by unit volume of solvent at experimental temperature and a total pressure of 1 atmosphere.

[b] Determined by static absorption method.

[c] Determined by flow method.

[d] $\alpha = \ell(1 - p_s^{\circ})$ where p_s° is the vapor pressure of solvent.

[e] Calculated by compiler.

AUXILIARY INFORMATION

METHOD/APPARATUS/PROCEDURE:

Two techniques were used: (a) static absorption method and (b) flow method. Static method consisted of equilibrating solvent and sulfur dioxide in a 50 ml flask and measuring the absorption volumetrically. The flow method consists of passing sulfur dioxide through a micro gas absorption tube and estimating sulfur dioxide by iodometric titration. The vapor pressure of the solvents were measured roughly and were assumed to be equal to the partial pressure of the solvent by the authors when calculating the Bunsen coefficient.

SOURCE AND PURITY OF MATERIALS:

1. Purity 99.0-99.8 mole per cent.

ESTIMATED ERROR:

$\delta T/K = \pm 0.5$; $\delta x_{SO_2} = \pm 5\%$

(estimated by compiler).

REFERENCES:

COMPONENTS:

1. Sulfur dioxide; SO_2; [7446-09-5]
2. Nitromethane; CH_3NO_2; [75-52-5]
   Nitrobenzene; $C_6H_5NO_2$; [98-95-3]

ORIGINAL MEASUREMENTS:

Sano, H.

Nippon Kagaku Zasshi

1968, *89*, 362-368.

VARIABLES:

PREPARED BY:

C. L. Young

EXPERIMENTAL VALUES:

T/°C = 30 T/K = 303

Vapor pressure of nitromethane = 0.042 atm

Absorption coefficient, S[a,b]	Bunsen coefficient,[b] α	Bunsen coefficient,[c,f] α	ℓ[c,d]	Mole ratio	Mole fraction[e]
	Nitromethane; CH_3NO_2; [75-52-5]				
150.5	142.0	149.3	143.5	0.531	0.347
	Nitrobenzene; $C_6H_5NO_2$; [98-95-3]				
70.4	64.1	65.0	65.0	0.500	0.333

[a] Volume of sulfur dioxide absorbed by unit volume of solvent at experimental temperature and a total pressure of 1 atmosphere.
[b] Determined by static absorption method.
[c] Determined by flow method.
[d] $\alpha = \ell(1 - p_s^\circ)$ where p_s° is the vapor pressure of solvent.
[e] Calculated by compiler.
[f] Assuming the vapor pressure of nitrobenzene is negligible.

AUXILIARY INFORMATION

METHOD/APPARATUS/PROCEDURE:

Two techniques were used: (a) static absorption method and (b) flow method. Static method consisted of equilibrating solvent and sulfur dioxide in a 50 ml flask and measuring the absorption volumetrically. The flow method consists of passing sulfur dioxide through a micro gas absorption tube and estimating sulfur dioxide by iodometric titration. The vapor pressure of the solvents were measured roughly and were assumed to be equal to the partial pressure of the solvent by the authors when calculating the Bunsen coefficient.

SOURCE AND PURITY OF MATERIALS:

1. Purity 99.0-99.8 mole per cent.

ESTIMATED ERROR:

$\delta T/K = \pm 0.5$; $\delta x_{SO_2} = \pm 5\%$

(estimated by compiler).

REFERENCES:

COMPONENTS:

1. Sulfur dioxide; SO_2; [7446-09-5]

2. Nitromethane; CH_3NO_2; [75-52-5]

ORIGINAL MEASUREMENTS:

Benoit, R.L.; Milanova, E.

Can. J. Chem., 1979, *59*, 1319-1323

VARIABLES:

PREPARED BY:

C.L. Young

EXPERIMENTAL VALUES:

T/K	Henry's law constant⁺ H/atm
298.15	2.33

⁺ Defined in the original as

$$H = \frac{\text{partial pressure of sulfur dioxide}}{\text{mole fraction of sulfur dioxide}}$$

AUXILIARY INFORMATION

METHOD/APPARATUS/PROCEDURE:

Few details given. Pressure measured with quartz spiral gauge. Partial pressure calculated assuming Raoult's law. Concentration of sulfur dioxide determined by iodometric titration. Concentration of sulfur dioxide ranged from 0.1 to 0.8 mol l^{-1}

SOURCE AND PURITY OF MATERIALS:

1. Anhydrous sample from Canadian Liquid Air, purity 99.98 mole per cent.

2. Dried over 4A molecular sieve

ESTIMATED ERROR:

$\delta T/K = \pm 0.1$; $\delta H/\text{atm} = \pm 4\%$ (estimated by compiler)

REFERENCES:

COMPONENTS:

1. Sulfur dioxide; SO_2; [7446-09-5]
2. Nitriles

ORIGINAL MEASUREMENTS:

Sano, H.

Nippon Kagaku Zasshi

1968, *89*, 362-368.

VARIABLES:

PREPARED BY:

C. L. Young

EXPERIMENTAL VALUES:

T/°C = 30 T/K = 303

Vapor pressure of benzonitrile = 0.001 atm

Absorption coefficient, $S^{a,b}$	Bunsen coefficient,[b] α	Bunsen coefficient,[c,f] α	$\ell^{c,d}$	Mole ratio	Mole fraction[e]
Acetonitrile; C_2H_3N; [75-05-8]					
-	-	175.7	-	0.920	0.479
Benzonitrile; C_7H_5N; [100-47-0]					
82.9	75.3	75.8	75.7	0.509	0.337
Benzyl cyanide (Benzeneacetonitrile); C_8H_7N; [140-29-4]					
98.3	89.3	90.2	90.2	0.781	0.439

AUXILIARY INFORMATION

METHOD/APPARATUS/PROCEDURE:

Two techniques were used: (a) static absorption method and (b) flow method. Static method consisted of equilibrating solvent and sulfur dioxide in a 50 ml flask and measuring the absorption volumetrically. The flow method consists of passing sulfur dioxide through a micro gas absorption tube and estimating sulfur dioxide by iodometric titration. The vapor pressure of the solvents were measured roughly and were assumed to be equal to the partial pressure of the solvent by the authors when calculating the Bunsen coefficient.

SOURCE AND PURITY OF MATERIALS:

1. Purity 99.0-99.8 mole per cent.

ESTIMATED ERROR:

$\delta T/K = \pm 0.5$; $\delta x_{SO_2} = \pm 5\%$

(estimated by compiler).

REFERENCES:

COMPONENTS:	ORIGINAL MEASUREMENTS:
1. Sulfur dioxide; SO_2; [7446-09-5]	Sano, H.
2. Nitriles	*Nippon Kagaku Zasshi*
	1968, *89*, 362-368.

[a] Volume of sulfur dioxide absorbed by unit volume of solvent at experimental temperature and a total pressure of 1 atmosphere.

[b] Determined by static absorption method.

[c] Determined by flow method.

[d] $\alpha = \ell(1 - p_s^\circ)$ where p_s° is the vapor pressure of solvent.

[e] Calculated by compiler.

[f] Assuming the vapor pressure of benzyl cyanide is negligible.

COMPONENTS:	ORIGINAL MEASUREMENTS:
1. Sulfur dioxide; SO_2; [7446-09-5] 2. Acetonitrile; C_2H_3N; [75-05-8]	Benoit, R.L.; Milanova, E. *Can. J. Chem.*, 1979, *57*, 1319-1323
VARIABLES:	PREPARED BY: C.L. Young

EXPERIMENTAL VALUES:

T/K	Henry's law constant $^+$ H/atm
298.15	1.91

$^+$ Defined in the original as

$$H = \frac{\text{partial pressure of sulfur dioxide}}{\text{mole fraction of sulfur dioxide}}$$

AUXILIARY INFORMATION

METHOD/APPARATUS/PROCEDURE:

Few details given. Pressure measured with quartz spiral gauge. Partial pressure calculated assuming Raoult's law. Concentration of sulfur dioxide determined by iodometric titration. Concentration of sulfur dioxide ranged from 0.2 to 1.4 mol l^{-1}

SOURCE AND PURITY OF MATERIALS:

1. Anhydrous sample from Canadian Liquid Air, purity 99.98 mole per cent.

2. Dried over 4A molecular sieve

ESTIMATED ERROR:

$\delta T/K = \pm 0.1$; $\delta H/\text{atm} = \pm 4\%$ (estimated by compiler)

REFERENCES:

COMPONENTS:

1. Sulfur dioxide; SO_2; [7446-09-5]
2. *N,N*-Dimethylformamide; C_3H_7NO; [68-12-2]

ORIGINAL MEASUREMENTS:

DuPont de Nemours and Co.

Chem. Eng. New. 1955, *33*, 2366.

VARIABLES:

PREPARED BY:

W. Gerrard

EXPERIMENTAL VALUES:

1 atm = 101.325 kPa

T/K	Volume of SO_2 (adjusted to 273.15 K and 1 atm) absorbed by 1 volume of *N,N*-dimethylformamide at 298.15 K and p_{SO_2} = 1 atm.	*Mole ratio SO_2/DMF	*Mole fraction x_{SO_2}
298.15	405	1.432	0.589

d_4^{25} taken to be 0.9445 g cm^{-3}

* Calculated by compiler.

Molar volume SO_2 at 273.15 K and 1 atm taken to be 21885 cm^3, based on the published density.

AUXILIARY INFORMATION

METHOD/APPARATUS/PROCEDURE:

No stated.

SOURCE AND PURITY OF MATERIALS:

Not stated.

ESTIMATED ERROR:

REFERENCES:

COMPONENTS:

1. Sulfur dioxide; SO_2; [7446-09-5]
2. *N,N*-Dimethylformamide; C_3H_7NO; [68-12-2]

ORIGINAL MEASUREMENTS:

Pfeifer, Gyula,

Magy. Kem. Folyoirat. 1963, *69*, No. 3. 138-141.

VARIABLES:

PREPARED BY:

W. Gerrard

EXPERIMENTAL VALUES:

1 atm = 760 mmHg = 101.325 kPa.

T/K	ml SO_2/g DMF	p_{SO_2}/mmHg	*M.R.	*M.F. x_{SO_2}
293.15	159	135	0.531	0.347
	177	155	0.591	0.372
	195	179	0.651	0.394
	231	225	0.772	0.435
	258	270	0.862	0.463
	284	314	0.949	0.487
	307	357	1.025	0.506
	332	401	1.109	0.526
	350	431	1.169	0.539
	382	512	1.276	0.561
	402	582	1.343	0.573
	414	660	1.383	0.580
	424	758	1.416	0.586
	428	842	1.430	0.588
	(424)	(760)	(1.416)	(0.586)
303.15	125	177	0.4175	0.295
	141	207	0.471	0.320
	160	235	0.534	0.348
	180	270	0.601	0.376
	216	334	0.721	0.419
	240	384	0.802	0.445
	262	425	0.875	0.467
	281	475	0.939	0.484
	308	641	1.029	0.507
	313	745	1.045	0.511

AUXILIARY INFORMATION

METHOD/APPARATUS/PROCEDURE:

A known volume of gas was added to the evacuated absorption vessel containing a known amount of component (2). After equilibrium, pressure and decrease in volume of gas were measured. The vapor pressure was measured by Smith-Menzies isoteniscopic method.

SOURCE AND PURITY OF MATERIALS:

1. Sulfur dioxide was probably of acceptable purity.
2. Dried and distilled; d^{20} = 0.950 g ml^{-1}

ESTIMATED ERROR:

REFERENCES:

COMPONENTS:	ORIGINAL MEASUREMENTS:
1. Sulfur dioxide; SO_2; [7446-09-5] 2. *N*,*N*-Dimethylformamide; C_3H_7NO; [68-12-2]	Pfeifer, Gyula, *Magy. Kem. Folyoirat*, 1963, *69*, No. 3, 138-141.

EXPERIMENTAL VALUES: 1 atm = 760 mmHg = 101.325 kPa

T/K	ml SO_2/g DMF	p_{SO_2}/mmHg	*M.R.	*M.F. x_{SO_2}
303.15	314	845	1.049	0.512
	(313)	(760)	(1.045)	(0.511)
313.15	124	210	0.414	0.293
	134	230	0.448	0.309
	164	292	0.548	0.354
	188	363	0.628	0.386
	214	423	0.715	0.417
	230	489	0.768	0.435
	237	511	0.792	0.442
	248	594	0.828	0.453
	254	715	0.848	0.459
	258	793	0.862	0.463
	(257)	(760)	(0.858)	(0.462)
323.15	86	225	0.287	0.223
	106	278	0.354	0.261
	117	315	0.391	0.281
	154	432	0.514	0.340
	157	454	0.524	0.344
	174	521	0.581	0.368
	192	610	0.641	0.391
	203	722	0.678	0.404
	208	795	0.695	0.410
	(205)	(760)	(0.685)	(0.406)
333.15	79	286	0.264	0.209
	94	343	0.314	0.239
	99	363	0.331	0.249
	116	444	0.387	0.279
	136	533	0.454	0.312
	149	612	0.498	0.332
	158	677	0.528	0.345
	164	719	0.548	0.354
	(167)	(760)	(0.558)	(0.358)

Vapor pressure, p_{DMF}, stated to be as follows:

T/K:	283.15	288.15	293.15	298.15	303.15	308.15	313.15	318.15	323.15	333.15
mmHg:	1.32	1.90	2.65	3.66	5.00	6.70	8.90	11.80	15.70	26.90

*M.R. Mole ratio SO_2/DMF, calculated by compiler.

*M.F. Mole fraction, x_{SO_2}, calculated by compiler.

COMPONENTS:

1. Sulfur dioxide; SO_2; [7446-09-5]
2. *N*,*N*-Dimethylformamide; C_3H_7NO; [68-12-2]

ORIGINAL MEASUREMENTS:

Albright, L.F.; Shannon, P.T.; Yu, S.-N.; Chueh, P.L.

Chem. Engng. Prog. Symp. Ser. 1963, *59*, (44), 66-74.

VARIABLES:

Temperature, pressure

PREPARED BY:

C.L. Young

EXPERIMENTAL VALUES:

T/K	P/bar	Mole fraction of sulfur dioxide in liquid, x_{SO_2}
298.15	0.096	0.180
	0.317	0.407
	1.331	0.626
	2.661	0.799
	3.089	0.851
310.93	0.129	0.180
	0.517	0.407
	1.999	0.625
	4.033	0.798
	4.654	0.850
	4.585	0.851
338.71	0.328	0.180
	1.489	0.406
	4.420	0.620
	8.494	0.793
	9.956	0.847
366.48	0.845	0.180
	3.296	0.401
	8.722	0.609
	15.975	0.787
	18.733	0.841

AUXILIARY INFORMATION

METHOD/APPARATUS/PROCEDURE:

Static equilibrium cell fitted with Bourdon pressure gauge. Temperature measured with mercury in glass thermometer. The equipment used in this investigation was stated to be similar to that used in previous studies, (1), (2).

SOURCE AND PURITY OF MATERIALS:

1. and 2. Commercial grade samples, no other details given.

ESTIMATED ERROR:

$\delta T/K = \pm 0.2$; $\delta P/\text{bar} = \pm 0.3$; $\delta x_{SO_2} = \pm 1\%$.

REFERENCES:

1. Albright, L. F.; Shannon, P. T.; Terrier, F.; Chueh, P. L. *Am. Inst. Chem. Engrs. J.* 1962, *8*, 668.
2. Albright, L. F.; Buclez, P. C.; Pluche, C. R.; Doody, T. C. *ASHRAE Trans.* 1960, *66*, 423.

COMPONENTS:

1. Sulfur dioxide; SO_2; [7446-09-5]
2. *N*,*N*-Dimethylformamide; C_3H_7NO; [68-12-2]

ORIGINAL MEASUREMENTS:

Gerrard, W.

J. Appl. Chem. Biotechnol. 1972, *22*, 623-650.

VARIABLES:

Temperature, pressure

PREPARED BY:

W. Gerrard

EXPERIMENTAL VALUES:

T/K	Total pressure /mmHg	Mole ratio SO_2/amide	Mole * fraction x_{SO_2}	T/K	Total pressure /mmHg	Mole ratio SO_2/amide	Mole * fraction x_{SO_2}
273.15	760	3.70	0.787	293.15	200	0.758	0.431
	709	3.47	0.776		127	0.590	0.371
	611	2.89	0.742	283.15	760	2.40	0.706
	527	2.48	0.712		733	2.30	0.697
	330	1.69	0.627		670	2.12	0.679
	167	1.13	0.531		614	1.97	0.663
	100	0.844	0.458		530	1.78	0.640
	32	0.504	0.335		479	1.60	0.615
	22	0.417	0.294		368	1.36	0.576
	7	0.230	0.187		275	1.13	0.531
					132	0.728	0.421
293.15	760	1.75	0.636		87	0.549	0.354
	695	1.66	0.624		65	0.367	0.268
	581	1.55	0.608		35	0.300	0.231
	491	1.37	0.578		26	0.275	0.216
	404	1.21	0.548				
	300	0.990	0.497	298.15	760	1.50	0.600

1 atm = 760 mmHg = 101.325 kPa.

* Calculated by compiler

<u>NOTE</u> The pressure/solubility values are the basis of the plots given in the original.

AUXILIARY INFORMATION

METHOD/APPARATUS/PROCEDURE:

Sulfur dioxide was bubbled into a weighed amount of component (2) in a bubbler tube as described in detail, diagram given, in the original paper. The amount of gas absorbed at equilibrium and at the observed temperature and pressure was weighed. By means of a manometer assembly, to which the bubbler tube was attached, the weight of gas absorbed at successively lower pressure was measured. Eventually the pressure was reduced to that of the component (2). The refractive index, and the infrared spectrum of the residue showed it to be essentially pure component (2).

SOURCE AND PURITY OF MATERIALS:

1. Sulfur dioxide was the best grade contained in a small cylinder. Its high purity was attested by the measurement of vapor pressure from 258 to 268 K.
2. The best obtainable specimen was carefully purified and rigorously attested.

ESTIMATED ERROR:

REFERENCES:

COMPONENTS:

1. Sulfur dioxide; SO_2; [7446-09-5]
2. Dimethylformamide; C_3H_7NO; [68-12-2]

ORIGINAL MEASUREMENTS:

Sano, H.

Nippon Kagaku Zasshi

1968, *89*, 362-368.

VARIABLES:

PREPARED BY:

C. L. Young

EXPERIMENTAL VALUES:

T/°C = 30 T/K = 303

Bunsen coefficient, α	Mole ratio	Mole fraction[a]
303	1.05	0.512

[a] Calculated by compiler.

AUXILIARY INFORMATION

METHOD/APPARATUS/PROCEDURE:

Static absorption method consisted of equilibrating solvent and sulfur dioxide in a 50 ml flask and measuring the absorption volumetrically. The vapor pressure of the solvents were measured roughly and were assumed to be equal to the partial pressure of the solvent by the authors when calculating the Bunsen coefficient.

SOURCE AND PURITY OF MATERIALS:

1. Purity 99.0-99.8 mole per cent.

ESTIMATED ERROR:

$\delta T/K = \pm 0.5$; $\delta x_{SO_2} = \pm 5\%$ (estimated by compiler).

REFERENCES:

COMPONENTS:	ORIGINAL MEASUREMENTS:
1. Sulfur dioxide; SO_2; [7446-09-5] 2. *N,N*-Dimethylformamide; C_3H_7NO; [68-12-2]	Benoit, R.L.; Milanova, E. *Can. J. Chem.*, 1979, *57*, 1319-23

VARIABLES:	PREPARED BY:
	C.L. Young

EXPERIMENTAL VALUES:

T/K	Henry's law constant+ *H*/atm
298.15	0.172

\+ Defined in the original as

$$H = \frac{\text{partial pressure of sulfur dioxide}}{\text{mole fraction of sulfur dioxide}}$$

AUXILIARY INFORMATION

METHOD/APPARATUS/PROCEDURE:

Few details given. Pressure measured with quartz spiral gauge. Partial pressure calculated assuming Raoult's law. Concentration of sulfur dioxide determined by iodometric titration. Concentration of sulfur dioxide ranged from 0.4 to 1.7 mol l^{-1}

SOURCE AND PURITY OF MATERIALS:

1. Anhydrous sample from Canadian Liquid Air, purity 99.98 mole per cent.

2. Dried over 4A molecular sieve

ESTIMATED ERROR:

$\delta T/K = \pm 0.1$; $\delta H/\text{atm} = \pm 4\%$ (estimated by compiler)

REFERENCES:

COMPONENTS:

1. Sulfur dioxide, SO_2; [7446-09-5]
2. *N,N*-Dimethylacetamide; C_4H_9NO; [127-19-5]

ORIGINAL MEASUREMENTS:

Albright, L.F.; Shannon, P.T.; Yu, S.-N. Chueh, P.L.

Chem.Engng. Prog. Symp.Ser. 1963, *59*, (44), 66-74.

VARIABLES:

Temperature, pressure

PREPARED BY:

C.L. Young

EXPERIMENTAL VALUES:

T/K	*P*/bar	Mole fraction of sulfur dioxide in liquid, x_{SO_2}
298.15	0.021	0.105
	0.061	0.166
	0.127	0.297
	0.248	0.391
	1.207	0.625
	2.413	0.771
310.93	0.041	0.105
	0.094	0.166
	0.228	0.297
	0.417	0.391
	1.875	0.624
	3.65	0.769
338.71	0.137	0.105
	0.254	0.166
	0.614	0.294
	1.155	0.389
	4.247	0.619
	7.708	0.763
366.48	0.399	0.105
	0.665	0.166
	1.517	0.291
	1.551	0.293
	2.613	0.383
	8.239	0.611
	14.445	0.756

AUXILIARY INFORMATION

METHOD/APPARATUS/PROCEDURE:

Static equilibrium cell fitted with Bourdon pressure gauge. Temperature measured with mercury in glass thermometer. The equipment used in this investigation was stated to be similar to that used in previous studies, (1), (2).

SOURCE AND PURITY OF MATERIALS:

1. and 2. Commercial grade samples, no other details given.

ESTIMATED ERROR:

$\delta T/K = \pm 0.2$; $\delta P/\text{bar} = \pm 0.3$; $\delta x_{SO_2} = \pm 1\%$.

REFERENCES:

1. Albright, L. F.; Shannon, P. T.; Terrier, F.; Chueh, P. L. *Am. Inst. Chem. Engrs. J.* 1962, *8*, 668.
2. Albright, L. F.; Buclez, P. C.; Pluche, C. R.; Doody, T. C. *ASHRAE Trans.* 1960, *66*, 423.

COMPONENTS:	ORIGINAL MEASUREMENTS:
1. Sulfur dioxide; SO_2; [7446-09-5] 2. Amines	Sano, H.; Nakamoto, Y. *Nippon Kagaku Zasshi* 1968, *89*, 369-373.
VARIABLES:	PREPARED BY: C. L. Young

EXPERIMENTAL VALUES: T/°C = 30 T/K = 303

Absorption coefficient, S^a	Bunsen coefficient, α	Mole ratio	Mole fraction[b]
1-Butanamine (*butylamine*); $C_4H_{11}N$; [109-73-9]			
-	350	1.6	0.615
N,N-Dibutylbutanamine (*tributylamine*); $C_{12}H_{27}N$; [102-82-9]			
206.1	187.5	2.02	0.669
2-Aminoethanol (*monoethanolamine*); C_2H_7NO; [141-43-5]			
-	380	1.1	0.52

[a] Volume of sulfur dioxide absorbed by unit volume of solvent at experimental temperature and a total pressure of 1 atmosphere.

[b] Calculated by compiler.

AUXILIARY INFORMATION

METHOD/APPARATUS/PROCEDURE:

Static absorption method consisted of equilibrating solvent and sulfur dioxide in a 50 ml flask and measuring the absorption volumetrically. The vapor pressure of the solvents were measured roughly and were assumed to be equal to the partial pressure of the solvent by the authors when calculating the Bunsen coefficient.

SOURCE AND PURITY OF MATERIALS:

1. Purity 99.0-99.8 mole per cent.

ESTIMATED ERROR:

$\delta T/K = \pm 0.5$; $\delta x_{SO_2} = \pm 5\%$
(estimated by compiler).

REFERENCES:

COMPONENTS:

1. Sulfur dioxide; SO_2; [7446-09-5]
2. Pyridine; C_5H_5N; [110-86-1]
   Methylpyridine (*Picoline*); C_6H_7N; [1333-41-1]
   Quinoline; C_9H_7N; [91-22-5]

ORIGINAL MEASUREMENTS:

Sano, H.; Nakamoto, Y.
Nippon Kagaku Zasshi
1968, *89*, 369-373.

VARIABLES:

PREPARED BY:

C. L. Young

EXPERIMENTAL VALUES:

T/°C = 30 T/K = 303

Bunsen coefficient, α	Mole ratio	Mole fraction[a]
	Pyridine	
424	1.54	0.606
	Methylpyridine	
335.5	1.47	0.595
	Quinoline	
229.5	1.21	0.549

[a] Calculated by compiler.

AUXILIARY INFORMATION

METHOD/APPARATUS/PROCEDURE:

Static absorption method consisted or equilibrating solvent and sulfur dioxide in a 50 ml flask and measuring the absorption volumetrically. The vapor pressure of the solvents were measured roughly and were assumed to be equal to the partial pressure of the solvent by the authors when calculating the Bunsen coefficient.

SOURCE AND PURITY OF MATERIALS:

1. Purity 99.0-99.8 mole per cent.

ESTIMATED ERROR:

$\delta T/K = \pm 0.5$; $\delta x_{SO_2} = \pm 5\%$ (estimated by compiler).

REFERENCES:

COMPONENTS:

1. Sulfur dioxide; SO_2; [7446-09-5]
2. Pyridine; C_5H_5N; [110-86-1]

ORIGINAL MEASUREMENTS:

Gerrard, W.

J. Appl. Chem. Biotechnol. 1972, *22*, 623-650.

VARIABLES:

Temperature, pressure

PREPARED BY:

W. Gerrard

EXPERIMENTAL VALUES:

T/K	Total pressure /mmHg	Mole ratio SO_2/C_5H_5N	*Mole fraction x_{SO_2}
273.15	**760	3.40	0.773
	735	3.35	0.770
	658	3.02	0.751
	580	2.62	0.724
	520	2.37	0.703
	312	1.70	0.630
	261	1.56	0.609
	102	1.13	0.531
283.15	757	2.26	0.693
293.15	757	1.70	0.630
298.15	757	1.56	0.609

** The plot of pressure vs solubility in the original paper is based on these data.

760 mmHg = 1 atm = 101.325 kPa.

* Calculated by compiler

AUXILIARY INFORMATION

METHOD/APPARATUS/PROCEDURE:

Sulfur dioxide was bubbled into a weighed amount of component (2) in a bubbler tube as described in detail, diagram given, in the original paper. The amount of gas absorbed at equilibrium and at the observed temperature and pressure was weighed. By means of a manometer assembly, to which the bubbler tube was attached, the weight of gas absorbed at successively lower pressure was measured. Eventually the pressure was reduced to that of the component (2). The refractive index, and the infrared spectrum of the residue showed it to be essentially pure component (2).

SOURCE AND PURITY OF MATERIALS:

1. Sulfur dioxide was the best grade contained in a small cylinder. Its high purity was attested by the measurement of vapor pressure from 258 to 268 K.
2. The best obtainable specimen was carefully purified and rigorously attested.

ESTIMATED ERROR:

REFERENCES:

COMPONENTS:

1. Sulfur dioxide; SO_2; [7446-09-5]
2. Pyridine; C_5H_5N; [110-86-1]

ORIGINAL MEASUREMENTS:

Benoit, R.L.; Milanova, E.

Can. J. Chem., 1979, *57*, 1319-23

VARIABLES:

PREPARED BY:

C.L. Young

EXPERIMENTAL VALUES:

T/K	Henry's law constant+ H/atm
298.15	0.088

\+ Defined in the original as

$$H = \frac{\text{partial pressure of sulfur dioxide}}{\text{mole fraction of sulfur dioxide}}$$

AUXILIARY INFORMATION

METHOD/APPARATUS/PROCEDURE:

Few details given. Pressure measured with quartz spiral gauge. Partial pressure calculated assuming Raoult's law. Concentration of sulfur dioxide determined by iodometric titration. Concentration of sulfur dioxide range from 0.5 to 1.2 mol l^{-1}

SOURCE AND PURITY OF MATERIALS:

1. Anhydrous sample from Canadian Liquid Air, purity 99.98 mole per cent.

2. Dried over 4A molecular sieve

ESTIMATED ERROR:

$\delta T/K = \pm 0.1$; $\delta H/\text{atm} = \pm 4\%$
(estimated by compiler)

REFERENCES:

COMPONENTS:

1. Sulfur dioxide; SO_2; [7446-09-5]
2. 1-Methyl-2-pyrrolidinone (*N-Methylpyrrolidone*); C_5H_9NO; [872-50-4]

ORIGINAL MEASUREMENTS:

Sano, H.
Nippon Kagaku Zasshi
1968, *89*, 362-368.

VARIABLES:

PREPARED BY:

C. L. Young

EXPERIMENTAL VALUES:

T/°C = 30 T/K = 303

Absorption coefficient, S^a	Bunsen coefficient, α	Mole ratio	Mole fraction[b]
340.3	309.6	1.52	0.603

[a] Volume of sulfur dioxide absorbed by unit volume of solvent at experimental temperature and a total pressure of 1 atmosphere.

[b] Calculated by compiler.

AUXILIARY INFORMATION

METHOD/APPARATUS/PROCEDURE:

Static absorption method consisted of equilibrating solvent and sulfur dioxide in a 50 ml flask and measuring the absorption volumetrically. The vapor pressure of the solvents were measured roughly and were assumed to be equal to the partial pressure of the solvent by the authors when calculating the Bunsen coefficient.

SOURCE AND PURITY OF MATERIALS:

1. Purity 99.0-99.8 mole per cent.

ESTIMATED ERROR:

$\delta T/K = \pm 0.5$; $\delta x_{SO_2} = \pm 5\%$ (estimated by compiler).

REFERENCES:

COMPONENTS:

1. Sulfur dioxide; SO_2; [7446-09-5]
2. Nitrobenzene; $C_6H_5NO_2$; [98-95-3]

ORIGINAL MEASUREMENTS:

Lloyd, S.J.

J. Phys. Chem. 1918, *22*, 300-302

VARIABLES:

PREPARED BY:

W. Gerrard

EXPERIMENTAL VALUES:

T/K	Weight of SO_2 per dm^3 of solution /g.	*Speculative values *Mole ratio, SO_2/Comp. (2)	*Mole fraction (as for 1 atm) x_{SO_2}
288.15	311.4	0.668	0.400
293.15	267.4	0.549	0.355
298.15	227.9	0.451	0.311
303.15	190.0	0.364	0.267
313.15	132.0	0.241	0.194
323.15	98.7	0.177	0.150
333.15	78.6	0.139	0.122

760 mmHg = 1 atm = 101.325 kPa

* Calculated by compiler by assuming that the volume of solution is the same as the volume of the component (2) before absorption of gas.

AUXILIARY INFORMATION

METHOD:/APPARATUS/PROCEDURE:

Gas was bubbled through component (2) in a long graduated tube of small diameter. After equilibrium had been attained at a total pressure of about 1 atm (barometric pressure, 756-760 mmHg), the volume of the solution (about $5 cm^3$) was measured, and the weight of absorbed sulfur dioxide was determined by an iodometric titration.

SOURCE AND PURITY OF MATERIALS:

(1) Prepared from concentrated sulfuric acid and a saturated aqueous solution of sodium bisulfite. Dried by sulfuric acid and phosphorus pentoxide.

(2) Distilled, dried by calcium chloride, and redistilled before use.

ESTIMATED ERROR:

REFERENCES:

COMPONENTS:

1. Sulfur dioxide; SO_2; [7446-09-5]
2. Nitrobenzene; $C_6H_5NO_2$; [98-95-3]

ORIGINAL MEASUREMENTS:

Albright, L.F.; Shannon, P.T.; Yu, S.-N.; Chueh, P.L.

Chem. Engng. Prog. Symp. Ser. 1963, *59*, (44), 66-74.

VARIABLES:

Temperature, pressure

PREPARED BY:

C.L. Young

EXPERIMENTAL VALUES:

T/K	P/bar	Mole fraction of sulfur dioxide in liquid, x_{SO_2}
298.15	1.32	0.371
	2.16	0.569
	3.30	0.842
	3.39	0.843
310.93	1.97	0.367
	3.25	0.565
	5.05	0.841
338.71	4.03	0.355
	6.65	0.549
	10.65	0.837
366.48	7.12	0.340
	11.68	0.515
	19.87	0.828

AUXILIARY INFORMATION

METHOD/APPARATUS/PROCEDURE:

Static equilibrium cell fitted with Bourdon pressure gauge. Temperature measured with mercury in glass thermometer. The equipment used in this investigation was stated to be similar to that used in previous studies, (1), (2).

SOURCE AND PURITY OF MATERIALS:

1. and 2. Commercial grade samples, no other details given.

ESTIMATED ERROR:

$\delta T/K = \pm 0.2$; $\delta P/\mathrm{bar} = \pm 0.3$; $\delta x_{SO_2} = \pm 1\%$.

REFERENCES:

1. Albright, L. F.; Shannon, P. T.; Terrier, F.; Chueh, P. L. *Am. Inst. Chem. Engrs. J.* 1962, *8*, 668.
2. Albright, L. F.; Buclez, P. C.; Pluche, C. R.; Doody, T. C. *ASHRAE Trans.* 1960, *66*, 423.

COMPONENTS:

1. Sulfur dioxide, SO_2; [7446-09-5]
2. Nitrobenzene; $C_6H_5NO_2$; [98-95-3]

ORIGINAL MEASUREMENTS:

Lenoir, J-Y.; Renault, P.; Renon, H.

J. Chem. Eng. Data. 1971, *16*, 340-342

VARIABLES:

PREPARED BY:

W. Gerrard

EXPERIMENTAL VALUES:

T/K	Henry's constant H_{SO_2}/atm	Mole fraction* at 1 atm
298	1.64	0.610

* Calculated by compiler assuming a linear function of p_{SO_2} vs x_{SO_2}, i.e. x_{SO_2} (1 atm) = $1/H_{SO_2}$.

The authors gave "Henry's constant" as $H_{SO_2} = \lim_{x_{SO_2}\to 0} \frac{f_{SO_2}(P,T)}{x_{SO_2}}$

x_{SO_2} is the mole fraction. f_{SO_2} is the fugacity of SO_2 "in the equilibrium condition" of pressure and temperature. H_{SO_2} is related to experimental chromatographic parameters based on the number of moles of solvent in the column, and the absolute retention time of the solute. Helium was the carrier gas.

AUXILIARY INFORMATION

METHOD/APPARATUS/PROCEDURE:

A conventional gas-liquid chromatographic unit was used. The chromatograph was an IGC 12 M Intersmat model, equipped with a thermal conductivity detector. The temperature control was stated to be within ±0.1° C. The pressure drop in the column was measured by a mercury manometer. Helium was the carrier gas.

There is serious uncertainty in the estimation of accuracy.

SOURCE AND PURITY OF MATERIALS:

1. Sulfur dioxide was provided by l'Air Liquide, and had a stated purity of 99.9%. It was used as provided.
2. Provided by Touzart & Matignon and Serlabo, and stated to have a minimum purity of 99%. Distillation was carried out when necessary, and the refractive index was measured.

ESTIMATED ERROR:

REFERENCES:

COMPONENTS:

1. Sulfur dioxide; SO_2; [7446-09-5]

2. Nitrobenzene; $C_6H_5NO_2$; [98-95-3]

ORIGINAL MEASUREMENTS:

Gerrard, W.

J. Appl. Chem. Biotechnol. 1972, 22, 623-650.

VARIABLES:

Temperature, pressure

PREPARED BY:

W. Gerrard

EXPERIMENTAL VALUES:

T/K	Pressure, total /mmHg	Mole ratio SO_2/Comp. (2)	*Mole fraction x_{SO_2}
273.15	772	2.0	0.667
283.15	772	0.878	0.468
280.15	760	1.105	0.525
	** 733	1.040	0.510
	670	0.901	0.474
	593	0.745	0.427
	427	0.487	0.328
	285	0.291	0.225
	186	0.177	0.150
	51	0.042	0.040
293.15	772	0.545	0.353
298.15	772	0.45	0.310

** The plot of pressure vs solubility given in the original paper is based on these data.

760 mmHg = 1 atm = 101.325 kPa.

* Calculated by compiler.

AUXILIARY INFORMATION

METHOD/APPARATUS/PROCEDURE:

Sulfur dioxide was bubbled into a weighed amount of component (2) in a bubbler tube as described in detail, diagram given, in the original paper. The amount of gas absorbed at equilibrium and at the observed temperature and pressure was weighed. By means of a manometer assembly, to which the bubbler tube was attached, the weight of gas absorbed at successively lower pressure was measured. Eventually the pressure was reduced to that of the component (2). The refractive index, and the infrared spectrum of the residue showed it to be essentially pure component (2).

SOURCE AND PURITY OF MATERIALS:

1. Sulfur dioxide was the best grade contained in a small cylinder. Its high purity was attested by the measurement of vapor pressure from 258 to 268 K.

2. The best obtainable specimen was carefully purified and rigorously attested.

ESTIMATED ERROR:

REFERENCES:

COMPONENTS:

1. Sulfur dioxide; SO_2; [7446-09-5]
2. Nitrobenzene; $C_6H_5NO_2$; [98-95-3]

ORIGINAL MEASUREMENTS:

Benoit, R.L.; Milanova, E.

Can. J. Chem., 1979, *57*, 1319-1323

VARIABLES:

PREPARED BY:

C.L. Young

EXPERIMENTAL VALUES:

T/K	Henry's law constant+ H/atm
298.15	3.18

+ Defined in the original as

$$H = \frac{\text{partial pressure of sulfur dioxide}}{\text{mole fraction of sulfur dioxide}}$$

AUXILIARY INFORMATION

METHOD/APPARATUS/PROCEDURE:

Few details given. Pressure measured with quartz spiral gauge. Partial pressure calculated assuming Raoult's law. Concentration of sulfur dioxide determined by iodometric titration. Concentration of sulfur dioxide ranged from 0.1 to 0.5 mol l^{-1}

SOURCE AND PURITY OF MATERIALS:

1. Anhydrous sample from Canadian Liquid Air, purity 99.98 mole per cent.
2. Dried over 4A molecular sieve

ESTIMATED ERROR:

$\delta T/K = \pm 0.1$; $\delta H/\text{atm} = \pm 4\%$ (estimated by compiler)

REFERENCES:

COMPONENTS:

1. Sulfur dioxide; SO_2; [7446-09-5]
2. Benzenamine (*aniline*); C_6H_7N; [62-53-3]
   ar-Methylbenzenamine (*toluidine (mixed)*); C_7H_9N; [26915-12-8]
   ar,*ar*-Dimethylbenzenamine (*xylidine (mixed)*); $C_8H_{11}N$; [1300-73-8]

ORIGINAL MEASUREMENTS:

Sano, H.; Nakamoto, Y.

Nippon Kagaku Zasshi

1968, *89*, 369-373.

VARIABLES:

PREPARED BY:

C. L. Young

EXPERIMENTAL VALUES:

T/°C = 30 T/K = 303

Bunsen coefficient, α	Mole ratio	Mole fraction[a]
Benzeneamine		
274	1.14	0.534
ar-Methylbenzenamine (mixture of isomers)		
241	-	-
ar,*ar*-Dimethylbenzenamine (mixture of isomers)		
234	-	-

[a] Calculated by compiler.

AUXILIARY INFORMATION

METHOD/APPARATUS/PROCEDURE:

Static absorption method consisted of equilibrating solvent and sulfur dioxide in a 50 ml flask and measuring the absorption volumetrically. The vapor pressure of the solvents were measured roughly and were assumed to be equal to the partial pressure of the solvent by the authors when calculating the Bunsen coefficient.

SOURCE AND PURITY OF MATERIALS:

1. Purity 99.0-99.8 mole per cent.

ESTIMATED ERROR:

$\delta T/K = \pm 0.5$; $\delta x_{SO_2} = \pm 5\%$ (estimated by compiler).

REFERENCES:

COMPONENTS:

1. Sulfur dioxide; SO_2; [7446-09-5]
2. Benzonitrile; C_7H_5N; [100-47-0]

ORIGINAL MEASUREMENTS:

Gerrard, W.

J. Appl. Chem. Biotechnol. 1972, *22*, 623-650.

VARIABLES:

Temperature, pressure

PREPARED BY:

W. Gerrard

EXPERIMENTAL VALUES:

T/K	Pressure, total /mmHg	Mole ratio SO_2/Comp.(2)	*Mole fraction x_{SO_2}
273.15	754	2.19	0.687
	** 736	2.07	0.674
	712	1.93	0.659
	660	1.64	0.621
	577	1.30	0.565
	469	0.951	0.487
	343	0.633	0.388
	176	0.288	0.224
	71	0.0972	0.0886
283.15	754	1.01	0.502
293.15	754	0.629	0.386
298.15	754	0.50	0.333

760 = 1 atm = 101.325 kPa.

* Calculated by compiler

** Data on which the graph in the original paper is based.

AUXILIARY INFORMATION

METHOD/APPARATUS/PROCEDURE:

Sulfur dioxide was bubbled into a weighed amount of component (2) in a bubbler tube as described in detail, diagram given, in the original paper. The amount of gas absorbed at equilibirium and at the observed temperature and pressure was weighed. By means of a manometer assembly, to which the bubbler tube was attached, the weight of gas absorbed at successively lower pressure was measured. Eventually the pressure was reduced to that of the component (2). The refractive index, and the infrared spectrum of the residue showed it to be essentially pure component (2).

SOURCE AND PURITY OF MATERIALS:

1. Sulfur dioxide was the best grade contained in a small cylinder. Its high purity was attested by the measurement of vapor pressure from 258 to 268 K.
2. The best obtainable specimen was carefully purified and rigorously attested.

ESTIMATED ERROR:

REFERENCES:

COMPONENTS:

1. Sulfur dioxide; SO_2; [7446-09-5]

2. 1-Methyl-2-nitrobenzene, (2-nitrotoluene); $C_7H_7NO_2$; [88-72-2]

ORIGINAL MEASUREMENTS:

Lloyd, S.J.

J. Phys. Chem. 1918, *22*, 300-302

VARIABLES:

PREPARED BY:

W. Gerrard

EXPERIMENTAL VALUES:

T/K	Weight of SO_2 per dm^3 of solution /g.	*Speculative values *Mole ratio, SO_2/ Comp.(2)	*Mole fraction (as for 1 atm) x_{SO_2}
288.15	290.8	0.711	0.415
293.15	236.0	0.545	0.353
298.15	192.2	0.426	0.299
303.15	160.7	0.347	0.258
313.15	118.5	0.248	0.199
323.15	87.0	0.178	0.151
333.15	68.8	0.140	0.123

760 mmHg = 1 atm = 101.325 kPa

* Calculated by compiler by assuming that the volume of solution is the same as the volume of the component (2) before absorption of gas.

AUXILIARY INFORMATION

METHOD:/APPARATUS/PROCEDURE:

Gas was bubbled through component (2) in a long graduated tube of small diameter. After equilibrium had been attained at a total pressure of about 1 atm (barometric pressure, 756-760 mmHg), the volume of the solution (about 5 cm^3) was measured, and the weight of absorbed sulfur dioxide was determined by an iodometric titration.

SOURCE AND PURITY OF MATERIALS:

(1) Prepared from concentrated sulfuric acid and a saturated aqueous solution of sodium bisulfite. Dried by sulfuric acid and phosphorus pentoxide.

(2) Distilled, dried by calcium chloride, and redistilled before use.

ESTIMATED ERROR:

REFERENCES:

COMPONENTS:

1. Sulfur dioxide; SO_2; [7446-09-5]
2. *N*-Methylbenzenamine, (methylaniline); C_7H_9N; [100-61-8]

ORIGINAL MEASUREMENTS:

Hill, A.E.; Fitzgerald, T.B.

J. Am. Chem. Soc. 1935, *57*, 250-254.

VARIABLES:

PREPARED BY:

W. Gerrard

EXPERIMENTAL VALUES:

T/K	Pressure, total, mmHg	Mole ratio SO_2/ component (2)	*Mole fraction, x_{SO_2}
298.15	2.0	0.00	
	19.5	0.124	0.110
	42.0	0.258	0.205
	72.5	0.396	0.284
	114.4	0.525	0.344
	166.1	0.652	0.395
	237.0	0.772	0.436
	329.0	0.896	0.473
	436.0	1.010	0.502
	552.5	1.115	0.527
	(760.0)		(0.560) Extrapolated by compiler.

760 mmHg = 1 atm = 101.325 kPa

* Calculated by compiler.

AUXILIARY INFORMATION

METHOD:/APPARATUS/PROCEDURE:

The apparatus had been previously described (1).

A weighed amount of liquid was treated with a measured volume of sulfur dioxide at a constant temperature and at a measured total pressure. A gas buret and a mercury manometer were used.

SOURCE AND PURITY OF MATERIALS:

(1) Not specified.

(2) Purified by distillation.

ESTIMATED ERROR:

REFERENCES:

1. Hill, A.E.

J. Am. Chem. Soc. 1931, *53*, 2598.

COMPONENTS:

1. Sulfur dioxide; SO_2; [7446-09-5]

2. *N,N*-Dimethylbenzenamine, (*N,N*-dimethylaniline); $C_8H_{11}N$; [121-69-7]

ORIGINAL MEASUREMENTS:

Hill, A.E.; Fitzgerald, T.B.

J. Am. Chem. Soc. 1935, *57*, 250-254.

VARIABLES:

PREPARED BY:

W. Gerrard

EXPERIMENTAL VALUES:

T/K	Pressure, total mmHg	Mole ratio SO_2/ component (2)	*Mole fraction, x_{SO_2}
298.15	2.0	0.00	
	14.6	0.185	0.156
	30.0	0.362	0.266
	56.5	0.564	0.361
	106.0	0.770	0.435
	189.1	0.950	0.487
	290.2	1.090	0.522
	449.5	1.271	0.560
	575.0	1.400	0.583
	(760.0)		(0.610) Extrapolated by compiler.

760 mmHg = 1 atm = 101.325 kPa

* Calculated by compiler.

AUXILIARY INFORMATION

METHOD:/APPARATUS/PROCEDURE:

The apparatus had been previously described (1).

A weighed amount of liquid was treated with a measured volume of sulfur dioxide at a constant temperature and at a measured total pressure. A gas buret and a mercury manometer were used.

SOURCE AND PURITY OF MATERIALS:

(1) Not specified.

(2) Purified by distillation.

ESTIMATED ERROR:

REFERENCES:

1. Hill, A.E.
J. Am. Chem. Soc. 1931, *53*, 2598.

COMPONENTS:

1. Sulfur dioxide; SO_2; [7446-09-5]
2. *N,N*-Dimethylbenzenamine, (*N,N*-dimethylaniline); $C_8H_{11}N$; [121-69-7]

ORIGINAL MEASUREMENTS:

Balej, J.; Regner, A.
Chem. Listy, 1956, *50*, 1374-1380

Collection Czech. Chem. Communs. 1956, *21*, 1545-1552.

VARIABLES:

PREPARED BY:

W. Gerrard

EXPERIMENTAL VALUES: 760 mmHg = 101.325 kPa.

T/K	p_{SO_2} /mmHg	Mole ratio SO /comp.(2)	* x_{SO_2}	T/K	p_{SO_2} /mmHg	Mole ratio SO /Comp.(2)	* x_{SO_2}
288.15	5.27	0.2238	0.1829	298.15	13.10	0.225	0.184
	12.30	0.3913	0.2812		13.78	0.230	0.187
	14.19	0.4171	0.2943		25.72	0.363	0.266
	21.51	0.5320	0.3473		48.0	0.520	0.342
	30.30	0.6225	0.3837		49.7	0.524	0.344
	48.70	0.7610	0.4321		56.0	0.562	0.360
	63.4	0.8550	0.4609		89.2	0.707	0.414
	78.6	0.8980	0.4731		119.5	0.792	0.442
	83.6	0.9150	0.4778		162.0	0.896	0.473
	95.0	0.9575	0.4891		232.5	1.028	0.507
	121.0	1.029	0.507				
	169.5	1.150	0.535	313.15	8.98	0.0703	0.0657
	220.5	1.243	0.554		13.30	0.0910	0.0834
293.15	12.25	0.3031	0.2326		25.80	0.1595	0.1376
	13.25	0.3122	0.2379		36.30	0.2219	0.1816
	21.23	0.4240	0.2978		39.10	0.2346	0.1900
	28.08	0.5075	0.3366		43.00	0.2460	0.1974
	39.90	0.5838	0.3686		58.5	0.3302	0.2482
	42.20	0.5975	0.3740		93.5	0.4320	0.3017
	89.0	0.829	0.453		103.95	0.4606	0.3153
	137.3	0.954	0.488		154.9	0.5850	0.3691
	150.0	0.983	0.496		158.0	0.5972	0.3739
	235.5	1.145	0.534		246.0	0.7212	0.4190

x_{SO_2} = mole fraction of SO_2. * Calculated by compiler.

AUXILIARY INFORMATION

METHOD/APPARATUS/PROCEDURE:

Diagram given. See Hill and Fitzgerald (1). Separate streams of purified nitrogen and sulfur dioxide were mixed in a vessel containing glass wool covered with phosphorus pentoxide. The gaseous mixture was then passed into the absorption vessel and then to a sampling tap to enable the SO_2 content of the gas phase to be determined by an iodometric titration. The SO_2 content of the solution was likewise determined. The method was described as a dynamic one.

SOURCE AND PURITY OF MATERIALS:

1. Probably of satisfactory purity.
2. B.p. 192-193.5°C, probably at atmospheric pressure.

ESTIMATED ERROR:

REFERENCES:

1. Hill, A.E.; Fitzgerald, T.B. *J. Am. Chem. Soc.* 1935, *57*, 250.

COMPONENTS:

1. Sulfur dioxide; SO_2; [7446-09-5]
2. *N,N*-Dimethylbenzenamine (*N,N-dimethylaniline*); $C_8H_{11}N$; [121-69-7]
   Benzenemethanamine (*Benzylamine*); C_7H_9N; [100-46-9]

ORIGINAL MEASUREMENTS:

Sano, H.; Nakamoto, Y.
Nippon Kagaku Zasshi
1968, *89*, 369-373.

VARIABLES:

PREPARED BY:

C. L. Young

EXPERIMENTAL VALUES:

T/°C = 30 T/K = 303

Absorption coefficient, S^a	Bunsen coefficient, α	Mole ratio	Mole fraction[b]
	N,N-Dimethylbenzenamine		
280	254	1.44	0.591
	Benzylamine		
-	288	1.40	0.583

[a] Volume of sulfur dioxide absorbed by unit volume of solvent at experimental temperature and a total pressure of 1 atmosphere.

[b] Calculated by compiler.

AUXILIARY INFORMATION

METHOD/APPARATUS/PROCEDURE:

Static absorption method consisted of equilibrating solvent and sulfur dioxide in a 50 ml flask and measuring the absorption volumetrically. The vapor pressure of the solvents were measured roughly and were assumed to be equal to the partial pressure of the solvent by the authors when calculating the Bunsen coefficient.

SOURCE AND PURITY OF MATERIALS:

1. Purity 99.0-99.8 mole per cent.

ESTIMATED ERROR:

$\delta T/K = \pm 0.5$; $\delta x_{SO_2} = \pm 5\%$ (estimated by compiler).

REFERENCES:

COMPONENTS:

1. Sulfur dioxide; SO_2; [7446-09-5]

2. *N*-Ethylbenzenamine, (ethylaniline); $C_8H_{11}N$; [103-69-5]

ORIGINAL MEASUREMENTS:

Hill, A.E.; Fitzgerald, T.B.

J. Am. Chem. Soc. 1935, *57*, 250-254.

VARIABLES:

PREPARED BY:

W. Gerrard

EXPERIMENTAL VALUES:

T/K	Pressure, total, mmHg	Mole ratio SO_2 / component (2)	*Mole fraction, x_{SO_2}
298.15	1.5	0.00	
	46.5	0.182	0.154
	101.0	0.361	0.265
	172.5	0.536	0.349
	264.5	0.706	0.414
	378.0	0.870	0.465
	510.0	1.027	0.507
	655.0	1.183	0.542
	(760.0)		(0.560) Extrapolated by compiler.

760 mmHg = 1 atm = 101.325 kPa.

*Calculated by compiler.

AUXILIARY INFORMATION

METHOD:/APPARATUS/PROCEDURE:

The apparatus had been previously described (1).

A weighed amount of liquid was treated with a measured volume of sulfur dioxide at a constant temperature and at a measured total pressure. A gas buret and a mercury manometer were used.

SOURCE AND PURITY OF MATERIALS:

(1) Not specified.

(2) Purified by distillation.

ESTIMATED ERROR:

REFERENCES:

1. Hill, A.E.
J. Am. Chem. Soc. 1931, *53*, 2598.

COMPONENTS:

1. Sulfur dioxide; SO_2; [7446-09-5]
2. *N,N*-Diethylbenzenamine, [diethylaniline); $C_{10}H_{15}N$; [91-66-7]

ORIGINAL MEASUREMENTS:

Foote, H.W.; Fleischer, J.

J. Am. Chem. Soc. 1934, *56*, 870-873.

VARIABLES:

PREPARED BY:

W. Gerrard

EXPERIMENTAL VALUES:

T/K	p_{SO_2} mmHg	Mole fraction x_{SO_2}
273	1008	0.891
	946	0.862
	845	0.817
	784	0.793
	733	0.774
	673	0.751
	584	0.716
	472	0.671
	359	0.618
	276	0.570
	225	0.532
	149	0.459
	81	0.347
	44	0.252
	13	0.092

NOTE : A mole fraction of 0.891 is equivalent to a mole ratio of 8.17 moles of SO_2 to 1 of N, N-diethyl aniline.

760 mmHg = 1 atm = 101.325 kPa. Observed pressure was taken to be p_{SO_2}.

AUXILIARY INFORMATION

METHOD /APPARATUS/PROCEDURE:

The apparatus and method had been previously described (1).

A glass bulb of 29.3 cm^3 capacity was connected via a tap to an open manometer, a vacuum flask of 2 dm^3 capacity, and a source of sulfur dioxide. Component (2) was weighed in the bulb. Sulfur dioxide was condensed in the bulb and 253 K. Equilibrium at 273.15 K and at the observed pressure, total, but assumed to be equal to p_{SO_2}, was attained.

SOURCE AND PURITY OF MATERIALS:

(1) Sulfur dioxide: Dried by sulfuric acid.

(2) "The compounds used were either of Eastman C.P. grade or were purified by customary methods."

ESTIMATED ERROR:

REFERENCES:

1. Foote, H.W.; Fleischer, J. *J. Am. Chem. Soc.* 1931, *53*, 1752

COMPONENTS:	ORIGINAL MEASUREMENTS:
1. Sulfur dioxide; SO_2; [7446-09-5] 2. *N*,*N*-Diethylbenzenamine, (diethylaniline); $C_{10}H_{15}N$; [91-66-7]	Hill, A.E.; Fitzgerald, T.B. *J. Am. Chem. Soc.* 1935, *57*, 250-254.
VARIABLES:	PREPARED BY: W. Gerrard

EXPERIMENTAL VALUES:

T/K	Pressure, total, mmHg	Mole ratio SO_2/ component (2)	*Mole fraction, x_{SO_2}
298.15	1.5	0.000	
	120.8	0.204	0.169
	249.0	0.410	0.291
	393.1	0.617	0.382
	529.5	0.803	0.445
	679.5	1.010	0.502
	(760.0)		(0.530) Extrapolated by compiler.

760 mmHg = 1 atm = 101.325 kPa

* Calculated by compiler.

AUXILIARY INFORMATION

METHOD:/APPARATUS/PROCEDURE:

The apparatus had been previously described (1).

A weighed amount of liquid was treated with a measured volume of sulfur dioxide at a constant temperature and at a measured total pressure. A gas buret and a mercury manometer were used.

SOURCE AND PURITY OF MATERIALS:

(1) Not specified.

(2) Purified by distillation.

ESTIMATED ERROR:

REFERENCES:

1. Hill, A.E.

J. Am. Chem. Soc. 1931, *53*, 2598

COMPONENTS:	ORIGINAL MEASUREMENTS:
1. Sulfur dioxide; SO_2; [7446-09-5] 2. Miscellaneous solvents	Bogeatzes, A. S.; Tassios, D. P. *Ind. Eng. Chem. Process Des. Develop.* 1973, *12*, 274-278.
VARIABLES:	PREPARED BY: C. L. Young

EXPERIMENTAL VALUES:

Solvent	Activity coefficient at infinite dilution, γ	T/K
N,N-Dimethylaniline (*N,N*-Dimethylbenzenamine); $C_8H_{11}N$; [121-69-7]	0.027	314.2
N,N-Dimethylformamide; C_3H_7NO; [68-12-2]	0.081	313.7
N,N-Dimethylacetamide; C_4H_9NO; [127-19-5]	0.036	314.7
	0.070	339.2
	0.077	366.2
Nitrobenzene; $C_6H_5NO_2$; [98-95-3]	0.910	298.7
Trimethylpyridine; $C_8H_{11}N$; [29611-84-5]	0.123	312.2
Xylidine (ar,ar-dimethylbenzenamine); $C_8H_{11}N$; [1300-73-8]	0.140	313.2
Benzonitrile; C_7H_5N; [100-47-0]	0.532	312.2
3-Methylsulfolane (tetrahydro-3-methylthiophene-1,1-dioxide); $C_5H_{10}O_2S$; [872-93-5]	0.609	313.2
3-Methoxysulfolane (tetrahydro-3-methoxythiophene-1,1-dioxide); $C_5H_{10}O_3S$; [20627-66-1]	1.090	312.2
N-Methylmethanamine (Dimethylamine); C_2H_7N; [124-40-3]	1.590	315.2
N,N'-Bis(2-aminoethyl)-1,2-Ethanediamine (Triethylenetetramine); $C_6H_{18}N_4$; [112-24-3]	1.580	314.2
Equimolar mixture of xylidine and *N,N*-dimethylaniline	0.040	313.2

AUXILIARY INFORMATION

METHOD/APPARATUS/PROCEDURE:

Gas chromatographic method. Activity coefficients determined from retention volumes of sulfur dioxide on solvents as stationary phases. It is probable that the activity coefficients are in error because of absorption at the carrier gas-solvent interface.

SOURCE AND PURITY OF MATERIALS:

No details given except purity better than 99 mole per cent.

ESTIMATED ERROR:

$\delta T/K = \pm 0.2$; $\delta\gamma/\gamma = \pm 0.05$.

REFERENCES:

COMPONENTS:

1. Sulfur dioxide; SO_2; [7446-09-5]
2. Organic compounds containing halogens

EVALUATOR:

Colin L. Young,
School of Chemistry,
University of Melbourne,
Parkville, Victoria 3052,
Australia.

December 1982

CRITICAL EVALUATION:

Trichloromethane; $CHCl_3$; [67-66-3]

Although this system has been investigated by three groups (1,2,3) the ranges of temperature and pressure of the measurements do not overlap. The measurements are broadly consistent but no precise comparison is possible.

Tetrachloromethane; CCl_4; [56-23-5]

This system has been investigated by three groups. The data of Sobolev *et al.* (4) is classified as doubtful, the mole fraction solubilities being considerably smaller than those of Horiuti (5) and Sano (2). The datum of Sano (2) is several per cent higher than the value obtained by interpolation of Horiuti values.

Chlorobenzene; C_6H_5Cl; [108-90-7]

The data of Horiuti (5) and Gerrard (6) for this system are in reasonable agreement. The isolated value of Sano (2) appears to be a few percent low when compared with the values of Gerrard at low temperatures.

References

1. Lindner, J. *Monatsh.* 1912, *33*, 613.
2. Sano, H. *Nippon Kagaku Zasshi* 1968, *89*, 362.
3. Lorimer, J. W.; Smith, B. C.; Smith, G. H. *J.C.S. Faraday I* 1975, *71*, 2232.
4. Sobolev, I. A.; Kukarin, V. A.; Dzhagatspanyan, R. V.; Kosorotov, V. I.; Zogarets, P. A.; Popov, A. I. *Khim. Prom.* 1970, *46*, 668.
5. Horiuti, J. *Sci. Pap. Inst. Phys. Chem. Res. (Jpn.)* 1931/32, *17*, 125.
6. Gerrard, W. *J. Appl. Chem. Biotechnol.* 1972, *22*, 623.

COMPONENTS:	ORIGINAL MEASUREMENTS:
1. Sulfur dioxide; SO_2; [7446-09-5] 2. Trichloromethane, (chloroform); $CHCl_3$; [67-66-3]	Lindner, J. *Monatsh.* 1912, *33*, 613-672.
VARIABLES:	PREPARED BY: W. Gerrard.

EXPERIMENTAL VALUES:

760 mmHg = 1 atm = 101.325 kPa.

T/K	Weight of SO_2 g	Weight of solution g (sp.gr.)	Pressure, SO_2/mmHg	*Mole ratio SO_2/ Comp.(2).	*Mole fraction x_{SO_2}
273.15	0.0738	160.7 (1.526)	(2.7?)	0.000857	-
	0.1879	160.3 (1.526)	5.6	0.00218	-
	0.9068	198.1 (1.525)	22.0	0.00858	-
	3.4395	169.1 (1.523)	90.2	0.0387	0.0373
	7.222	133.0 (1.514)	219.6	0.107	0.0967
	**Rational extrapolation		(760)	(1.174)	(0.540)
298.15	0.0727	160.7 (1.480)	5.7	0.000844	-
	0.1853	160.3 (1.480)	12.9	0.00216	0.00215
	0.9008	198.1 (1.479)	48.0	0.00852	0.00845
	3.3943	169.1 (1.472?)	200.2	0.0382	0.0368
	7.0866	133.0 (1.472)	488.8	0.105	0.0950
	**Rational extrapolation		(760)	(0.202)	(0.168)

**By compiler's plot of x_{SO_2} vs p_{SO_2} on the reference line diagram of Gerrard (1)

*Calculated by compiler.

AUXILIARY INFORMATION

METHOD/APPARATUS/PROCEDURE:

Somewhat complex arrangement of six glass vessels containing the liquid. A small sealed tube containing a weighed amount of sulfur dioxide was fitted onto each tube assembly by a flexible joint. The previously scratched tip of the tube was broken to release sulfur dioxide, and the absorption of this was encouraged by a twirling motion of the apparatus. The pressure was measured manometrically. To obtain the amount absorbed, an allowance was made for the amount in the gas phase. The description of the procedure for deciding the partial pressure of the sulfur dioxide is somewhat involved, and cannot be briefly reported. One of the bulbs contained pure chloroform so that its vapor pressure at the operational temperature could be indicated.

SOURCE AND PURITY OF MATERIALS:

1. Sulfur dioxide may be taken as pure.

2. May be taken as pure, density compared with the literature value.

ESTIMATED ERROR:

REFERENCES:

Gerrard, W. "*Solubility of Gases and Liquids*," Plenum Press, New York, 1976.

COMPONENTS:

1. Sulfur dioxide; SO_2; [7446-09-5]
2. Trichloromethane (*chloroform*); $CHCl_3$; [67-66-3]

   Tetrachloromethane (*carbon tetrachloride*); CCl_4; [56-23-5]

ORIGINAL MEASUREMENTS:

Sano, H.

Nippon Kagaku Zasshi

1968, *89*, 362-368.

VARIABLES:

PREPARED BY:

C. L. Young

EXPERIMENTAL VALUES:

T/°C = 30 T/K = 303

Vapor pressure of trichloromethane, $p^{\circ}_{CHCl_3}$ = 0.272 atm

Vapor pressure of tetrachloromethane, $p^{\circ}_{CCl_4}$ = 0.200 atm

Absorption coefficient, $S^{a,b}$	Bunsen coefficient,[b] α	Bunsen coefficient,[c] α	$\ell^{c,d}$	Mole ratio	Mole fraction[e]
		Trichloromethane			
30.1	37.6	-	-	0.134	0.118
		Tetrachloromethane			
12.3	14.0	12.2	9.8	0.061	0.0575

[a] Volume of sulfur dioxide absorbed by unit volume of solvent at experimental temperature and a total pressure of 1 atmosphere.

[b] Determined by static absorption method.

[c] Determined by flow method.

[d] $\alpha = \ell(1 - p^{\circ}_{s})$ where p°_{s} is the vapor pressure of solvent.

[e] Calculated by compiler.

AUXILIARY INFORMATION

METHOD/APPARATUS/PROCEDURE:

Two techniques were used: (a) static absorption method and (b) flow method. Static method consisted of equilibrating solvent and sulfur dioxide in a 50 ml flask and measuring the absorption volumetrically. The flow method consists of passing sulfur dioxide through a micro gas absorption tube and estimating sulfur dioxide by iodometric titration. The vapor pressure of the solvents were measured roughly and were assumed to be equal to the partial pressure of the solvent by the authors when calculating the Bunsen coefficient.

SOURCE AND PURITY OF MATERIALS:

1. Purity 99.0-99.8 mole per cent.

ESTIMATED ERROR:

$\delta T/K = \pm 0.5$; $\delta x_{SO_2} = \pm 5\%$

(estimated by compiler).

REFERENCES:

COMPONENTS:

1. Sulfur dioxide; SO_2; [7446-09-5]
2. Trichloromethane, (chloroform); $CHCl_3$; [67-66-3]

ORIGINAL MEASUREMENTS:

Lorimer, J.W.; Smith, B.C.; Smith, G.H.

J.C.S. Faraday I, 1975, *71*, 2232-50.

VARIABLES:

Temperature, pressure

PREPARED BY:

C.L. Young

EXPERIMENTAL VALUES:

T/K	p/mmHg	p/kPa	Mole fraction of sulfur dioxide in liquid, x_{SO_2}	in vapor, y_{SO_2}
227.6	3.44	0.459	0.0000	0.000
	30.58	4.077	0.0920	0.897
	52.83	7.043	0.2091	0.946
	67.28	8.970	0.3154	0.961
	73.73	9.830	0.3812	0.966
	82.05	10.939	0.4878	0.972
	89.25	11.899	0.5928	0.977
	91.65	12.219	0.6335	0.979
	98.81	13.174	0.7674	0.984
	106.48	14.196	0.8885	0.990
	116.49	15.531	1.0000	1.000
237.4	7.20	0.960	0.0000	0.000
	51.50	6.866	0.0824	0.871
	89.35	11.912	0.1922	0.932
	116.32	15.508	0.3018	0.952
	126.75	16.899	0.3590	0.958
	142.74	19.030	0.4701	0.966
	157.11	20.946	0.5803	0.972
	162.44	21.657	0.6120	0.974
	176.86	23.579	0.7577	0.981
	191.17	25.487	0.8861	0.989
	208.74	27.830	1.0000	1.000

AUXILIARY INFORMATION

METHOD/APPARATUS/PROCEDURE:

Static vapor-liquid equilibrium apparatus. Total pressure measured using mercury manometers and cathetometer. Temperature measured using a sulfur dioxide vapor pressure thermometer. Some details in source.

SOURCE AND PURITY OF MATERIALS:

No details given.

ESTIMATED ERROR:

$\delta T/K = \pm 0.1$; $\delta p/kPa = \pm 0.01$; δx_{SO_2}, $\delta y_{SO_2} = \pm 1\%$.

REFERENCES:

COMPONENTS:

1. Sulfur dioxide; SO_2; [7446-09-5]

2. Trichloromethane, (chloroform); $CHCl_3$; [67-66-3]

ORIGINAL MEASUREMENTS:

Lorimer, J.W.; Smith, B.C.; Smith, G.H.

J.C.S. Faraday I, 1975, *71*, 2232-50.

EXPERIMENTAL VALUES:

T/K	p/mmHg	p/kPa	Mole fraction of sulfur dioxide in liquid, x_{SO_2}	in vapor, y_{SO_2}
249.8	16.31	2.174	0.0000	0.000
	88.56	11.807	0.0650	0.827
	156.27	20.834	0.1656	0.909
	211.30	28.171	0.3019	0.940
	226.36	30.179	0.3093	0.944
	266.04	35.469	0.4274	0.957
	296.86	39.578	0.5493	0.966
	303.07	40.406	0.5708	0.968
	338.45	45.123	0.7395	0.977
	372.26	49.631	0.8816	0.987
	405.40	54.049	1.0000	1.000

COMPONENTS:	ORIGINAL MEASUREMENTS:
1. Sulfur dioxide; SO_2; [7446-09-5] 2. Tetrachloromethane, (Carbon tetrachloride); CCl_4; [56-23-5]	Horiuti, J. *Sci. Pap. Inst. Phys. Chem. Res. (Jpn.)*, 1931/32, *17*, 125-256.
VARIABLES:	PREPARED BY:
Temperature, pressure	C. L. Young

EXPERIMENTAL VALUES:

T/K	$P^{\S}$/bar	Mole fraction of sulfur dioxide in liquid, x_{SO_2}	T/K	$P^{\dagger}$/bar	Mole fraction of sulfur dioxide in liquid, x_{SO_2}	Ostwald coefficient, L
298.15	0.1497	0.0000	283.15	0.9338	0.1111	30.96
	0.3537	0.0126	298.15	0.8585	0.0604	18.45
	0.4765	0.0208	313.15	0.7285	0.0342	12.52
	0.6776	0.0348				
	0.7090	0.0371				
	0.8610	0.0490				
	0.9611	0.0567				
	1.085	0.0673				
	1.172	0.0748				
	1.260	0.0827				
	1.321	0.0880				

§ total pressure.

† partial pressure, total pressure = 1 atmosphere = 1.01325×10^5 Pa.

Ostwald coefficient (as $x \to 0$, T/K = 298.15) = 15.1.

Henry's law constant, dp/dx (as $x \to 0$, T/K = 298.15) = 12700 mmHg = 16.93 bar

AUXILIARY INFORMATION

METHOD/APPARATUS/PROCEDURE:

Apparatus consisted of a gas buret, a solvent reservoir and an absorption pipet. Dry gas introduced into degassed solvent. System mixed using a magnetic stirrer until saturation occurred. Care was taken to prevent solvent vapor from mixing with gas in the gas buret. Volume of gas determined from gas buret reading; volume of liquid determined from height of meniscus in absorption pipet.

SOURCE AND PURITY OF MATERIALS:

1. Sample obtained from Nippon Sulphuric Acid Manufacturing Co. Dried and fractionated several times.
2. Kahlbaum sample, dried and distilled. B. pt. 76.74 °C.

ESTIMATED ERROR:

$\delta T/K = \pm 0.1$; $\delta P/\text{bar} = \pm 0.001$;

$\delta x_{SO_2} = \pm 0.001$ (estimated by compiler).

REFERENCES:

COMPONENTS:	ORIGINAL MEASUREMENTS:
1. Sulfur dioxide; SO_2; [7446-09-5] 2. Tetrachloromethane, (Carbon tetrachloride); CCl_4; [56-23-5]	Sobolev, I.A.; Kukarin, V.A.; Dzhagatspanyan, R.V.; Kosorotov, V.I.; Zogorets, P.A.; Popov, A.I. *Khim. Prom.* 1970, *46*, 668-70.
VARIABLES: Temperature	PREPARED BY: C.L. Young

EXPERIMENTAL VALUES:

T/K	Solubility*, S /mol l^{-1}	Mole fraction of+ sulfur dioxide in liquid, x_{SO_2}
283.15	0.67	0.061
288.15	0.536	0.049
293.15	0.435	0.040
298.15	0.365	0.034
303.15	0.296	0.028

\+ calculated by compiler

* at atmospheric pressure

$\ln S = 1528/(T/K) - 8.47$ (where S is in units of mol l^{-1} atm^{-1}).

AUXILIARY INFORMATION

METHOD/APPARATUS/PROCEDURE:

Bubbler method. Samples of saturated liquid taken and added to sodium hydroxide solution and then excess alkali back titrated.

SOURCE AND PURITY OF MATERIALS:

1. Purity at least 95 mole per cent.
2. Chemically pure. Dried.

ESTIMATED ERROR:

$\delta T/K = \pm 0.1$; δ(Solubility) = ±2%; $\delta x_{SO_2} = \pm 0.001$; (estimated by compiler).

REFERENCES:

COMPONENTS:

1. Sulfur dioxide; SO_2; [7446-09-5]
2. 2,2,2-Trichloroethanol; $C_2H_3Cl_3O$; [115-20-8]

ORIGINAL MEASUREMENTS:

Gerrard, W.

J. Appl. Chem. Biotechnol. 1972, *22*, 623-650.

VARIABLES:

Temperature

PREPARED BY:

W. Gerrard.

EXPERIMENTAL VALUES:

T/K	P/kPa	Mole ratio SO_2 /ROH	*Mole fraction x_{SO_2}
273.15	99.1	0.70	0.412
283.15		0.377	0.274
293.15		0.240	0.194
298.15		0.200	0.167

* Calculated by compiler.

P is the total pressure.

AUXILIARY INFORMATION

METHOD/APPARATUS/PROCEDURE:

Sulfur dioxide was bubbled into a weighed amount of component (2) in a bubbler tube as described in detail, diagram given, in the original paper. The amount of gas absorbed at equilibrium and at the observed temperature and pressure was weighed. By means of a manometer assembly, to which the bubbler tube was attached, the weight of gas absorbed at successively lower pressure was measured. Eventually the pressure was reduced to that of the component (2). The refractive index, and the infrared spectrum of the residue showed it to be essentially pure component (2).

SOURCE AND PURITY OF MATERIALS:

1. Sulfur dioxide was the best grade contained in a small cylinder. Its high purity was attested by the measurement of vapor pressure from 258 to 268 K.
2. The best obtainable specimen was carefully purified and rigorously attested.

ESTIMATED ERROR:

REFERENCES:

COMPONENTS:

1. Sulfur dioxide; SO_2; [7446-09-5]
2. 1,2-Dichloroethane; $C_2H_4Cl_2$; [107-06-2]
   1,2-Dichloropropane; $C_3H_6Cl_2$; [78-87-5]
   Trichloroethene (*trichloroethylene*); C_2HCl_3; [79-01-6]

ORIGINAL MEASUREMENTS:

Sano, H.
Nippon Kagaku Zasshi
1968, *89*, 362-368.

VARIABLES:

PREPARED BY:

C. L. Young

EXPERIMENTAL VALUES:

T/°C = 30 T/K = 303

Vapor pressure of 1,2-dichloroethane, p_s° = 0.110 atm

Vapor pressure of 1,2-dichloropropane, p_s° = 0.065 atm

Vapor pressure of trichloroethene, p_s° = 0.120 atm

Absorption coefficient, S[a,b]	Bunsen coefficient,[b] α	Bunsen coefficient,[c] α	ℓ[c,d]	Mole ratio	Mole fraction[e]
		1,2-Dichloroethane			
67.0	68.4	72.9	62.0	0.296	0.228
		1,2-Dichloropropane			
47.6	46.0	47.0	43.5	0.200	0.167
		Trichloroethene			
22.0	22.5	22.9	20.1	0.091	0.0834

[a] Volume of sulfur dioxide absorbed by unit volume of solvent at experimental temperature and a total pressure of 1 atmosphere.

[b] Determined by static absorption method.

[c] Determined by flow method.

[d] $\alpha = \ell(1 - p_s^\circ)$ where p_s° is the vapor pressure of solvent.

[e] Calculated by compiler.

AUXILIARY INFORMATION

METHOD/APPARATUS/PROCEDURE:

Two techniques were used: (a) static absorption method and (b) flow method. Static method consisted of equilibrating solvent and sulfur dioxide in a 50 ml flask and measuring the absorption volumetrically. The flow method consists of passing sulfur dioxide through a micro gas absorption tube and estimating sulfur dioxide by iodometric titration. The vapor pressure of the solvents were measured roughly and were assumed to be equal to the partial pressure of the solvent by the authors when calculating the Bunsen coefficient.

SOURCE AND PURITY OF MATERIALS:

1. Purity 99.0-99.8 mole per cent.

ESTIMATED ERROR:

$\delta T/K = \pm 0.5$; $\delta x_{SO_2} = \pm 5\%$

(estimated by compiler).

REFERENCES:

COMPONENTS:	ORIGINAL MEASUREMENTS:
1. Sulfur dioxide; SO_2; [7446-09-5] 2. 1,2-Dichloroethane; $C_2H_4Cl_2$; [107-06-2]	Benoit, R.L.; Milanova, E. *Can. J. Chem.*, 1979, *59*, 1319-1323
VARIABLES:	PREPARED BY: C.L. Young

EXPERIMENTAL VALUES:

T/K	Henry's law constant[+] H/atm
298.15	4.23

[+] Defined in the original as

$$H = \frac{\text{partial pressure of sulfur dioxide}}{\text{mole fraction of sulfur dioxide}}$$

AUXILIARY INFORMATION

METHOD/APPARATUS/PROCEDURE:

Few details given. Pressure measured with quartz spiral gauge. Partial pressure calculated assuming Raoult's law. Concentration of sulfur dioxide determined by iodometric titration. Concentration of sulfur dioxide ranged from 0.05 to 0.2 mol l^{-1}

SOURCE AND PURITY OF MATERIALS:

1. Anhydrous sample from Canadian Liquid Air, purity 99.98 mole per cent.

2. Dried over 4A molecular sieve

ESTIMATED ERROR:

$\delta T/K = \pm 0.1$; $\delta H/\text{atm} = \pm 4\%$ (estimated by compiler)

REFERENCES:

COMPONENTS:

1. Sulfur dioxide; SO_2; [7446-09-5]

2. Chlorobenzene; C_6H_5Cl; [108-90-7]

ORIGINAL MEASUREMENTS:

Horiuti, J.

Sci. Pap. Inst. Phys. Chem. Res. (Jpn.), 1931/32, *17*, 125-256.

VARIABLES:

Temperature, pressure

PREPARED BY:

C. L. Young

EXPERIMENTAL VALUES:

T/K	$P^{\S}$/bar	Mole fraction of sulfur dioxide in liquid, x_{SO_2}	T/K	$P^{\dagger}$/bar	Mole fraction of sulfur dioxide in liquid, x_{SO_2}	Ostwald coefficient, L
298.15	0.0155	0.000	273.15	1.011	0.5310	169.3
	0.3662	0.0602	283.15	1.008	0.3415	97.4
	0.4058	0.0660	293.15	1.003	0.2181	59.14
	0.6745	0.1163	303.15	0.9950	0.1532	41.17
	0.7626	0.1328	313.15	0.9816	0.1102	29.92
	0.9318	0.1666	323.15	0.9612	0.0821	22.88
	1.0369	0.1876	333.15	0.9306	0.0614	17.83
	1.193	0.2202	343.15	0.8882	0.0466	14.33
	1.229	0.2271	353.15	0.8265	0.0365	12.23
	1.373	0.2590				
	1.397	0.2635				

§ total pressure.

† partial pressure, total pressure = 1 atmosphere = 1.01325×10^5 Pa.

Ostwald coefficient (as $x \to 0$, T/K = 298.15) = 40.4.

Henry's law constant, dp/dx (as $x \to 0$, T/K = 298.15) = 4502 mmHg = 6.002 bar.

AUXILIARY INFORMATION

METHOD/APPARATUS/PROCEDURE:

Apparatus consisted of a bas buret, a solvent reservoir and an absorption pipet. Dry gas introduced into degassed solvent. System mixed using a magnetic stirrer until saturation occurred. Care was taken to prevent solvent vapor from mixing with gas in the gas buret. Volume of gas determined from gas buret reading; volume of liquid determined from height of meniscus in absorption pipet.

SOURCE AND PURITY OF MATERIALS:

1. Sample obtained from Nippon Sulphuric Acid Manufacturing Co. Dried and fractionated several times.

2. Sample from Kahlbaum, dried and distilled. Boiling point at atmospheric pressure, 131.96 °C.

ESTIMATED ERROR:

$\delta T/K = \pm 0.1$; $\delta P/\text{bar} = \pm 0.001$;

$\delta x_{SO_2} = \pm 0.001$ (estimated by compiler).

REFERENCES:

COMPONENTS:

1. Sulfur dioxide; SO_2; [7446-09-5]
2. Chlorobenzene; C_6H_5Cl; [108-90-7]

   1,2-Dichlorobenzene; $C_6H_4Cl_2$; [95-50-1]

ORIGINAL MEASUREMENTS:

Sano, H.

Nippon Kagaku Zasshi

1968, *89*, 362-368.

VARIABLES:

PREPARED BY:

C. L. Young

EXPERIMENTAL VALUES:

T/°C = 30 T/K = 303

Vapor pressure of chlorobenzene, p_s° = 0.020 atm

Absorption coefficient, S[a,b]	Bunsen coefficient,[b] α	Bunsen coefficient,[c,f] α	ℓ[c,d]	Mole ratio	Mole fraction[e]
		Chlorobenzene			
40.1	37.2	38.0	37.1	0.170	0.145
		1,2-Dichlorobenzene			
25.7	23.6	24.1	24.1	0.117	0.105

[a] Volume of sulfur dioxide absorbed by unit volume of solvent at experimental temperature and a total pressure of 1 atmosphere.

[b] Determined by static absorption method.

[c] Determined by flow method.

[d] $\alpha = \ell(1 - p_s^\circ)$ where p_s° is the vapor pressure of solvent.

[e] Calculated by compiler.

[f] Assuming vapor pressure of 1,2-dichlorobenzene to be negligible.

AUXILIARY INFORMATION

METHOD/APPARATUS/PROCEDURE:

Two techniques were used: (a) static absorption method and (b) flow method. Static method consisted of equilibrating solvent and sulfur dioxide in a 50 ml flask and measuring the absorption volumetrically. The flow method consists of passing sulfur dioxide through a micro gas absorption tube and estimating sulfur dioxide by iodometric titration. The vapor pressure of the solvents were measured roughly and were assumed to be equal to the partial pressure of the solvent by the authors when calculating the Bunsen coefficient.

SOURCE AND PURITY OF MATERIALS:

1. Purity 99.0-99.8 mole per cent.

ESTIMATED ERROR:

$\delta T/K = \pm 0.5$; $\delta x_{SO_2} = \pm 5\%$

(estimated by compiler).

REFERENCES:

COMPONENTS:

1. Sulfur dioxide; SO_2; [7446-09-5]
2. Chlorobenzene; C_6H_5Cl; [108-90-7]

ORIGINAL MEASUREMENTS:

Gerrard, W.

J. Appl. Chem. Biotechnol. 1972, *22*, 623-650.

VARIABLES:

Temperature, pressure

PREPARED BY:

W. Gerrard.

EXPERIMENTAL VALUES:

T/K	Pressure, total /mmHg	Mole ratio SO_2/Comp.(2)	*Mole fraction x_{SO_2}
273.15	760	1.20	0.545
	** 715	1.06	0.516
	694	0.962	0.490
	636	0.783	0.439
	530	0.546	0.353
283.15	760	0.510	0.338
293.15	760	0.28	0.219
298.15	760	0.23	0.187

760 mmHg = 1 atm = 101.325 kPa.

* Calculated by compiler

** The plot of pressure vs solubility given in the original paper is based on these data.

AUXILIARY INFORMATION

METHOD/APPARATUS/PROCEDURE:

Sulfur dioxide was bubbled into a weighed amount of component (2) in a bubbler tube as described in detail, diagram given, in the original paper. The amount of gas absorbed at equilibrium and at the observed temperature and pressure was weighed. By means of a manometer assembly, to which the bubbler tube was attached, the weight of gas absorbed at successively lower pressure was measured. Eventually the pressure was reduced to that of the component (2). The refractive index, and the infrared spectrum of the residue showed it to be essentially pure component (2).

SOURCE AND PURITY OF MATERIALS:

1. Sulfur dioxide was the best grade contained in a small cylinder. Its high purity was attested by the measurement of vapor pressure from 258 to 268 K.
2. The best obtainable specimen was carefully purified and rigorously attested.

ESTIMATED ERROR:

REFERENCES:

COMPONENTS:

1. Sulfur dioxide; SO_2; [7446-09-5]

2. Bromobenzene; C_6H_5Br; [108-86-1]

ORIGINAL MEASUREMENTS:

Gerrard, W.

J. Appl. Chem. Biotechnol. 1972, *22*, 623-650

VARIABLES:

Temperature

PREPARED BY:

W. Gerrard

EXPERIMENTAL VALUES:

T/K	Pressure,total /mmHg	Mole ratio SO_2/Comp.(2)	*Mole fraction x_{SO_2}
273.15	760	0.91	0.476
283.15	760	0.41	0.291
293.15	760	0.245	0.197
298.15	760	0.21	0.174

760 = mmHg = 1 atm = 101.325 kPa

* Calculated by compiler.

AUXILIARY INFORMATION

METHOD/APPARATUS/PROCEDURE:

Sulfur dioxide was bubbled into a weighed amount of component (2) in a bubbler tube as described in detail, diagram given, in the original paper. The amount of gas absorbed at equilibrium and at the observed temperature and pressure was weighed. By means of a manometer assembly, to which the bubbler tube was attached, the weight of gas absorbed at successively lower pressure was measured. Eventually the pressure was reduced to that of the component (2). The refractive index, and the infrared spectrum of the residue showed it to be essentially pure component (2).

SOURCE AND PURITY OF MATERIALS:

1. Sulfur dioxide was the best grade contained in a small cylinder. Its high purity was attested by the measurement of vapor pressure from 258 to 268 K.

2. The best obtainable specimen was carefully purified and rigorously attested.

ESTIMATED ERROR:

REFERENCES:

COMPONENTS:

1. Sulfur dioxide; SO_2; [7446-09-5]
2. Iodobenzene; C_6H_5I; [591-50-4]

ORIGINAL MEASUREMENTS:

Gerrard, W.

J. Appl. Chem. Biotechnol. 1972, *22*, 623-650

VARIABLES:

Temperature,

PREPARED BY:

W. Gerrard.

EXPERIMENTAL VALUES:

T/K	Pressure, total /mmHg	Mole ratio SO_2/Comp.(2)	*Mole fraction x_{SO_2}
273.15	760	0.81	0.448
283.15	760	0.335	0.251
293.15	760	0.21	0.174
298.15	760	0.18	0.153

760 mmHg = 1 atm = 101.325 kPa.

* calculated by compiler

AUXILIARY INFORMATION

METHOD/APPARATUS/PROCEDURE:

Sulfur dioxide was bubbled into a weighed amount of component (2) in a bubbler tube as described in detail, diagram given, in the original paper. The amount of gas absorbed at equilibrium and at the observed temperature and pressure was weighed. By means of a manometer assembly, to which the bubbler tube was attached, the weight of gas absorbed at successively lower pressure was measured. Eventually the pressure was reduced to that of the component (2). The refractive index, and the infrared spectrum of the residue showed it to be essentially pure component (2).

SOURCE AND PURITY OF MATERIALS:

1. Sulfur dioxide was the best grade contained in a small cylinder. Its high purity was attested by the measurement of vapor pressure from 258 to 268 K.
2. The best obtainable specimen was carefully purified and rigorously attested.

ESTIMATED ERROR:

REFERENCES:

COMPONENTS:

1. Sulfur dioxide; SO_2; [7446-09-5]
2. 2-Chlorophenol; C_6H_5ClO; [95-57-8]

ORIGINAL MEASUREMENTS:

Sano, H.; Nakamoto, Y.

Nippon Kagaku Zasshi

1968, *89*, 369-373.

VARIABLES:

PREPARED BY:

C. L. Young

EXPERIMENTAL VALUES:

T/°C = 30 T/K = 303

Absorption coefficient, S^a	Bunsen coefficient, α	Mole ratio	Mole fraction[b]
76.2	69.4	0.323	0.244

[a] Volume of sulfur dioxide absorbed by unit volume of solvent at experimental temperature and a total pressure of 1 atmosphere.

[b] Calculated by compiler.

AUXILIARY INFORMATION

METHOD/APPARATUS/PROCEDURE:

Static absorption method consisted of equilibrating solvent and sulfur dioxide in a 50 ml flask and measuring the absorption volumetrically. The vapor pressure of the solvents were measured roughly and were assumed to be equal to the partial pressure of the solvent by the authors when calculating the Bunsen coefficient.

SOURCE AND PURITY OF MATERIALS:

1. Purity 99.0-99.8 mole per cent.

ESTIMATED ERROR:

$\delta T/K = \pm 0.5$; $\delta x_{SO_2} = \pm 5\%$ (estimated by compiler).

REFERENCES:

COMPONENTS:

1. Sulfur dioxide; SO_2; [7446-09-5]
2. 1-Chloro-octane; (octyl chloride); $C_8H_{17}Cl$; [111-85-3]

ORIGINAL MEASUREMENTS:

Gerrard, W. *J. Appl. Chem. Biotechnol.* 1972, *22*, 623-650.

VARIABLES:

Temperature

PREPARED BY:

W. Gerrard

EXPERIMENTAL VALUES:

T/K	Pressure, total /mmHg	Mole ratio SO_2/Comp.(2)	*Mole fraction x_{SO_2}
273.15	748	0.75	0.429
283.15	748	0.40	0.286
293.15	748	0.25	0.200
298.15	748	0.21	0.174

760 mmHg = 1 atm = 101.325 kPa.

* Calculated by compiler.

AUXILIARY INFORMATION

METHOD/APPARATUS/PROCEDURE:

Sulfur dioxide was bubbled into a weighed amount of component (2) in a bubbler tube as described in detail, diagram given, in the original paper. The amount of gas absorbed at equilibrium and at the observed temperature and pressure was weighed. By means of a manometer assembly, to which the bubbler tube was attached, the weight of gas absorbed at successively lower pressure was measured. Eventually the pressure was reduced to that of the component (2). The refractive index, and the infrared spectrum of the residue showed it to be essentially pure component (2).

SOURCE AND PURITY OF MATERIALS:

1. Sulfur dioxide was the best grade contained in a small cylinder. Its high purity was attested by the measurement of vapor pressure from 258 to 268 K.
2. The best obtainable specimen was carefully purified and rigorously attested.

ESTIMATED ERROR:

REFERENCES:

COMPONENTS:

1. Sulfur dioxide; SO_2; [7446-09-5]

2. 1-Bromooctane, (octyl bromide); $C_8H_{17}Br$; [111-83-1]

ORIGINAL MEASUREMENTS:

Gerrard. W.

J. Appl. Chem. Biotechnol. 1972, *22*, 623-650.

VARIABLES:

Temperature

PREPARED BY:

W. Gerrard

EXPERIMENTAL VALUES:

T/K	Pressure, total /mmHg	Mole ratio SO_2/ Comp. (2)	*Mole fraction x_{SO_2}
273.15	757	0.66	0.398
283.15	757	0.365	0.267
293.15	757	0.235	0.190
298.15	757	0.195	0.163

760 mmHg = 1 atm = 101.325 kPa.

* Calculated by compiler.

AUXILIARY INFORMATION

METHOD/APPARATUS/PROCEDURE:

Sulfur dioxide was bubbled into a weighed amount of component (2) in a bubbler tube as described in detail, diagram given, in the original paper. The amount of gas absorbed at equilibrium and at the observed temperature and pressure was weighed. By means of a manometer assembly, to which the bubbler tube was attached, the weight of gas absorbed at successively lower pressure was measured. Eventually the pressure was reduced to that of the component (2). The refractive index, and the infrared spectrum of the residue showed it to be essentially pure component (2).

SOURCE AND PURITY OF MATERIALS:

1. Sulfur dioxide was the best grade contained in a small cylinder. Its high purity was attested by the measurement of vapor pressure from 258 to 268 K.

2. The best obtainable specimen was carefully purified and rigorously attested.

ESTIMATED ERROR:

REFERENCES:

COMPONENTS:	ORIGINAL MEASUREMENTS:
1. Sulfur dioxide; SO_2; [7446-09-5] 2. Heptadecane; $C_{17}H_{36}$; [629-78-7] 3. Tetrachloromethane, (Carbon tetrachloride); CCl_4; [56-23-5]	Sobolev, I. A.; Kurkarin, V. A.; Dzhogatspanyan, R. V.; Kosorotov, V. I.; Zogorets, P. A.; Popov, A. I. *Khim. Prom.* 1970, *46*, 668-70.
VARIABLES:	PREPARED BY:
Temperature, concentration	C. L. Young

EXPERIMENTAL VALUES:

T/K	Percentage of heptadecane by volume	Solubility*, S /mol ℓ^{-1}
283.15	25	0.530
288.15		0.440
293.15		0.350
298.15		0.299
303.15		0.248
283.15	50	0.421
288.15		0.350
293.15		0.292
298.15		0.248
303.15		0.211
283.15	75	0.313
288.15		0.266
293.15		0.226
298.15		0.196
303.15		0.172

* at atmospheric pressure

Authors claim solubility is given by

$$\ln S = -\left((967x - 1582)/(T/K) + 8.47 - 2.71x\right)$$

where x is the mole fraction of heptadecane and S is in units of mol l^{-1} atm^{-1}.

AUXILIARY INFORMATION

METHOD/APPARATUS/PROCEDURE:	SOURCE AND PURITY OF MATERIALS:
Bubbler method. Samples of saturated liquid taken and added to excess sodium hydroxide solution and back titrated.	1. Purity at least 95 mole per cent. 2 and 3. Chemically pure, dried.
	ESTIMATED ERROR:
	$\delta T/K = \pm 0.1$; δ(Solubility) = ±2% (estimated by compiler).
	REFERENCES:

COMPONENTS:

1. Sulfur dioxide; SO_2; [7446-09-5]

2. Soft paraffin (see note).

3. Tetrachloromethane, (carbon tetrachloride); CCl_4; [56-23-5]

ORIGINAL MEASUREMENTS:

Sobolev, I.A.; Kukarin, V.A.; Dzhagatspanyan, R.V.; Kosorotov, V.I.; Zogorets, P.A; Popov, A.I.

Khim. Prom. 1970, *46*, 668-70.

VARIABLES:

Temperature, composition

PREPARED BY:

C.L. Young

EXPERIMENTAL VALUES:

T/K	Percentage of soft paraffin (by volume)	Solubility*, S /mol l^{-1}
283.15	25	0.533
288.15		0.419
293.15		0.344
298.15		0.294
303.15		0.246
283.15	50	0.412
288.15		0.343
293.15		0.283
298.15		0.238
303.15		0.208
283.15	75	0.317
288.15		0.265
293.15		0.221
298.15		0.195
303.15		0.167
288.15	90	0.227
293.15		0.204
298.15		0.180
303.15		0.157

* at atmospheric pressure.

Author claim solubility is given by $\ln S = -((838x - 1582)/(T/K) + 8.27 - 2.49x)$ where x is the mole fraction of hydrocarbon (appears to be based on assuming a molecular weight of about 250) and S is in units of mol l^{-1} atm^{-1}

AUXILIARY INFORMATION

METHOD/APPARATUS/PROCEDURE:

Bubbler method. Sample of saturated liquid taken and added to excess sodium hydroxide solution and back titrated.

NOTE: Mixture of C_{12}-C_{22} hydrocarbons (86.7% normal paraffins, 11.3% branched paraffins, 0.4% aromatic and 1.6% naphthenic hydrocarbons).

SOURCE AND PURITY OF MATERIALS:

1. Purity at least 95 mole per cent.

2. See note.

3. Chemically pure. Dried.

ESTIMATED ERROR:

$\delta T/K = \pm 0.1$; δ(Solubility) = ±2%; $\delta x_{SO_2} = \pm 0.001$; (estimated by compiler.

REFERENCES:

COMPONENTS:	EVALUATOR:
1. Sulfur dioxide; SO_2; [7446-09-5] 2. Sulfinylbismethane; C_2H_6OS; [67-68-5]	Colin L. Young, School of Chemistry, University of Melbourne, Parkville, Victoria 3052, Australia. December 1982

CRITICAL EVALUATION:

The isolated value of Sano (1) is in fair agreement with the data of Smedslund (2). Both the value of Lenoir *et al.* (3) and Benoit and Milanova (4) were determined at low partial pressures of sulfur dioxide and extrapolation to 1 atmosphere partial pressure assuming Henry's law, is not valid. However, the two values, one obtained using chemical analysis and the other using gas chromatographic retention datum, are in fair agreement, which suggests the values are essentially correct but refer to very low partial pressures of sulfur dioxide.

References

1. Sano, H. *Nippon Kagaku Zasshi* 1968, *89*, 362.
2. Smedslund, T. H. *Finska Kemisteamfundets Medd.* 1950, *59*, 40.
3. Lenoir, J.-Y.; Renault, P.; Renon, H. *J. Chem. Engng. Data* 1971, *16*, 340.
4. Benoit, R. L.; Milanova, E. *Can. J. Chem.* 1979, *57*, 1319.

COMPONENTS:

1. Sulfur dioxide; SO_2; [7446-09-5]
2. Sulfinylbismethane (Dimethylsulfoxide); C_2H_6OS; [67-68-5]

ORIGINAL MEASUREMENTS:

Smedslund, T.H.
Finska Kemistsamfundets Medd. 1950, *59*, 40-43.

VARIABLES:

Temperature

PREPARED BY:

W. Gerrard

EXPERIMENTAL VALUES:

T/K	SO_2, Weight %	*Mole ratio $SO_2/(CH_3)_2SO$	*Mole fraction x_{SO_2}
291	60.13	1.839	0.648
293	59.12	1.763	0.638
373	17.9	0.266	0.210
434	3.2	0.0403	0.039

T/K	SO_2, Weight %	Partial Pressure SO_2, mmHg	* Mole ratio $SO_2/(CH_3)_2SO$	Mole fraction * x_{SO_2}
"293"	4.9	5.2	0.0628	0.0590
Probably	14.6	21	0.208	0.172
	27.6	71	0.465	0.317

See also general statement (data not given) in U.S. Patent No. 2,539,871, 1951.

* Calculated by compiler.

AUXILIARY INFORMATION

METHOD/APPARATUS/PROCEDURE:

The first set of data are for "the saturation of dimethylsulfoxide at a pressure of 1 atm." Details are not available for the second set.

SOURCE AND PURITY OF MATERIALS:

1. Information not available.
2. Dimethylsulfoxide was prepared by the catalytic oxidation of dimethyl sulfide in the vapor state.

   The purified product had b.p. 63°C/6 mmHg. (1).

ESTIMATED ERROR:

REFERENCES:

1. Smedslund, T.H. *U.S. Patent*, 2,581,050, 1952.

COMPONENTS:

1. Sulfur dioxide; SO_2; [7446-09-5]
2. Sulfinylbismethane (*dimethyl-sulfoxide*); C_2H_6OS; [67-68-5]
   Sulfuric acid, dimethyl ester (*dimethyl sulfate*); $C_2H_6O_4S$; [77-78-1]
   Tetramethylene sulfone (*tetra-hydro thiophene, 1,1-dioxide*); $C_4H_8O_2S$; [126-33-0]

ORIGINAL MEASUREMENTS:

Sano, H.

Nippon Kagaku Zasshi

1968, *89*, 362-368.

VARIABLES:

PREPARED BY:

C. L. Young

EXPERIMENTAL VALUES:

T/°C = 30 T/K = 303

Vapor pressure of dimethyl sulfate = 0.001 atm

Absorption coefficient, S[a]	Bunsen coefficient, α	Mole ratio	Mole fraction[b]
	Sulfinylbismethane		
-	447	1.42	0.587
	Dimethyl sulfate		
124.2	113.3	0.482	0.325
	Tetrahydrothiophene 1,1-dioxide		
186.4	169.6	0.929	0.482

[a] Volume of sulfur dioxide absorbed by unit volume of solvent at experimental temperature and a total pressure of 1 atmosphere.

[b] Calculated by compiler.

AUXILIARY INFORMATION

METHOD/APPARATUS/PROCEDURE:

Static absorption method consisted of equilibrating solvent and sulfur dioxide in a 50 ml flask and measuring the absorption volumetrically. The vapor pressure of the solvents were measured roughly and were assumed to be equal to the partial pressure of the solvent by the authors when calculating the Bunsen coefficient.

SOURCE AND PURITY OF MATERIALS:

1. Purity 99.0-99.8 mole per cent.

ESTIMATED ERROR:

$\delta T/K = \pm 0.5$; $\delta x_{SO_2} = \pm 5\%$

(estimated by compiler).

REFERENCES:

COMPONENTS:

1. Sulfur dioxide, SO_2; [7446-09-5]
2. Sulfinylbismethane, (dimethylsulfoxide); C_2H_6OS; [67-68-5]

ORIGINAL MEASUREMENTS:

Lenoir, J-Y.; Renault, P.; Renon, H.

J. Chem. Eng. Data 1971, *16*, 340-342.

VARIABLES:

PREPARED BY:

W. Gerrard

EXPERIMENTAL VALUES: 760 mmHg = 1 atm = 101.325 kPa.

T/K	Henry's constant H_{SO_2}/atm	*Mole fraction at 1 atm,
298	0.117	8.55 (An impossible condition)

* Calculated by compiler assuming a linear function of p_{SO_2} vs x_{SO_2}, i.e. x_{SO_2} (1 atm) = $1/H_{SO_2}$.

The authors gave "Henry's constant" as

$$H_{SO_2} = \lim_{x_{SO_2} \to 0} \frac{f_{SO_2}(P,T)}{x_{SO_2}}$$

x_{SO_2} is the mole fraction. f_{SO_2} is the fugacity of SO_2 "in the equilibrium condition"of pressure and temperature. H_{SO_2} is related to experimental chromatographic parameters based on the number of moles of solvent in the column, and the absolute retention time of the solute. Helium was the carrier gas.

AUXILIARY INFORMATION

METHOD/APPARATUS/PROCEDURE:

A conventional gas-liquid chromatographic unit was used. The chromatograph was an IGC 12 M Intersmat model, equipped with a thermal conductivity detector. The temperature control was stated to be within ±0.1 °C. The pressure drop in the column was measured by a mercury manometer. Helium was the carrier gas.

There is serious uncertainty in the estimation of accuracy.

SOURCE AND PURITY OF MATERIALS:

1. Sulfur dioxide was provided by l'Air Liquide, and had a stated purity of 99.9%. It was used as provided.
2. Provided by Touzart & Matignon and Serlabo, and stated to have a minimum purity of 99%. Distillation was carried out when necessary and the refractive index was measured.

ESTIMATED ERROR:

REFERENCES:

COMPONENTS:

1. Sulfur dioxide; SO_2; [7446-09-5]

2. Sulfinylbismethane (Dimethylsulfoxide); C_2H_6OS; [67-68-5]

ORIGINAL MEASUREMENTS:

Benoit, R.L.; Milanova, E.

Can. J. Chem., 1979, *57*, 1319-1323

VARIABLES:

PREPARED BY:

C.L. Young

EXPERIMENTAL VALUES:

T/K	Henry's law constant[+] H/atm
298.15	0.109

[+] Defined in the original as

$$H = \frac{\text{partial pressure of sulfur dioxide}}{\text{mole fraction of sulfur dioxide}}$$

AUXILIARY INFORMATION

METHOD/APPARATUS/PROCEDURE:

Few details given. Pressure measured with quartz spiral gauge. Partial pressure calculated assuming Raoult's law. Concentration of sulfur dioxide determined by iodometric titration. Concentration of sulfur dioxide ranged from 0.3 to 1.5 mol l^{-1}

SOURCE AND PURITY OF MATERIALS:

1. Anhydrous sample from Canadian Liquid Air, purity 99.98 mole per cent.

2. Dried over 4A molecular sieve.

ESTIMATED ERROR:

$\delta T/K = \pm 0.1$; $\delta H/\text{atm} = \pm 4\%$ (estimated by compiler)

REFERENCES:

COMPONENTS:	ORIGINAL MEASUREMENTS:
1. Sulfur dioxide; SO_2; [7446-09-5] 2. Tetrahydrothiophene, 1,1-dioxide, (sulfolane); $C_4H_8O_2S$; [126-33-0]	Benoit, R.L.; Milanova, E. *Can. J. Chem.*, 1979, *57*, 1319-1323
VARIABLES:	PREPARED BY: C.L. Young

EXPERIMENTAL VALUES:

T/K	Henry's law constant[+] H/atm
303.15	1.01

[+] Defined in the original as

$$H = \frac{\text{partial pressure of sulfur dioxide}}{\text{mole fraction of sulfur dioxide}}$$

AUXILIARY INFORMATION

METHOD/APPARATUS/PROCEDURE:

Few details given. Pressure measured with quartz spiral gauge. Partial pressure calculated assuming Raoult's law. Concentration of sulfur dioxide determined by iodometric titration. Concentration of sulfur dioxide ranged from 0.1 to 1.0 mol l^{-1}

SOURCE AND PURITY OF MATERIALS:

1. Anhydrous sample from Canadian Liquid Air, purity 99.98 mole per cent.

2. Dried over 4A molecular sieve

ESTIMATED ERROR:

$\delta T/K = \pm 0.1$; $\delta H/\text{atm} = \pm 4\%$ (estimated by compiler)

REFERENCES:

COMPONENTS:

1. Sulfur dioxide; SO_2; [7446-09-5]

2. Esters of phosphoric acid

ORIGINAL MEASUREMENTS:

Sano, H.

Nippon Kagaku Zasshi

1968, *89*, 362-368.

VARIABLES:

PREPARED BY:

C. L. Young

EXPERIMENTAL VALUES:

T/°C = 30 T/K = 303

Absorption coefficient, S[a,b]	Bunsen coefficient,[b] α	Bunsen coefficient,[c,f] α	ℓ[c,d]	Mole ratio	Mole fraction[e]
Phosphoric acid, tributyl ester (*Tributyl phosphate*); $C_{12}H_{27}PO_4$; [126-73-8]					
84.9	77.3	75.5	75.5	0.942	0.485
Phosphoric acid, tris(methylphenyl) ester (*Tritolyl phosphate*); $C_{21}H_{21}PO_4$; [1330-78-5]					
61.2	55.7	56.0	56.0	0.790	0.441

[a] Volume of sulfur dioxide absorbed by unit volume of solvent at experimental temperature and a total pressure of 1 atmosphere.

[b] Determined by static absorption method.

[c] Determined by flow method.

[d] $\alpha = \ell(1 - p_s^\circ)$ where p_s° is the vapor pressure of solvent.

[e] Calculated by compiler.

[f] Assuming vapor pressure of solvent to be negligible.

AUXILIARY INFORMATION

METHOD/APPARATUS/PROCEDURE:

Two techniques were used: (a) static absorption method and (b) flow method. Static method consisted of equilibrating solvent and sulfur dioxide in a 50 ml flask and measuring the absorption volumetrically. The flow method consists of passing sulfur dioxide through a micro gas absorption tube and estimating sulfur dioxide by iodometric titration. The vapor pressure of the solvents were measured roughly and were assumed to be equal to the partial pressure of the solvent by the authors when calculating the Bunsen coefficient.

SOURCE AND PURITY OF MATERIALS:

1. Purity 99.0-99.8 mole per cent.

ESTIMATED ERROR:

$\delta T/K = \pm 0.5$; $\delta x_{SO_2} = \pm 5\%$

(estimated by compiler).

REFERENCES:

COMPONENTS:

1. Sulfur dioxide; SO_2; [7746-09-5]
2. Esters of phosphoric acid.

ORIGINAL MEASUREMENTS:

Lenoir, J-Y,; Renault, P.; Renon, H.

J. Chem. Eng. Data. 1971, *16*, 340-342

VARIABLES:

PREPARED BY:

W. Gerrard.

EXPERIMENTAL VALUES:

T/K	Henry's constant H_{SO_2}/atm	Mole fraction at 1 atm *
Phosphoric acid, triethyl ester; $C_6H_{15}O_4P$; [78-40-0]		
325	1.41	0.709
Phosphoric acid, tripropyl ester; $C_9H_{21}O_4P$; [513-08-6]		
323	0.621	(1.61)
343	1.45	0.690
Phosphoric acid, tris (2-methyl propyl) ester; $C_{12}H_{27}O_4P$; [126-71-6]		
325	0.695	(1.44)

* Calculated by compiler assuming a linear function p_{SO_2} vs x_{SO_2}, i.e. x_{SO_2} (1 atm) = $1/H_{SO_2}$.

The authors gave "Henry's constant" as

$$H_{SO_2} = \lim_{x_{SO_2}\to 0} \frac{f_{SO_2}(P,T)}{x_{SO_2}}$$

x_{SO_2} is the mole fraction. f_{SO_2} is the fugacity of SO_2 "in the equilibrium condition" of pressure and temperature. H_{SO_2} is related to experimental chromatographic parameters based on the number of moles of solvent in the column, and the absolute retention time of the solute. Helium was the carrier gas.

AUXILIARY INFORMATION

METHOD/APPARATUS/PROCEDURE:

A conventional gas-liquid chromatographic unit was used. The chromatograph was an IGC 12 M Intersmat model, equipped with a thermal conductivity detector. The temperature control was stated to be within ±0.1° C. The pressure drop in the column was measured by a mercury manometer. Helium was the carrier gas.

There is serious uncertainty in the estimation of accuracy.

SOURCE AND PURITY OF MATERIALS:

1. Sulfur dioxide was provided by l'Air Liquide, and had a stated purity of 99.9%. It was used as provided.

2. Provided by Touzart & Matignon and Serlabo, and stated to have a minimum purity of 99%. Distillation was carried out when necessary, and the refractive index was measured.

ESTIMATED ERROR:

REFERENCES:

COMPONENTS:

1. Sulfur dioxide; SO_2; [7446-09-5]
2. Tributylphosphate, (Phosphoric acid, tributyl ester); $C_{12}H_{27}O_4P$; [126-73-8]

ORIGINAL MEASUREMENTS:

Cooper, D.F.; Smith, J.W.

J. Chem. Eng. Data, 1974, *19*, 133-6

VARIABLES:

PREPARED BY:

C. L. Young

EXPERIMENTAL VALUES:

T/K	p_{SO_2}/mmHg [+]	p_{SO_2}/kPa [+]	Wt ratio $SO_2/C_{12}H_{27}O_4P$	Mole fraction of sulfur dioxide in liquid, x_{SO_2} [*]
279.2	6.9	0.92	0.0239	0.0904
	8.7	1.16	0.0291	0.1079
	12.4	1.65	0.0389	0.1392
	19.3	2.57	0.0584	0.1954
	30.4	4.05	0.0807	0.2512
	43.7	5.83	0.1045	0.3029
	64.0	8.53	0.1315	0.3535
	72.8	9.71	0.1426	0.3722
	105	14.0	0.1731	0.4185
	162	21.6	0.2156	0.4727
	219	29.2	0.2538	0.5134
	299	39.9	0.3116	0.5644
	394	52.5	0.3771	0.6106
	492	65.6	0.4256	0.6389
290.2	33.8	4.51	0.0608	0.2018
	44.9	5.99	0.0730	0.2328
	57.8	7.71	0.0871	0.2658
	73.9	9.85	0.1016	0.2970
	94.6	12.6	0.1192	0.3314
	130	17.3	0.1458	0.3774
	179	23.9	0.1764	0.4231
	267	35.6	0.2203	0.4780

AUXILIARY INFORMATION

METHOD/APPARATUS/PROCEDURE:

Glass equilibrium cell. After equilibrium established samples of liquid removed and analysed by GC. Partial pressure of sulfur dioxide obtained by mixing with air to a total pressure of atmospheric pressure (757 ± 3 mmHg)

SOURCE AND PURITY OF MATERIALS:

1. Matheson sample, purity 99.9 mole per cent.
2. BDH sample containing 0.099 weight per cent water.

ESTIMATED ERROR:

$\delta T/K = \pm 0.1$; $\delta p/kPa = \pm 2\%$
$\delta x_{SO_2} = \pm 2\%$ (estimated by compiler)

REFERENCES:

COMPONENTS:	ORIGINAL MEASUREMENTS:
1. Sulfur dioxide; SO_2; [7446-09-5] 2. Tributylphosphate (Phosphoric acid, tributyl ester); $C_{12}H_{27}O_4P$; [126-73-8]	Cooper, D. F.; Smith, J. W. *J. Chem. Eng. Data* 1974, *19*, 133-6.

T/K	p_{SO_2}/mmHg +	p_{SO_2}/kPa +	Wt ratio $SO_2/C_{12}H_{27}O_4P$	Mole fraction of sulfur dioxide in liquid, x_{SO_2} *
290.2	362	48.3	0.2663	0.5254
	534	71.2	0.3279	0.5768
	661	88.1	0.3809	0.6129
307.0	7.4	0.99	0.0073	0.0295
	11.0	1.47	0.0106	0.0422
	17.6	2.35	0.0167	0.0649
	31.6	4.21	0.0288	0.1069
	38.8	5.17	0.0345	0.1254
	44.4	5.92	0.0388	0.1389
	55.1	7.35	0.0467	0.1626
	70.8	9.44	0.0575	0.1929
	105	14.0	0.0785	0.2461
	177	23.6	0.1131	0.3198
	269	35.9	0.1478	0.3806
	326	43.5	0.1655	0.4076
	458	61.1	0.2009	0.4551
	486	64.8	0.2078	0.4635
	570	76.0	0.2279	0.4865
327.3	13.6	1.81	0.0062	0.0251
	23.7	3.16	0.0106	0.0422
	37.2	4.96	0.0155	0.0605
	54.4	7.25	0.0222	0.0845
	69.6	9.28	0.0285	0.1059
	115	15.3	0.0427	0.1508
	115	15.3	0.0429	0.1514
	139	18.5	0.0486	0.1681
	176	23.5	0.0611	0.2026
	226	30.1	0.0733	0.2336
	320	42.7	0.0995	0.2926
	405	54.0	0.1160	0.3254
	429	57.2	0.1192	0.3314
	449	59.9	0.1220	0.3365
	480	64.0	0.1320	0.3543
	511	68.1	0.1367	0.3624
	590	78.7	0.1458	0.3774
	677	90.3	0.1610	0.4010
342.7	21.7	2.89	0.0056	0.0228
	58.7	7.83	0.0137	0.0539
	142	18.9	0.0315	0.1158
	200	26.7	0.0449	0.1573
	262	34.9	0.0548	0.1856
	331	44.1	0.0630	0.2076
	365	48.7	0.0710	0.2279
	430	57.3	0.0850	0.2611
	587	78.3	0.1028	0.2994
	663	88.4	0.1148	0.3231

+ p_{SO_2} = partial pressure of SO_2

* Calculated by compiler

COMPONENTS:

1. Sulfur dioxide; SO_2; [7446-09-5]
2. Carbon disulfide; CS_2; [75-15-0]

ORIGINAL MEASUREMENTS:

Sano, H.

Nippon Kagaku Zasshi

1968, *89*, 362-368.

VARIABLES:

PREPARED BY:

C. L. Young

EXPERIMENTAL VALUES:

T/°C = 30 T/K = 303

Bunsen coefficient, α	Mole ratio	Mole fraction[a]
10.1	0.027	0.0263

[a] Calculated by compiler.

AUXILIARY INFORMATION

METHOD/APPARATUS/PROCEDURE:

Static absorption method consisted of equilibrating solvent and sulfur dioxide in a 50 ml flask and measuring the absorption volumetrically. The vapor pressure of the solvents were measured roughly and were assumed to be equal to the partial pressure of the solvent by the authors when calculating the Bunsen coefficient.

SOURCE AND PURITY OF MATERIALS:

1. Purity 99.0-99.8 mole per cent.

ESTIMATED ERROR:

$\delta T/K = \pm 0.5$; $\delta x_{SO_2} = \pm 5\%$

(estimated by compiler).

REFERENCES:

COMPONENTS:	ORIGINAL MEASUREMENTS:
1. Sulfur dioxide; SO_2; [7446-09-5] 2. Sulfuryl chloride; O_2Cl_2S; [7791-25-5]	Schulze, H. *J. Prakt. Chem.* 1881, *24*, 168-183
VARIABLES:	PREPARED BY: W. Gerrard.

EXPERIMENTAL VALUES: 760 mmHg = 1 atm = 101.325 kPa.

T/K	Weight of SO_2 /1 g of comp.(2)	*Mole ratio SO_2/Comp.(2)	*Mole fraction x_{SO_2}
273.15	0.323	0.681	0.405

*Calculated by compiler.

Pressure of SO_2 stated to be 725 mmHg.

1 Volume of sulfuryl chloride absorbed 187 volumes of SO_2

AUXILIARY INFORMATION

METHOD/APPARATUS/PROCEDURE:

The gas was passed into a weighed amount of sulfuryl chloride in a thin-walled test-tube fitted with a rubber stopper carrying an inlet, and an outlet tube. The test tube was cooled in melting snow. The amount of gas absorbed was determined by weighing.

SOURCE AND PURITY OF MATERIALS:

1. Probably of satisfactory purity.
2. Probably of satisfactory purity. The paper gave details on the preparation of sulfuryl chloride.

ESTIMATED ERROR:

REFERENCES:

COMPONENTS:	ORIGINAL MEASUREMENTS:
(1) Sulfur dioxide; SO_2; [7446-09-5] (2) Lithium tetrachloro aluminate(1-); $AlCl_4.Li$; [14024-11-4] (3) Thionyl chloride; $SOCl_2$; [7719-09-7]	Domeniconi, M.; Klinedinst, K.; Marincic, N.; Schlaikjer, C.; Staniewicz, R.; Swette, L. ONR Contract # N00014-76-C-0524 Interim Report Jan. 1976 - Oct. 1977.
VARIABLES: T/K: 273 - 333 P: Atmospheric	PREPARED BY: H. L. Clever

EXPERIMENTAL VALUES:

Fifteen measurements of the solubility of sulfur dioxide, SO_2, in a 1.8 mole $LiAlCl_4$ dm^{-3} solution in thionyl chloride were made between 273 and 333 K. The mole fraction solubility values are shown on a graph (Figure 10 of the report). The data are not tabulated, but they were fitted to the following equation by the authors.

$$\ln x_{SO_2} = [(2.417 \pm 0.184) \times 10^3]/T - (9.863 \pm 0.613)$$

The equation can be used to calculate the mole fraction solubility of sulfur dioxide at a partial pressure of 1 atm between temperatures of 273 and 333 K in 1.8 molar $LiAlCl_4$ in $SOCl_2$.

The authors imply that they also measured the solubility of sulfur dioxide in pure thionyl chloride. However, they give neither data nor an equation. They do show what may be a smoothed curve for the sulfur dioxide solubility in thionyl chloride (Figure 10 of report).

Figure 10 of the report compares the ln mole fraction solubility *vs.* 1/T and Figure 11 compares the molar solubility *vs.* temperature in pure thionyl chloride and in the 1.8 M $LiAlCl_4$ solution. In both solubility units the solubility appears to be equal at a temperature near 288 K. At lower temperatures sulfur dioxide is more soluble in pure thionyl chloride than in the 1.8 M $LiAlCl_4$ solution, above 288K the sulfur dioxide is more soluble in the solution than in the pure solvent.

In the original report the equation intercept is given erroneously as 0.863 instead of 9.863.

AUXILIARY INFORMATION

METHOD/APPARATUS/PROCEDURE:	SOURCE AND PURITY OF MATERIALS:
The sulfur dioxide was determined gravimetrically. A special Erlenmeyer flask fitted with a side arm bubbler was dried and weighed. The solvent was added and the flask weighed again The flask was placed in either an oil or a water bath. A glass wool-packed water-jacketed condenser was attached to the flask. The sulfur dioxide was bubbled in until saturation. The flask was weighed. Dry argon was bubb bubbled through the solution as it was heated to reflux,thus driving off the previously dissolved SO_2. The flask was again weighed. The authors do not explain their reason for weighing the SO_2 gain on saturation, and the weight loss on degassing. It is not clear whether this is a check on the measurement, or the degassing procedure for the next determination at another temperature.	(1) Sulfur dioxide. (2) Lithium tetrachloroaluminate(1-). (3) Thionyl chloride. Source not given. Distilled twice before use. ESTIMATED ERROR: $\delta X_1/X_1 = 0.20$ (Compiler) REFERENCES:

COMPONENTS:

(1) Sulfur dioxide; SO_2; [7446-09-5]

(2) Lithium tetrachloro aluminate(1-); $AlCl_4.Li$; [14024-11-4]

(3) Thionyl chloride; $SOCl_2$; [7719-09-7]

ORIGINAL MEASUREMENTS:

Driscoll, J. R.; Brummer, S. B.; Gudrais, P.; Holleck, G. L.; Toland, D. E.; Turchan, M. J.

Report ECOM-74-0030-F
March 1978

VARIABLES:

T/K: 273 , 298
P: atmosphere

PREPARED BY:

H. L. Clever

EXPERIMENTAL VALUES:

T/K	Mol Fraction X_1	Molarity mol SO_2 dm^{-3} solution[a]
Thionyl chloride; $SOCl_2$; 7719-09-7		
273	0.465	8.0
298	0.160	2.3
Lithium tetrachloro aluminate(1-) + thionyl chloride; 1.5 mol $LiAlCl_4$ dm^{-3} solution		
273	>0.498	8.4
	0.512	8.7
298	0.192	2.7
	0.187	2.6

[a] The mole fraction values were calculated directly from the weights of the components in the saturated solutions. The molarities were calculated assuming a solution density of 1.60 g cm^{-3}. The solution densities were calculated from $\rho(\text{solution}) = \rho(SO_2)\cdot x(SO_2) + \rho(\text{solvent})\cdot x(\text{solvent})$. All of the calculated densities were within ± 0.03 of 1.60 g cm^{-3}. The 1.60 value was chosen to calculate all of the molarity values.

The present values confirm the authors earlier values by gas chromatography. They are much greater than values reported by Spandau and Brunneck (1). The present results are in substantial agreement of those of Domeniccui *et al.* (2).

AUXILIARY INFORMATION

METHOD/APPARATUS/PROCEDURE:

The solubility was determined gravimetrically. The sufur dioxide gas cylinder was connected to three gas bubblers joined in series. The first bubbler contained glass wool, the second bubbler contained a weighed portion of the solvent, and the third bubbler contained concentrated H_2SO_4. The third bubbler both isolated the system from the atmosphere, and acted as a flow monitor. The flow of SO_2 was set to minimize thionyl chloride loss by allowing little or no exit bubbling. Bubbling was continued until there was no more weight gain.

The sample bubbler was maintained at 25 °C in air or at 0 °C in a ice bath

SOURCE AND PURITY OF MATERIALS:

(1) Sulfur dioxide. Matheson Co., Inc. Anhydrous, minimum purity 99.98 per cent.

(2) Lithium tetrachloro aluminate(1-). Source and purity not given.

(3) Thionyl chloride. Eastman Co. Number 246, used as received.

ESTIMATED ERROR:

$\delta X_1/X_1 = 0.03 - 0.05$ (Compiler)

REFERENCES:

1. Spandau, H.; Brunneck, E. *Z.Anorg.Allgem.Chem.* 1952, *270*, 201.

2. Domeniccui, M.; Kliaedinst, K.; Marincic, N.; Schlaikjer, C.; Staniewicz, R.; Swette, L. Interim Report, Jan.1976-Oct. 1977, ONR Contract # N00014-76-C-0524.

COMPONENTS:	ORIGINAL MEASUREMENTS:
1. Sulfur dioxide; SO_2; [7446-09-5] 2. Rubber.	Venable, C.S.; Fuwa, T. *Ind. Eng. Chem.* 1922, *14*, 139-142.
VARIABLES:	PREPARED BY: W. Gerrard

EXPERIMENTAL VALUES:

The data were expressed as quantity of gas measured as S = cm^3 gas (273.15K, 1 atm) absorbed by 100 cm^3 of rubber.

At 293-295 K, S = 1950. The pressure appears to be the prevailing barometric one.

The compiler assumes that the isoprene unit, C_5H_8, may be taken to calculate an equivalent mole ratio (and hence mole fraction) as follows:

Equivalent mole ratio SO_2/C_5H_8 = (19.50/21868) x (68/0.75) = 0.0808; x_{SO_2} = 0.075.

The gram mole volume of sulfur dioxide is taken to be 21868 cm^3 at 273.15 K and 1 atm.); 0.75 is the assumed density of the rubber.

The x_{SO_2} value of 0.075 obtained in this way fits the value for the pure hydrocarbons, C_5 onwards, obtained by Makranczy *et al.* (1).

AUXILIARY INFORMATION

METHOD/APPARATUS/PROCEDURE:	SOURCE AND PURITY OF MATERIALS:
The rubber was held under high vacuum for 48 hours. A weighed amount of the degassed rubber, as thin strips, was left in contact with the gas until equilibrium had been attained, as indicated by the constancy of pressure in the bulb. The gas was removed by a pump and measured over mercury. Temperature control was by a large bath of water.	1. Dried over phosphorus pentoxide. From a cylinder. ESTIMATED ERROR: REFERENCES: 1. Makranczy, J.; Magyery-Balog, K.; Rusz, L.; Patyi, L. *Hung. J. Ind. Chem.* 1976, *4*, 269.

COMPONENTS:	EVALUATOR:
1. Chlorine; Cl_2; [7782-50-5] 2. Water; H_2O; [7732-18-5]	Rubin Battino, Department of Chemistry, Wright State University, Dayton, Ohio 45435, U.S.A. July 1981

CRITICAL EVALUATION:

The solubility of chlorine in pure water is complicated by two factors: (a) it reacts chemically with the water hydrolyzing to form Cl^- and HClO; and (b) at relatively low pressures (near atmospheric) it forms a hydrate ($Cl_2.8H_2O$) which separates out. The hydrolysis has been studied in great detail by Yakovkin (1) whose work was analyzed by Arkadiev (2). Unfortunately, neither of these papers was available to the Evaluator at the time of writing this critical evaluation. However, a summary of both of these papers was provided by Adams and Edmonds (3). Formation of the hydrate was specifically studied by Bakhuis Roozeboom (4) and Stackelberg and Müller (5).

There were a number of papers reporting on the solubility of chlorine in water. We take these solubilities to be bulk solubilities, i.e., we assume that all of the chlorine containing species present are present as Cl_2(aq). The work reported by Bunsen (6) and cited in other places was apparently carried out by Schönfeld (7) using a chemical method for determining the amount of dissolved chlorine. Roscoe (8) repeated Schönfeld's work at one temperature while studying gaseous mixtures of chlorine and hydrogen in aqueous solutions. Bakhuis Roozeboom (4) determined the solubility of chlorine in water at three temperatures (and extrapolated to two lower temperatures). Mellor (9) measured the solubility of chlorine in water at one temperature by chemical means while studying hydrogen and chlorine reactions. Winkler's results (10) are reported in the International Critical Tables for a wide range of temperature (283-383 K). Finally, Whitney and Vivian (11) reported on solubilities they determined while studying the hydrolysis constants of chlorine in water.

The solubility results fall into two main groups with that of Bakhuis Roozeboom falling roughly in the middle. Roscoe, Mellor, and Bunsen/Schönfeld's values are in quite good agreement with each other. Winkler, Whitney and Vivian, and Yakovkin's results (the latter as analyzed by Arkadiev and by Adams and Edmonds) are in good agreement with each other, but higher than the other group. The recommended values below were calculated from Table 1 in Adams and Edmonds' paper (3). They should be reliable to about ±2 per cent. The mole fractions are calculated for 101.325 kPa partial pressure of the gas.

T/K	$10^3 x_1$	T/K	$10^3 x_1$
283.15	2.48	343.15	0.849
293.15	1.88	353.15	0.784
303.15	1.50	363.15	0.737
313.15	1.23	373.15	0.697
323.15	1.06	383.15	0.668
333.15	0.939		

(cont.)

COMPONENTS:	EVALUATOR:
1. Chlorine; Cl_2; [7782-50-5] 2. Water; H_2O; [7732-18-5]	Rubin Battino, Department of Chemistry, Wright State University, Dayton, Ohio 45435, U.S.A. July 1981

CRITICAL EVALUATION:

References:

1. Yakovkin, A. *J. Russ. Phys.-Chem.* 1900, *32*, 673.
2. Arkadiev, V. *J. Russ. Phys.-Chem.* 1918, *50*, 205.
3. Adams, F. W.; Edmonds, R. G. *Ind. Eng. Chem.* 1937, *29*, 447-50.
4. Bakhuis Roozeboom, H. W. *Recl. Trav. Chim. Pays-Bas* 1884, *3*, 59-72.
5. Stackelberg, von M.; Müller, H. R. *Z. Elektrochem.* 1954, *58*, 25-39.
6. Bunsen, R. *Phil. Mag.* 1855, *9*, 181-201.
7. Schönfeld, M. *Ann.* 1855, *93*, 26.
8. Roscoe, H. E. *J. Chem. Soc.* 1856, *8*, 14-26.
9. Mellor, W. *J. Chem. Soc.* 1901, *79*, 216-38.
10. Winkler, L. W. *Mathematikai es Termeszettudomanyi Ertesitö Budapest* 1907, *25*, 86.
11. Whitney, R. P.; Vivian, J. E. *Ind. Eng. Chem.* 1941, *33*, 741-4; *Tech. Assoc. Papers* 1941, *24*, 435-6; *Paper Trade J.* 1941, *113*, 31-2. (Note that the last two papers are essentially identical.)

COMPONENTS:

1. Chlorine; Cl_2; [7782-50-5]

2. Water; H_2O; [7732-18-5]

ORIGINAL MEASUREMENTS:

Bunsen, R.

Phil. Mag.

1855, *9*, 181-201.

VARIABLES:

T/K = 283-313

PREPARED BY:

R. Battino

EXPERIMENTAL VALUES:

t/°C	T^a/K	$10^3 x_1$ [a,b]	$L^{a,c}$	α^d/cm³ (STP) cm⁻³ atm⁻¹
10	283	2.074	2.680	2.5852
11	284	2.039	2.644	2.5413
12	285	2.005	2.607	2.4977
13	286	1.970	2.571	2.5453
14	287	1.936	2.535	2.4111
15	288	1.901	2.498	2.3681
16	289	1.867	2.462	2.3253
17	290	1.834	2.425	2.2828
18	291	1.800	2.388	2.2405
19	292	1.767	2.351	2.1984
20	293	1.733	2.314	2.1565
21	294	1.700	2.277	2.1148
22	295	1.667	2.240	2.0734
23	296	1.635	2.203	2.0322
24	297	1.602	2.166	1.9912
25	298	1.570	2.129	1.9504
26	299	1.538	2.092	1.9099
27	300	1.506	2.054	1.8695
28	301	1.474	2.017	1.8295
29	302	1.442	1.979	1.7895
30	303	1.411	1.942	1.7499
31	304	1.379	1.905	1.7104
32	305	1.348	1.867	1.6712
33	306	1.317	1.829	1.6322

(cont.)

AUXILIARY INFORMATION

METHOD/APPARATUS/PROCEDURE:

These data are values reported in several places as having been determined by a chemical method by M. Schönfeld (1) of Dusseldorf.

SOURCE AND PURITY OF MATERIALS:

No details given.

ESTIMATED ERROR:

$\delta\alpha/\alpha = \pm 0.02$ (compiler's estimate).

REFERENCES:

1. Schönfeld, M.
Annalen
1855, *93*, 26.

COMPONENTS:	ORIGINAL MEASUREMENTS:
1. Chlorine; Cl_2; [7782-50-5]	Bunsen, R.
	Phil. Mag.
2. Water; H_2O; [7732-18-5]	1855, *9*, 181-201.

EXPERIMENTAL VALUES:

t/°C	T^a/K	$10^3 x_1$ [a,b]	$L^{a,c}$	α^d/cm³ (STP) cm⁻³ atm⁻¹
34	307	1.286	1.792	1.5934
35	308	1.256	1.754	1.5550
36	309	1.225	1.716	1.5166
37	310	1.195	1.679	1.4785
38	311	1.165	1.641	1.4406
39	312	1.135	1.603	1.4029
40	313	1.105	1.566	1.3655

[a] Calculated by compiler.

[b] Mole fraction solubility at 101.325 kPa partial pressure of gas.

[c] Ostwald coefficient.

[d] Bunsen coefficient.

COMPONENTS:	ORIGINAL MEASUREMENTS:
1. Chlorine; Cl_2; [7782-50-5] 2. Water; H_2O; [7732-18-5]	Roscoe, H. E. *J. Chem. Soc.* 1856, *8*, 14-26.
VARIABLES: T/K = 287	PREPARED BY: R. Battino

EXPERIMENTAL VALUES:

t/°C	T^a/K	$10^3\ x_1{}^{a,b}$	$L^{a,c}$	α^d/cm³ (STP) cm⁻³ atm⁻¹
14.0	287.2	1.920	2.514	2.3911

[a] Calculated by compiler.

[b] Mole fraction solubility at 101.325 kPa partial pressure of gas.

[c] Ostwald coefficient.

[d] Bunsen coefficient.

AUXILIARY INFORMATION

METHOD/APPARATUS/PROCEDURE:

The amount of chlorine dissolved appears to have been determined chemically using Schönfeld's method.

SOURCE AND PURITY OF MATERIALS:

1. From electrolysis of HCl solutions.

ESTIMATED ERROR:

$\delta\alpha/\alpha = \pm 0.02$ (compiler's estimate).

REFERENCES:

1. Schönfeld, M. *Annalen* 1855, *93*, 26.

COMPONENTS:	ORIGINAL MEASUREMENTS:
1. Chlorine; Cl_2; [7782-50-5] 2. Water; H_2O; [7732-18-5]	Bakhuis Roozeboom, H. W., *Recl. Trav. Chim. Pays-Bas* 1884, *3*, 59-72.
VARIABLES:	PREPARED BY:
T/K = 273-285	R. Battino

EXPERIMENTAL VALUES:

t/°C	T^{a}/K	$10^3\ x_1^{a,b}$	$L^{a,c}$	$\alpha^{a,d}$	β^{e}
0	273	3.72	4.64	4.64	4.61[f]
3	276	3.17	4.00	3.96	3.93[f]
6	279	2.77	3.53	3.45	3.42
9	282	2.46	3.17	3.07	3.03
12	285	2.26	2.93	2.81	2.77

[a] Calculated by compiler.

[b] Mole fraction solubility at 101.325 kPa partial pressure of gas.

[c] Ostwald coefficient.

[d] Bunsen coefficient. Units of cm^3(STP) cm^{-3} atm^{-1}.

[e] Solubility coefficient similar to Bunsen coefficient but at 1 atm total pressure. $\alpha = \beta \times 760/(760 - P_{H_2O})$.

[f] Points interpolated by author. The original values are presented as 1.07% Cl_2 (6 °C), 0.95% Cl_2 (9 °C), and 0.87% Cl_2 (12 °C) at a total pressure of 1 atm.

AUXILIARY INFORMATION

METHOD/APPARATUS/PROCEDURE:

The author was investigating the hydrate formed between chlorine and water and incidentally determined the solubility of chlorine in water at 6, 9 and 12 °C at a total pressure of 1 atm. The β values given above are those appearing in the *International Critical Tables* (1).

SOURCE AND PURITY OF MATERIALS:

ESTIMATED ERROR:

REFERENCES:

1. *International Critical Tables*, Vol. III, p.256.

COMPONENTS:	ORIGINAL MEASUREMENTS:
1. Chlorine; Cl_2; [7782-50-5] 2. Water; H_2O; [7732-18-5]	Mellor, W. *J. Chem. Soc.* 1901, *79*, 216-38.

VARIABLES:	PREPARED BY:
T/K = 294.2	R. Battino

EXPERIMENTAL VALUES:

t/°C	T^a/K	$10^3\ x_1^{a,b}$	$L^{a,c}$	α^d/cm^3 (STP) cm^{-3} atm^{-1}
21.0	294.2	1.701	2.278	2.1157

[a] Calculated by compiler.

[b] Mole fraction solubility at 101.325 kPa partial pressure of gas.

[c] Ostwald coefficient.

[d] Bunsen coefficient.

AUXILIARY INFORMATION

METHOD/APPARATUS/PROCEDURE:

Chlorine bubbled through solution. Amount determined by reaction with potassium iodide and titration of iodine with sodium thiosulfate soln.

SOURCE AND PURITY OF MATERIALS:

1. Prepared by reaction of hydrochloric acid with chromic acid.

ESTIMATED ERROR:

$\delta T/K = \pm 0.2$; $\delta\alpha/\alpha = \pm 0.02$ (compiler's estimate).

REFERENCES:

COMPONENTS:

1. Chlorine; Cl_2; [7782-50-5]

2. Water; H_2O; [7732-18-5]

ORIGINAL MEASUREMENTS:

Winkler, L. W.

Mathematikai es Termeszettudomanyi Ertesitö, Budapest

1907, *25*, 86. (*International Critical Tables*, Vol. III, p.256)

VARIABLES:

T/K = 283-373

PREPARED BY:

R. Battino

EXPERIMENTAL VALUES:

t/°C	T^a/K	$10^3\, x_1^{a,b}$	$L^{a,c}$	$\alpha^{a,d}$	β^e
10	283	2.556	3.304	3.187	3.148
11	284	2.476	3.211	3.087	3.047
12	285	2.399	3.112	2.991	2.950
13	286	2.326	3.037	2.899	2.856
14	287	2.256	2.955	2.811	2.767
15	288	2.188	2.876	2.726	2.680
16	289	2.123	2.799	2.644	2.597
17	290	2.061	2.726	2.566	2.517
18	291	2.009	2.655	2.491	2.440
19	292	1.944	2.588	2.420	2.368
20	293	1.891	2.525	2.353	2.299
21	294	1.844	2.470	2.294	2.238
22	295	1.800	2.418	2.238	2.180
23	296	1.757	2.368	2.184	2.123
24	297	1.716	2.320	2.133	2.070
25	298	1.677	2.275	2.084	2.019
26	299	1.641	2.232	2.038	1.970
27	300	1.605	2.190	1.993	1.923
28	301	1.573	2.153	1.953	1.880
29	302	1.543	2.118	1.915	1.839
30	303	1.514	2.084	1.878	1.799
35	308	1.374	1.920	1.702	1.608

(cont.)

AUXILIARY INFORMATION

METHOD/APPARATUS/PROCEDURE:

As the original article was not available, the above values were obtained from the compilation in the *International Critical Tables*.

SOURCE AND PURITY OF MATERIALS:

ESTIMATED ERROR:

REFERENCES:

COMPONENTS:	ORIGINAL MEASUREMENTS:
1. Chlorine; Cl_2; [7782-50-5] 2. Water; H_2O; [7732-18-5]	Winkler, L. W. *Mathematikai es Termeszettudomanyi Ertesitö, Budapest* 1907, *25*, 86. (*International Critical Tables*, Vol. III, p.256)

t/°C	T^a/L	$10^3\ x_1^{a,b}$	$L^{a,c}$	$\alpha^{a,d}$	β^e
40	313	1.265	1.793	1.564	1.450
45	318	1.182	1.698	1.458	1.320
50	323	1.125	1.639	1.385	1.216
60	333	1.042	1.556	1.276	1.025
70	343	1.022	1.564	1.245	0.862
80	353	1.059^f	1.657^f	1.282^f	0.683^f
90	363	1.052^f	1.682^f	1.265^f	0.39^f

[a] Calculated by compiler.

[b] Mole fraction solubility at 101.325 kPa partial pressure of gas.

[c] Ostwald coefficient.

[d] Bunsen coefficient in units of cm^3(STP) cm^{-3} atm^{-1}.

[e] Solubility coefficient similar to the Bunsen coefficient but at 1 atm total pressure. $\alpha = \beta \times 760/(760 - P_{H_2O})$.

[f] Doubtful points due to very high vapor pressure of water at these temperatures.

COMPONENTS:

1. Chlorine; Cl_2; [7782-50-5]

2. Water; H_2O; [7732-18-5]

ORIGINAL MEASUREMENTS:

Whitney, R.P.; Vivian, J.E.

Ind. Eng. Chem. 1941, *33*, 741-4

VARIABLES:

Temperature, pressure

PREPARED BY:

C.L. Young

EXPERIMENTAL VALUES:

T/K	Partial Pressure p_{Cl_2}/atm	p_{Cl_2}/kPa	Solubility + g/10^2g H_2O	Mole fraction of chlorine, x_{Cl_2}
283.2	0.0601	6.09	0.124	0.000315
	0.0660	6.69	0.131	0.000333
	0.155	15.7	0.228	0.000579
	0.157	15.9	0.237	0.000602
	0.292	29.6	0.368	0.000934
	0.293	29.7	0.366	0.000929
	0.498	50.5	0.541	0.001373
	0.498	50.5	0.555	0.001408
288.2	0.0581	5.89	0.111	0.000282
	0.0611	6.19	0.115	0.000292
	0.140	14.2	0.195	0.000495
	0.145	14.7	0.201	0.000510
	0.291	29.5	0.330	0.000838
	0.297	30.1	0.330	0.000838
	0.488	49.4	0.477	0.00121
	0.490	49.6	0.480	0.00128
	0.951	96.4	0.825	0.00209
	0.951	96.4	0.825	0.00209
293.2	0.0665	6.74	0.112	0.000284
	0.1195	12.11	0.160	0.000406
	0.260	26.3	0.262	0.000665
	0.260	26.3	0.265	0.000673
	0.487	49.3	0.415	0.00105

AUXILIARY INFORMATION

METHOD/APPARATUS/PROCEDURE:

Single pass flow apparatus. Chlorine and nitrogen at a total pressure of atmospheric pressure was bubbled through water in an "equilibrium" cell. After one hour samples of saturated liquid analysed by adding to excess potassium iodide and titrating the liberated iodine with sodium thiosulfate. Partial pressure of chlorine varied by adjusting proportion in nitrogen chlorine mixture. Also see references (1) and (2) (which are identical) for additional calculations on this system.

SOURCE AND PURITY OF MATERIALS:

1. Purity 98.5 mole per cent.

2. Distilled.

ESTIMATED ERROR:

$\delta T/K = \pm 0.05$; $\delta x_{Cl_2}/x_{Cl_2} = \pm 0.02$

(estimated by compiler)

REFERENCES:

1. Whitney, R. P.; Vivian, J. E. *Tech. Assoc. Papers* 1941, *24*, 435-6.

2. Whitney, R. P.; Vivian, J. E. *Paper Trade J.* 1941, *113*, 31-2.

COMPONENTS:	ORIGINAL MEASUREMENTS:
1. Chlorine; Cl_2; [7782-50-5]	Whitney, R.P.; Vivian, J.E.
2. Water; H_2O; [7732-18-5]	*Ind. Eng. Chem.* 1941, *33*, 741-4.

EXPERIMENTAL VALUES:

T/K	Partial pressure p_{Cl_2}/atm	p_{Cl_2}/kPa	Solubility[+], g/10^5g H_2O	Mole fraction of chlorine, x_{Cl_2}
293.2	0.490	49.6	0.422	0.00107
	0.965	97.8	0.725	0.00184
298.2	0.0621	6.29	0.104	0.000264
	0.0655	6.64	0.106	0.000269
	0.104	10.5	0.139	0.000353
	0.109	11.0	0.143	0.000363
	0.174	17.6	0.189	0.000480
	0.270	27.4	0.246	0.000625
	0.551	55.8	0.391	0.000992
	0.589	59.7	0.418	0.00106
	0.961	97.4	0.628	0.00159

T/K	$10^4\ x_1$ [a]
283.2	26.25
288.2	22.11
293.2	19.20
298.2	16.19

[a] Extrapolated values at 101.325 kPa partial pressure of gas calculated by compiler from a least squares smoothing of the data at each temperature.

COMPONENTS:	EVALUATOR:
1. Chlorine; Cl_2; [7782-50-5] 2. Water; H_2O; [7732-18-5] 3. Hydrochloric acid; HCl; [7647-01-0]	Colin L. Young, School of Chemistry, University of Melbourne, Parkville, Victoria 3052, Australia. March 1982

EVALUATION:

Much of the solubility data for chlorine in aqueous hydrochloric acid solutions has been determined in order to study the various chemical equilibria established in solution. Jakowkin (Yakovkin) (1) found that the solubility of chlorine in water was decreased by small additions of hydrogen chloride, but is increased by larger additions. Other early investigations were those of Roscoe (2), Berthelot (3) and Goodwin (4). None of these workers data are considered accurate enough for inclusion in the compiled tables. The early data of Mellor (5) are compiled but are classified as doubtful. The data of Hine and Inuta (6) are classified as tentative but are considered inferior to the data of Sherrill and Izard (7), (8) but because of the different temperatures employed, the two sets of data are not directly comparable.

References

1. Jakowkin, A. A.
Z. physik. Chem., 1899, *29*, 613.

2. Roscoe, H. E.
J. Chem. Soc., 1956, *8*, 14.

3. Berthelot, M.
Ann. Chim. Phys., 1881, *22*, 462 and *Compt. rend.*, 1880, *91*, 194.

4. Goodwin, W. L.
Trans. Roy. Soc. Edin., 1882, *30*, 597 or *Ber.*, 1882, *15*, 3039.

5. Mellor, W.
J. Chem. Soc., 1901, *79*, 216.

6. Hine, F.; Inuta, S.
Bull. Chem. Soc. Japan, 1968, *41*, 71.

7. Sherrill, M. S.; Izard, E. F.
J. Am. Chem. Soc., 1928, *50*, 1665.

8. Sherrill, M. S.; Izard, E. F.
J. Am. Chem. Soc., 1931, *53*, 1667.

COMPONENTS:	ORIGINAL MEASUREMENTS:
1. Chlorine; Cl_2; [7782-50-5]	Mellor, W.
2. Water; H_2O; [7732-18-5]	*J. Chem. Soc.*
3. Hydrochloric acid; HCl; [7647-01-0]	1901, *79*, 216.

VARIABLES:	PREPARED BY:
	C. L. Young

EXPERIMENTAL VALUES:

t/°C	T/K	Conc. of HCl /g dm^{-3}	*p*/mmHg	*p*/kPa	Conc. of Cl_2 † /g dm^{-3}	Bunsen Coefficient α	Ostwald Coefficient
21.0	294.2	313.401	761	101.5	12.03	3.5492	3.8224
21.0	294.2	282.060	761	101.5	11.87	3.3278	3.5859
21.0	294.2	250.720	759	101.2	10.68	3.1272	3.3677
21.0	294.2	219.380	759	101.2	9.93	2.9243	3.1312
20.2	293.4	188.040	761	101.5	9.23	2.7020	2.9117
20.2	293.4	156.700	761	101.5	8.58	2.5095	2.7043
20.5	293.7	125.360	761	101.5	7.76	2.2711	2.4473
20.5	293.7	94.020	760	101.3	7.19	2.1044	2.2677
20.0	293.2	62.680	760	101.3	6.38	1.8682	2.0131
21.0	294.2	31.340	762	101.6	5.81	1.6736	1.8033
21.0	294.2	15.670	759	101.2	5.10	1.4933	1.6092
21.0	294.2	12.540	759	101.2	4.85	1.4200	1.5292
21.0	294.2	9.402	760	101.3	4.76	1.3942	1.5013
20.0	293.2	6.248	760	101.3	4.94	1.4483	1.5607
21.0	294.2	3.134	760	101.3	5.30	1.5496	1.6698
21.0	294.2	0.00	760	101.3	7.23	2.1157	2.2799

† volume measured after saturation with chlorine.

AUXILIARY INFORMATION

METHOD/APPARATUS/PROCEDURE:

Chlorine bubbled through solution. Amount determined by reaction with potassium iodide and titration of iodine with sodium thiosulfate solution.

SOURCE AND PURITY OF MATERIALS:

No details given.

ESTIMATED ERROR:

$\delta T/K = \pm 0.2$; $\delta\alpha/\alpha = \pm 0.02$ (compiler's estimate).

REFERENCES:

COMPONENTS:	ORIGINAL MEASUREMENTS:
1. Chlorine; Cl_2; [7782-50-5] 2. Water; H_2O; [7732-18-5] 3. Hydrochloric acid; HCl; [7647-01-0]	Sherrill, M. S.; Izard, E. F. *J. Am. Chem. Soc.* <u>1928</u>, *50*, 1665-1675 and *J. Am. Chem. Soc.* <u>1931</u>, *53*, 1667-1674.
VARIABLES:	**PREPARED BY:**
Concentration of acid	C. L. Young

EXPERIMENTAL VALUES:

T/K	Concentration of acid /mol kg^{-1} (H_2O)	Concentration of chlorine /mol kg^{-1} (H_2O)	Mole fraction of chlorine,[§] x_{Cl_2}
298.15	0.000	0.0923	0.001660
	0.010	0.08264	0.001486
	0.100	0.06394	0.001146
	0.200	0.0619	0.00111
	0.496	0.0630	0.00112
	1.019	0.0665	0.00118
	1.991	0.0737	0.00128
	2.990	0.0824	0.00141
	3.989	0.0914	0.00153
	* 4.903	0.0978	0.00162
	5.180	0.1004	0.00165
	* 7.016	0.1150	0.00184
	* 8.163	0.1219	0.00191
	* 8.854	0.1281	0.00199
	* 9.805	0.1326	0.00203

§ at a partial pressure of chlorine of 101.3 kPa.

* taken from earlier reference at a partial pressure of chlorine of 101.3 kPa.

AUXILIARY INFORMATION

METHOD/APPARATUS/PROCEDURE:

Chlorine bubbled through solution. Absorption train painted black to prevent decomposition of chlorine by light. Before the liquid sample was analysed the saturated solution was allowed to equilibrate for at least 3 hours without chlorine being bubbled through. Amount of chlorine determined by reaction with potassium iodide and titrating liberated iodine with sodium thiosulfate.

SOURCE AND PURITY OF MATERIALS:

1. Chlorine prepared by action of hydrochloric acid on chromic acid.

ESTIMATED ERROR:

$\delta T/K = \pm 0.01$; $\delta x_{Cl_2} = \pm 2\%$ (estimated by compiler).

REFERENCES:

COMPONENTS:

1. Chlorine; Cl_2; [7782-50-5]
2. Water; H_2O; [7732-18-5]
3. Hydrochloric acid; HCl; [7647-01-0]

ORIGINAL MEASUREMENTS:

Hine, F.; Inuta, S.

Bull. Chem. Soc. Japan

1968, *41*, 71-75.

VARIABLES:

Temperature, pressure, concentration of component 3

PREPARED BY:

C. L. Young

EXPERIMENTAL VALUES:

These workers presented data in graphical form for the solubility of chlorine at partial pressures up to 10^5 Pa in hydrochloric acid of concentrations in the range 2 to 10 moles per litre at temperatures between 30 to 90 °C. They proposed the following smoothing equation for the solubility of chlorine

$$S = (\alpha C + \beta)p \qquad \ldots (1)$$

where p is the pressure in units of atmosphere. C is the concentration of hydrogen chloride in molality or g dm^{-3}. α and β were functions of temperature and when C was expressed in g dm^{-3}

$$\log \alpha = -1.21 \times 10^{-2}\, t - 1.603$$

$$\text{and} \quad \beta = 2.14 \times 10^{2}\, t^{-1.21}$$

$$\text{where} \quad t = \text{Temperature in °C.}$$

The constants for equation (1) are given below.

(cont.)

AUXILIARY INFORMATION

METHOD/APPARATUS/PROCEDURE:

Flow system similar to that used by Whitney and Vivian (1). Chlorine and nitrogen mixture of known composition and pressure bubbled through acid of known composition. After one hour, samples of saturated liquid analysed by adding to excess potassium iodide and titrating the liberated iodine with sodium thiosulfate. Partial pressure of chlorine varied by adjusting proportion in nitrogen chlorine mixture.

SOURCE AND PURITY OF MATERIALS:

No details given.

ESTIMATED ERROR:

REFERENCES:

1. Whitney, R. P.; Vivian, J. E. *Ind. Eng. Chem.* 1941, *33*, 741.

COMPONENTS:

1. Chlorine; Cl_2; [7782-50-5]

2. Water; H_2O; [7732-18-5]

3. Hydrochloric acid; HCl; [7647-01-0]

ORIGINAL MEASUREMENTS:

Hine, F.; Inuta, S.

Bull. Chem. Soc. Japan

1968, *41*, 71-75.

EXPERIMENTAL VALUES:

t/°C	(K/T) × 10^3	α		β	
		α(†) × 10^3	α(*) × 10^2	β(§) × 10^2	β(#)
30	3.300	6.57	1.08	4.85	3.42
40	3.195	5.11	0.844	3.49	2.46
50	3.096	3.65	0.592	2.72	1.92
70	2.915	2.20	0.350	1.76	1.22
90	2.755	1.38	0.204	1.32	0.916

† α/(mole Cl_2 per 1000 gm solvent)/(mole HCl per 1000 gm solvent)(atm))

* α/(g Cl_2 per dm^3 solvent)/(g HCl per dm^3 solvent)(atm))

§ β/(mole Cl_2 per 1000 gm solvent/atm)

\# β/(g Cl_2 per dm^3 solvent/atm)

COMPONENTS:	EVALUATOR:
1. Chlorine; Cl_2; [7782-50-5] 2. Electrolyte solutions	Colin L. Young, School of Chemistry, University of Melbourne, Parkville, Victoria 3052, Australia. June 1981.

CRITICAL EVALUATION:

Note on the data of Karapysh and coworkers

Karapysh and coworkers have made fairly extensive studies of the solubility of chlorine in aqueous solutions but the data are, in general, presented in a graphical form or the work was presented in journals which are not readily available. In many cases the information was presented in an ambiguous manner and the meaning of the tables and graphs was not clear.

The solubility of chlorine in perchloric acid has been investigated by Semchenko *et al.* (1), Loseva *et al.* (2), Levchenko and Karapysh (3) and Semchenko and Karapysh (4). In reference (1) a graph of solubility (g dm^{-3}) versus chlorine pressure was given.

In a later paper Smirnov *et al.* (5) presented the solubility of chlorine in hydrochloric acid. The data were presented only in graphical form.

Smirnov and Aliev (6) presented results for the solubility of chlorine in aqueous solutions of lithium chloride, sodium chloride and of potassium chloride. The results were presented in graphical form. A table was given in the paper but the values do not agree, even approximately, with the data presented in the graphs.

Aliev *et al.* (7) studied the solubility of chlorine in water and in aqueous sodium chloride solution. Blaznina *et al.* (8) studied the solubility of chlorine in aqueous barium chloride at concentrations in the range 0.1 to 3.0 mol dm^{-3} and temperatures in the range 10 to 50 °C. Karapysh and Semchenko (9) studied the solubility of chlorine in aqueous solutions of lithium chloride, strontium chloride and barium chloride. Karapysh and Semchenko (10) also studied the solubility of chlorine in aqueous solutions of lithium and sodium perchlorates. The papers (7) to (10) could not be obtained by us despite considerable effort.

References

1. Semchenko, D. P.; Smirnov, V. A.; Aliev, Z. M.; Karapysh, V. V. *Zh. Fiz. Khim.* 1974, *48*, 1002; *Russ. J. Phys. Chem.* 1974, *48*, 580-581.

2. Loseva, G. K.; Semchenko, D. P.; Il'in, K. G. *Tr. Novocherk. Politekh. Inst.* 1969, *197*, 37.

3. Levchenko, V. M.; Karapysh, V. V. *Gidrokhim. Mater.* 1969, *51*, 153.

4. Semchenko, D. P.; Karapysh, V. V. *Tr. Novocherk. Politekh. Inst.* 1956, *48*, 19.

5. Smirnov, V. A.; Aliev, Z. M.; Karapysh, V. V.; Gurchin, I. I. *Zh. Fiz. Khim.* 1974, *48*, 1241; *Russ. J. Phys. Chem.* 1974, *48*, 720.

6. Smirnov, V. A.; Aliev, Z. M. *Zh. Fiz. Khim.* 1976, *50*, 1132; *Russ. J. Phys. Chem.* 1976, *50*, 684.

7. Aliev, Z. M.; Karapysh, V. V.; Smirnov, V. A. *Tr. Novocherk. Politekh. Inst.* 1972, *266*, 155.

8. Blaznina, V. A.; Fedorina, V. F.; Shul'gina, G. A.; Karapysh, V. V. *Nauch. Raboty Stud. Khim.-Tekhnol.Fak., Novocherkasshii Politekh. Inst. im. S. Ordzhonikidze* 1959, *7*, 17.

9. Karapysh, V. V.; Semchenko, D. P. *Tr. Novocherk. Politekh. Inst.* 1960, *98*, 173.

10. Karapysh, V. V.; Semchenko, D. P. *Tr. Novocherk. Politekh. Inst.* 1959, *65*, 111.

COMPONENTS:

1. Chlorine; Cl_2; [7782-50-5]
2. Water; H_2O; [7732-18-5]
3. Sulfuric acid; H_2SO_4; [7664-93-9]

ORIGINAL MEASUREMENTS:

Sherrill, M. S.; Izard, E. F.

J. Am. Chem. Soc.

1931, *53*, 1667-1674.

VARIABLES:

Concentration of acid

PREPARED BY:

C. L. Young

EXPERIMENTAL VALUES:

T/K	Conc. of salt /mol kg^{-1} (H_2O)	Conc. of chlorine /mol kg^{-1} (H_2O)	Mole fraction of chlorine x_{Cl_2}
298.15	0.0000	0.0923	0.001660
	0.4995	0.06272	0.001119
	1.0240	0.05617	0.000993
	1.9686	0.04973	0.000865
	3.0090	0.04550	0.000777
	3.9934	0.04278	0.000718
	4.9901	0.04066	0.000672

AUXILIARY INFORMATION

METHOD/APPARATUS/PROCEDURE:

Chlorine bubbled through solution. Absorption train painted black to prevent decomposition of chlorine by light. Before the liquid sample was analysed the saturated solution was allowed to equilibrate for at least 3 hours without chlorine being bubbled through. Amount of chlorine determined by reaction with potassium iodide and titrating liberated iodine with sodium thiosulfate.

SOURCE AND PURITY OF MATERIALS:

1. Chlorine prepared by action of hydrochloric acid on chromic acid.

ESTIMATED ERROR:

$\delta T/K = \pm 0.01$; $\delta x_{Cl_2} = \pm 2\%$

(estimated by compiler).

REFERENCES:

COMPONENTS:

1. Chlorine; Cl_2; [7782-50-5]
2. Water; H_2O; [7732-18-5]
3. Barium chloride; $BaCl_2$; [10361-37-2]

ORIGINAL MEASUREMENTS:

Sherrill, M. S.; Izard, E. F.

J. Am. Chem. Soc.

1931, *53*, 1667-1674.

VARIABLES:

Concentration of salt

PREPARED BY:

C. L. Young

EXPERIMENTAL VALUES:

T/K	Conc. of salt /mol kg^{-1} (H_2O)	Conc. of chlorine /mol kg^{-1} (H_2O)	Mole fraction of chlorine x_{Cl_2}
298.15	0.000	0.0923	0.001660
	0.330	0.06216	0.001112
	0.667	0.05494	0.000977
	1.000	0.04924	0.000871
	1.333	0.04468	0.000785

AUXILIARY INFORMATION

METHOD/APPARATUS/PROCEDURE:

Chlorine bubbled through solution. Absorption train painted black to prevent decomposition of chlorine by light. Before the liquid sample was analysed the saturated solution was allowed to equilibrate for at least 3 hours without chlorine being bubbled through. Amount of chlorine determined by reaction with potassium iodide and titrating liberated iodine with sodium thiosulfate.

SOURCE AND PURITY OF MATERIALS:

1. Chlorine prepared by action of hydrochloric acid on chromic acid.

ESTIMATED ERROR:

$\delta T/K = \pm 0.01$; $\delta x_{Cl_2} = \pm 2\%$

(estimated by compiler).

REFERENCES:

COMPONENTS:

1. Chlorine; Cl_2; [7782-50-5]
2. Water; H_2O; [7732-18-5]
3. Sodium chloride; NaCl; [7647-14-5]

ORIGINAL MEASUREMENTS:

Sherrill, M. S.; Izard, E. F.

J. Am. Chem. Soc.

1931, *53*, 1667-1674.

VARIABLES:

Concentration of salt

PREPARED BY:

C. L. Young

EXPERIMENTAL VALUES:

T/K	Conc. of salt /mol kg^{-1} (H_2O)	Conc. of chlorine /mol kg^{-1} (H_2O)	Mole fraction of chlorine x_{Cl_2}
298.15	0.000	0.0923	0.001660
	0.501	0.06575	0.001183
	0.998	0.05795	0.001025
	2.991	0.04161	0.000711
	3.989	0.03601	0.000605
	4.989	0.03087	0.000510

AUXILIARY INFORMATION

METHOD/APPARATUS/PROCEDURE:

Chlorine bubbled through solution. Absorption train painted black to prevent decomposition of chlorine by light. Before the liquid sample was analysed the saturated solution was allowed to equilibrate for at least 3 hours without chlorine being bubbled through. Amount of chlorine determined by reaction with potassium iodide and titrating liberated iodine with sodium thiosulfate.

SOURCE AND PURITY OF MATERIALS:

1. Chlorine prepared by action of hydrochloric acid on chromic acid.

ESTIMATED ERROR:

$\delta T/K = \pm 0.01$; $\delta x_{Cl_2} = \pm 2\%$ (estimated by compiler).

REFERENCES:

COMPONENTS:

1. Chlorine; Cl_2; [7782-50-5]
2. Water; H_2O; [7732-18-5]
3. Potassium chloride; KCl; [7447-40-7]

ORIGINAL MEASUREMENTS:

Sherrill, M. S.; Izard, E. F.

J. Am. Chem. Soc.

1931, *53*, 1667-1674.

VARIABLES:

Concentration of salt

PREPARED BY:

C. L. Young

EXPERIMENTAL VALUES:

T/K	Conc. of salt /mol kg^{-1} (H_2O)	Conc. of chlorine /mol kg^{-1} (H_2O)	Mole fraction of chlorine x_{Cl_2}
298.15	0.000	0.0923	0.001660
	0.500	0.06610	0.001179
	1.000	0.06109	0.001080
	2.013	0.05388	0.000936
	3.014	0.04865	0.000831
	3.913	0.04724	0.000794

AUXILIARY INFORMATION

METHOD/APPARATUS/PROCEDURE:

Chlorine bubbled through solution. Absorption train painted black to prevent decomposition of chlorine by light. Before the liquid sample was analysed the saturated solution was allowed to equilibrate for at least 3 hours without chlorine being bubbled through. Amount of chlorine determined by reaction with potassium iodide and titrating liberated iodine with sodium thiosulfate.

SOURCE AND PURITY OF MATERIALS:

1. Chlorine prepared by action of hydrochloric acid on chromic acid.

ESTIMATED ERROR:

$\delta T/K = \pm 0.01$; $\delta x_{Cl_2} = \pm 2\%$ (estimated by compiler).

REFERENCES:

COMPONENTS:	EVALUATOR:
1. Chlorine; Cl_2; [7782-50-5] 2. Organic liquids	P. G. T. Fogg, Department of Chemistry, Polytechnic of North London, Holloway, London, U.K. September 1980

CRITICAL EVALUATION:

It is not possible, in general, to judge between the merits of different apparatus from the information which has been given by authors. In the descriptions of technique it has not always been clearly stated or implied whether or not allowance has been made for the partial pressure of the solvent in the calculation of the partial pressure of chlorine. In the case of solvents which have low volatility at the temperature of measurement such correction is of little importance. However, at a temperature of about 50 K below the normal boiling point of a typical solvent the partial vapor pressure of the solvent is about 0.08 atm and appropriate correction becomes increasingly important for measurements at higher temperatures. Uncertainty as to whether appropriate correction has been made by authors leads to corresponding uncertainty in the significance of the results which have been reported.

Where solubilities have been reported for a range of temperatures and the partial pressure of chlorine equal to 1 atm or a sum of partial pressures of chlorine and solvent equal to 1 atm, the evaluator has carried out a least squares analysis to fit values of $RT\ln x_{Cl_2}$ to equations of the form:

$$-RT\ln x_{Cl_2} = a + bT$$

where x_{Cl_2} is the mole fraction solubility of chlorine at a temperature T. In several cases solubilities were measured at an unstated barometric pressure. Any differences between a standard atmosphere and barometric pressure have been neglected for the purpose of general comparison of results.

If $\Delta\bar{G}°$, $\Delta\bar{H}°$ and $\Delta\bar{S}°$ are defined as the change in Gibbs free energy, enthalpy and entropy for the hypothetical process:

$$Cl_2\ (1\ \text{atm}) \rightarrow Cl_2\ (\text{solution},\ x_{Cl_2} = 1)$$

it follows that:

$$\Delta\bar{G}° \simeq -RT\ln x_{Cl_2}$$

$$\Delta\bar{H}° \simeq a$$

$$\Delta\bar{S}° \simeq -b$$

provided that the solubility is measured at a partial pressure of chlorine equal to 1 atm (1). Values of a and b are given in table I.

In the cases where it appears to the evaluator that no allowance has been made by the authors for the vapor pressure of a volatile solvent, "corrected" solubility values have been determined. These "corrected" solubilities have been calculated using published vapor pressure data (2). It has been assumed that dissolution of chlorine lowers the partial vapor pressure of the solvent in accordance with the relationship:

$$p_{\text{solvent}} = p°_{\text{solvent}}(1 - x_{Cl_2})$$

and that x_{Cl_2} (p_{Cl_2} = 1 atm) may be calculated from x_{Cl_2} ($p_{Cl_2} + p_{\text{Solvent}}$ = 1 atm) by assuming that x_{Cl_2} is directly proportional to p_{Cl_2} over a small pressure range. This procedure, used by Hildebrand, is subject to limitations (3), (4). The "corrected" solubilities have been subjected to least squares analysis and fitted to equations of the form:

$$-RT\ln x_{Cl_2(\text{corrected})} = a' + b'T.$$

COMPONENTS:	EVALUATOR:
1. Chlorine; Cl_2; [7782-50-5] 2. Organic liquids	P. G. T. Fogg, Department of Chemistry, Polytechnic of North London, Holloway, London, U.K. September 1980

CRITICAL EVALUATION:

Values of a' and of b' are also given in table I. In view of the uncertainties involved they are not necessarily better constants than values of a and b.

Solubility measurements made at a single temperature have been collected together in table II and values of x_{Cl_2}(corrected) are given where appropriate.

The average value of a is -18935 J mol^{-1}. The average is -17284 J mol^{-1} if the "corrected" values, a', are taken into account. The enthalpy of vaporization of liquid chlorine, calculated from the variation of vapor pressure with temperature (2), is 19912 J mol^{-1} at 298.16 K. The average value of b is 80.10 J K^{-1} mol^{-1} or 74.43 J K^{-1} mol^{-1} if "corrected" values of b' are taken into account. These values can be compared with the value of $\Delta S°$ for the process:

$$Cl_2(\text{liquid}) \rightarrow Cl_2(\text{vapor; 1 atm}) \text{ at } 298.16 \text{ K.}$$

The value of this entropy change, calculated by the evaluator from the variation of vapor pressure of chlorine with temperature (2), is 83.68 J K^{-1} mol^{-1}. It follows that, on average, for the solvents under consideration there is no great difference between the values of $\Delta\bar{G}°$, $\Delta\bar{H}°$, and $\Delta\bar{S}°$ for the hypothetical process:

$$Cl_2(\text{gas, 1 atm}) \rightarrow Cl_2(\text{solution, } x_{Cl_2} = 1)$$

and the corresponding values for the process:

$$Cl_2(\text{gas, 1 atm}) \rightarrow Cl_2(\text{liquid}).$$

Experimental measurements of solubilities of chlorine which have led to values of a or a' and b or b' which differ appreciably from mean values are not necessarily unreliable but ought to be used with caution. Measurements which have given large standard deviations from the smoothing equations should also be treated with similar caution.

Experimental values of x_{Cl_2}(1 atm) as a function of T for various systems are shown in figs. 1 and 2. On these graphs are also plotted reference lines corresponding to the equation:

$$x_{Cl_2} = p_{Cl_2}/p°_{Cl_2}$$

where $p°_{Cl_2}$ is the vapor pressure of liquid chlorine at temperature T. As explained by Gerrard (4), comparison of experimental results with reference lines can give an estimation of the reliability of the measurements.

Experimental curves showing the variation of the mole fraction solubility of a gas as a function of temperature at constant pressure are often similar in shape to the corresponding reference lines. Solvents of similar structure usually show similar behavior relative to the reference line for a particular gas.

The general considerations outlined above lead to the following detailed observations on the results which have been reported:

Trichloromethanesulfenyl chloride

Values of a and b are large compared with average values. There is nothing unusual in the graph of x_{Cl_2} against T and the measurements (5) are classified as tentative.

COMPONENTS:	EVALUATOR:
1. Chlorine; Cl_2; [7782-50-5] 2. Organic liquids	P. G. T. Fogg, Department of Chemistry, Polytechnic of North London, Holloway, London, U.K. September 1980

CRITICAL EVALUATION:

Carbon disulfide

Values of a and b are large compared with the average. "Corrected" values, a' and b' are a little closer in line with the average. "Uncorrected" values of x_{Cl_2} become increasingly lower than corresponding points on the reference line with increasing temperature. In contrast, "corrected" values become increasingly higher than the reference line with increasing temperature. The data of Raskina *et al.* (5) are classified as tentative.

Tetrachloromethane

The solubilities of chlorine in this liquid are evaluated on page 395.

Mixtures of tetrachloromethane, trichloromethanesulfenyl chloride and carbon disulfide

The solubilities of chlorine in six mixtures of the three solvents were measured by Raskina *et al.* (5). In the case of two of the mixtures there is a discrepancy between solubilities given by the authors as mole fractions and those given as weight fractions. It is not clear which figures are erroneous.

Each mole fraction solubility, x_{Cl_2}, reported by the authors is close to the sum of the solubilities, x'_{Cl_2}, in the separate portions of the pure solvents which were used to make up the mixture. (Coef. corr. 0.8731; $x_{Cl_2}/x'_{Cl_2} = 1.01 \pm 0.03$.) The agreement between weight fraction solubilities, w_{Cl_2}, given by the authors and the sum weight fraction solubilities, w'_{Cl_2}, in separate portions of the pure liquid is not as good. (Coef. corr. 0.3772; $w_{Cl_2}/w'_{Cl_2} = 0.97 \pm 0.07$.)

Tetrachloroethene

The standard deviation of "uncorrected" values of $RT\ln x_{Cl_2}$ is about 8%. Corrected values have a standard deviation of about 3%. Uncorrected values of a and b are above average whereas corrected values are about average. A plot of uncorrected values of x_{Cl_2} crosses the reference line, i.e., exhibits both positive and negative deviations from Raoult's law. Corrected values of x_{Cl_2} fall on a curve which approaches but does not cross the reference line as the temperature is increased. The "corrected" values are to be preferred (6).

1,1,2,2-Tetrachloroethane

Standard deviations of both "uncorrected" and "corrected" values of $RT\ln x_{Cl_2}$ are about 6%. Corrected and uncorrected values of a and b are clearly considerably smaller than in any other similar system. The measurements (7) must therefore be considered doubtful.

Acetic acid

Values of a and b based on measurements by Kumar and Gehlawat (8) are appreciably greater than mean values. A plot of "uncorrected" values of x_{Cl_2} (1 atm) against T is close to a straight line which would indicate a very unusual temperature dependence of solubility. "Corrected" values of solubility do not show such an unusual temperature dependence and are deemed more reliable. Jones (9) measured solubilities at 289.2 K and low pressures of chlorine. Extrapolation of these results to 1 atm gives a value close to that obtained by extrapolating the results of Kumar and Gehlawat (8) to 289.2 K by use of the "corrected" smoothing equation.

COMPONENTS:	EVALUATOR:
1. Chlorine; Cl_2; [7782-50-5] 2. Organic liquids	P. G. T. Fogg, Department of Chemistry, Polytechnic of North London, Holloway, London, U.K. September 1980

CRITICAL EVALUATION:

However, in view of uncertainty of the "correction" to Kumar and Gehlawat (8) values and the extrapolation from low pressure needed to compare Jones (9) values, both sets of data are classified as doubtful.

1,1,2,3,4,4-Hexachloro-1,3-butadiene

The measurements of Kogan *et al.* (10) indicate that x_{Cl_2} is proportional to p_{Cl_2} in the range 408 mmHg to 974 mmHg at 303 K with p_{Cl_2}/x_{Cl_2} equal to 4827 ± 17 mmHg. Values of x_{Cl_2}(1 atm) for temperatures in the range 253 K to 365 K are close to corresponding values for tetrachloromethane and are classified as tentative.

Chlorobenzene

Values of a and b based on the measurements by Cervinka (11) at three temperatures are well below average and differ appreciably from values based upon measurements by Krentsel *et al.* (12) and by Zetkin *et al.* (13). However, with the exception of a measurement at 298 K by Cervinka (11) which appears to be erroneous and which distorts the pattern of his results, solubilities reported by the different authors fall into a consistent pattern. "Corrected" smoothing equation constants, a' and b', from Zetkin's results agree more closely than "uncorrected" values with the corresponding constants based upon the work of Krentsel *et al.* (12). However, because of the low volatility of the solvent at the temperature of measurement, such correction for the vapor pressure of the solvent is of little significance. The following equation, based upon "corrected" measurements by Zetkin *et al.* (13) and measurements by Krentsel *et al.* (12), is recommended:

$$\Delta G°/\text{J mol}^{-1} = (-RT\ln x_{Cl_2})/\text{J mol}^{-1} = -19804 + 81.654T/K$$

$$\text{Std. dev. } \Delta G°/\text{J mol}^{-1} = 154; \quad \text{Coef. corr.} = 0.9951.$$

Benzene

There is appreciable disagreement between results over a range of temperatures which have been published by different authors. Corresponding values of a and b are inconsistent. However "corrected" solubilities based upon measurements by Zetkin *et al.* (13) are close to the values based upon the results of Cervinka (11). To estimate solubilities at 1 atm from the results given by Krivonos (14), it is assumed that Bunsen coefficients are independent of the pressure at which they are measured. This procedure may introduce considerable error and explain why these solubilities are appreciably greater than those based upon the work of Cervinka (11) and Zetkin *et al.* (13). Nevertheless, the measurement at 1 atm and 293 K by Pozin and Opykhtina (15) is close to the value based upon Krivonos' (14) work. The following equation is based upon a combination of "corrected" measurements by Zetkin *et al.* (13) and measurements by Cervinka (14):

$$\Delta G°/\text{J mol}^{-1} = (-RT\ln x_{Cl_2})/\text{J mol}^{-1} = -15043 + 66.633T/K$$

$$\text{Std. dev. } \Delta G°/\text{J mol}^{-1} = 192; \quad \text{Coef. corr.} = 0.9872.$$

Values of x_{Cl_2} measured by Pozin and Opykhtina (15) at 293 K for the range 49 mmHg to 763 mmHg give a standard deviation of 9.6% when expressed as a linear function of p_{Cl_2}.

COMPONENTS:	EVALUATOR:
1. Chlorine; Cl_2; [7782-50-5] 2. Organic liquids	P. G. T. Fogg, Department of Chemistry, Polytechnic of North London, Holloway, London, U.K. September 1980

CRITICAL EVALUATION:

Mixtures of trifluoromethylbenzene and its ring substituted chlorinated derivatives

Solubilities of chlorine in pure trifluoromethylbenzene and in mixtures with chlorinated derivatives were measured by Ushakov *et al.* (16). Solubilities increased with the proportion of chlorinated derivatives present. In each case plots of x_{Cl_2} against T fall on a smooth curve which tends to approach the reference line with increase in temperature. Data for the smoothing equation:

$$-RT\ln x_{Cl_2} = a + bT$$

are as follows:

Composition before passage of chlorine (wt-%)			a	b	Coef. corr.	(Std. dev.) /J mol^{-1} of $RT\ln x_{Cl_2}$
$C_7H_5F_3$	$C_7H_4ClF_3$	$C_7H_3Cl_2F_3$	/J mol^{-1}	/JK^{-1}mol^{-1}		
100	-	-	-22918	94.563	0.9972	184
84.84	15.16	-	-20233	84.933	0.9982	105
58.29	39.47	2.24	-20851	86.514	0.9994	65
22.82	60.86	16.32	-16565	71.346	0.9989	70

The proportions of the different isomers having the same empirical formulae were not determined.

Cyclohexane

Solubilities at three temperatures, 283 K, 293 K and 313 K, over a range of pressures were measured by Tsiklis and Svetlova (17). In each case a plot of x_{Cl_2} *vs.* p_{Cl_2} (T const.) is a curve which, when extrapolated, cuts the reference line for constant temperature at values of x_{Cl_2} ranging from about 0.1 to about 0.16. This behavior is unusual and the data ought to be used with caution. The data are classified as doubtful.

Methylbenzene

Values of a and b are high compared with the average. "Corrected" values, a' and b', while still above the average values appear to be reasonable. A plot of "uncorrected" values of x_{Cl_2} against T cuts the reference line. "Corrected" values fall on a curve which approaches but does not cut the reference line. These values seem to be more acceptable than the "uncorrected" values (18).

Heptane

Solubilities measured by Semb (19) are consistent with those measured by Zetkin *et al.* (20) over a shorter temperature range and with the measurement by Taylor and Hildebrand (3) at the single temperature of 273 K. Values above about 333 K should be used with caution because of uncertainties associated with the effect of the vapor pressure of the solvent and the difficulties of making satisfactory corrections as the

COMPONENTS:	EVALUATOR:
1. Chlorine; Cl_2; [7782-50-5] 2. Organic liquids	P. G. T. Fogg, Department of Chemistry, Polytechnic of North London, Holloway, London, U.K. September 1980

CRITICAL EVALUATION:

temperature approaches the b.p. of the solvent. The values of a and b based on Zetkin *et al.* results lie close to values for other hydrocarbons also based upon Zetkin *et al.* results. "Correction" of results leads to some inconsistencies between values of a' and b' for different hydrocarbons. It may therefore be preferable to use "uncorrected" values of solubilities of heptane and other hydrocarbons studied by Zetkin *et al.* (20).

Dichlorodioxochromium

A plot of x_{Cl_2} (1 atm) against T based upon measurements by Roozeboom (21) approaches the reference line with increase in temperature and values are acceptable. Values of x_{Cl_2} at 273.2 K for the pressure range 753 to 1302 mmHg differ from values given by the equation:

$$x_{Cl_2} = p_{Cl_2}/p°_{Cl_2}$$

by about 4% and can be fitted to a curve which cuts the reference line for a constant temperature of 273.2 K at $x_{Cl_2} = 0$ and $x_{Cl_2} = 1$. This is a good indication that the measurements are acceptable (4) and add support to the validity of the way in which the original measurements were interpreted when the data were compiled.

References

1. Battino, R.; Clever, H. L. *Chem. Rev.* 1966, *66*, 395.
2. *Lange's Handbook of Chemistry* (12th edition), McGraw-Hill, New York, 1979.
3. Taylor, N. W.; Hildebrand, J. H. *J. Amer. Chem. Soc.* 1923, *45*, 682.
4. Gerrard, W. "*Gas Solubilities - Widespread Applications*", Pergamon Press, Oxford, 1980.
5. Raskina, A. D.; Zetkin, V. I.; Zakharov, E. V.; Kolesnikov, I. M.; Kosorotov, V. I. *Zh. Prikl. Khim. (Leningrad)* 1972, *45*, 1333.
6. Curda, M.; Holas, J. *Chem. Prum.* 1964, *14*, 547.
7. Kalz, G.; Naumann, A. *Plaste Kautsch.* 1971, *18*, 500.
8. Kumar, S.; Gehlawat, J. K. *J. Chem. Technol. Biotechnol.* 1979, *29*, 353.
9. Jones, W. J. *J. Chem. Soc.* 1911, *99*, 392.
10. Kogan, L. M.; Kol'tsov, N. S.; Litvinov, N. D. *Zh. Fiz. Khim.* 1963, *37*, 1875.
11. Cervinka, M. *Chem. Prum.* 1960, *10*, 249.
12. Krentsel, L. B.; Litmanovich, A. D.; Shtern, V. Ya. *Zh. Prikl. Khim. (Leningrad)* 1972, *45*, 1875.

COMPONENTS:	EVALUATOR:
1. Chlorine; Cl_2; [7782-50-5] 2. Organic liquids	P. G. T. Fogg, Department of Chemistry, Polytechnic of North Soldon, Holloway, London, U.K. September 1980

CRITICAL EVALUATION:

13. Zetkin, V. I.; Kosorotov, V. I.; Stul, B. Ya.; Dzhagatspanyan, R. V. *Khim. Prom.* 1971, *47*, 102.

14. Krivonos, F. F. *J. Appl. Chem. USSR*, 1958, *31*, 487.

15. Pozin, M. E.; Opykhtina, M. A. *Zh. Prikl. Khim. (Leningrad)* 1947, *20*, 523.

16. Ushakov, A. A.; Kosorotov, V. I.; Stul', B. Ya.; Motsarev, G. V.; Dzhagatspanyan, R. V. *Zh. Prikl. Khim. (Leningrad)* 1977, *50*, 425.

17. Tsiklis, D. S.; Svetlova, G. M. *Zh. Fiz. Khim.* 1958, *32*, 1476.

18. Egunov, A. V.; Konobeev, B. I.; Ryabov, E. A.; Gubanova, T. I. *Zh. Prikl. Khim. (Leningrad)* 1973, *46*, 1855.

19. Semb, J. *J. Am. Pharm Assoc.* 1935, *24*, 547.

20. Zetkin, V. I.; Kosorotov, V. I.; Zakharov, E. V.; Martynov, Yu. M.; Dzhagatspanyan, R. V. *Zh. Fiz. Khim.* 1970, *44*, 830.

21. Roozeboom, H. W. B. *Rec. Trav. Chim.* 1885, *4*, 379.

Table I. Quantitative evaluation of measurements of the solubility of chlorine in liquids at 1 atm.

Solvent	b.p. /K	Authors	Temp. range /K	Allowance for v.p. of solvent	Smoothing equation: Constants a /J mol^{-1}	Smoothing equation: Constants b /JT^{-1} mol^{-1}	Coef. corr.	(Std.dev.) /J mol^{-1} of $RT\ln x_{Cl_2}$	Corrected smoothing equation: Constants a /J mol^{-1}	Corrected smoothing equation: Constants b /JT^{-1} mol^{-1}	Coef. corr.	(Std.dev.) /J mol^{-1} of $RT\ln x_{Cl_2}$
$CHCl_3$; Trichloromethane	335	VK	257-298	Yes	-19788	84.219	0.9973	108				
CH_2Cl_2; Dichloromethane	313	VK	254-298	Yes	-17147	76.348	0.9996	42				
CCl_4; Tetrachloromethane	349	†	253-353	Yes	-18728	78.052	0.9995	83				
CCl_4S; Trichloromethane-sulfinyl chloride	422	RZZKK	275-308	No	-23717	98.600	0.9906	159				
CS_2; Carbon disulfide	319	RZZKK	275-298	No	-24996	102.504	0.9995	30	-14570	63.636	0.9971	43
C_2Cl_4; Tetrachloroethene	394	CH	273-363	No	-22825	92.530	0.9920	387	-18750	78.377	0.9983	151
$C_2H_2Cl_4$; 1,1,2,2-Tetrachloro-ethane	419	KN	298-383	No	-12509	62.349	0.9721	502	-8674	50.056	0.9612	479
$C_2H_4Br_2$; 1,2-Dibromoethane	405	TH	293;313	Yes	-18257	75.999	-	-				
$C_2H_4O_2$; Acetic acid	391	KG	303-333	No	-26872	111.932	0.9926	176	-24136	102.76	0.9935	151
C_4Cl_6; 1,1,2,3,4,4-Hexa-chloro-1,3-butadiene	488	KKL	253-365	No	-17228	72.090	0.9997	66	-17082	71.570	0.9997	72
$C_6H_3Cl_3$; 1,2,4-Trichloro-benzene	487	ZKSD	288-333	No	-22450	89.500	0.9995	48				
$C_6H_3Cl_2NO_2$; 1,2-Dichloro-4-nitrobenzene		ZKZD	333-433	No	-17771	78.307	0.9999	30				

(cont.)

Table I (cont.)

Solvent	b.p. /K	Authors	Temp. range /K	Allowance for v.p. of solvent	Smoothing equation: Constants a /J mol^{-1}	Smoothing equation: Constants b /JT^{-1} mol^{-1}	Smoothing equation: Coef. corr.	Smoothing equation: (Std.dev.) /J mol^{-1} of RTlnx_{Cl_2}	Corrected smoothing equation: Constants a /J mol^{-1}	Corrected smoothing equation: Constants b /JT^{-1} mol^{-1}	Corrected smoothing equation: Coef. corr.	Corrected smoothing equation: (Std.dev.) /J mol^{-1} of RTlnx_{Cl_2}
$C_6H_3Cl_2NO_2$; 1,4-Dichloro-2-nitrobenzene	540	ZKZD	333-433	No	-18585	78.712	0.9999	31				
$C_6H_4ClNO_2$; 1-Chloro-2-nitrobenzene	519	ZKZD	313-433	No	-17700	78.794	0.9999	41				
$C_6H_4ClNO_2$; 1-Chloro-3-nitrobenzene	508	ZKZD	333-433	No	-19424	84.055	0.9999	36				
$C_6H_4ClNO_2$; 1-Chloro-4-nitrobenzene	515	ZKZD	373-433	No	-22866	91.459	1.0000	12				
$C_6H_4Cl_2$; 1,2-Dichlorobenzene	454	ZKSD	288-333	No	-20174	81.553	0.9967	117	-19930	80.718	0.9966	117
C_6H_5Cl; Chlorobenzene	405	C	298-338	Yes	-13327	62.404	0.9943	134				
C_6H_5Cl; Chlorobenzene	405	KLS	313-343	Yes	-17357	74.542	0.9999	10				
C_6H_5Cl; Chlorobenzene	405	ZKSD	283-333	No	-20090	82.631	0.9971	199	-18919	78.541	0.9960	131
$C_6H_5NO_2$; Nitrobenzene	484	ZKZD	313-433	No	-19405	82.342	0.9996	102	-16838	74.745	0.9978	213
C_6H_6; Benzene	353	C	298-341	Yes	-13448	62.072	0.9970	82				
C_6H_6; Benzene	353	K	283-323	Yes	-11182	50.557	0.9963	69				
C_6H_6; Benzene	353	ZKSD	283-333	No	-23857	97.081	0.9991	78	-14836	65.573	0.9916	160
$C_7H_5Cl_3$; Trichloromethyl benzene	494	EKRG	303-423	No	-19730	81.311	0.9955	287				
$C_7H_5F_3$; Trifluoromethyl benzene	375	UKSM	279-353	No	-22918	94.563	0.9972	184	-16158	71.402	0.9966	155

(cont.)

Table I (cont.)

Solvent	b.p. /K	Authors	Temp. range /K	Allowance for v.p. of solvent	Smoothing equation: Constants a /J mol^{-1}	Smoothing equation: Constants b /JT^{-1} mol^{-1}	Smoothing equation: Coef. corr.	Smoothing equation: (Std.dev.) /J mol^{-1} of RTlnx_{Cl_2}	Corrected smoothing equation: Constants a /J mol^{-1}	Corrected smoothing equation: Constants b /JT^{-1} mol^{-1}	Corrected smoothing equation: Coef. corr.	Corrected smoothing equation: (Std.dev.) /J mol^{-1} of RTlnx_{Cl_2}
$C_7H_6Cl_2$; Dichloromethyl benzene	478	EKRG	303-403	No	-18877	78.531	0.9957	235				
C_7H_7Cl; Chloromethyl benzene	452	EKRG	303-383	No	-18003	75.960	0.9958	191				
C_7H_8; Methyl benzene	384	EKRG	303-353	No	-26746	104.440	0.9967	158	-19709	81.343	0.9932	179
C_7H_{16}; Heptane	372	S	254-358	No	-18261	77.873	0.9947	278	-12108	55.750	0.9883	298
C_7H_{16}; Heptane		ZKZMD	283-323	No	-15768	69.270	0.9998	19	-12660	58.281	0.9990	40
C_7F_{16}; Hexadeca-fluoro heptane	356	GH	273-298	Yes	-13666	65.106	0.9964	24				
C_8H_{18}; Octane	399	ZKZMD	283-323	No	-15972	69.416	0.9994	38	-14882	65.579	0.9988	51
C_8H_{18}; 2,2,4-Trimethyl pentane	372	ZKZMD	283-323	No	-15022	66.962	0.9996	28	-11914	55.928	0.9976	61
$C_{10}H_{24}$; Decane	447	ZKZMD	283-323	No	-16770	71.116	0.9994	39	-16613	70.582	0.9993	43
$C_{14}H_{30}$; Tetradecane	527	ZKZMD	283-323	No	-16899	69.183	0.9994	37	-16899	69.183	0.9994	37
Cl_2CrO_2; Dichloro-dioxo chromium *	390	R	247-273	No	-19362	82.241	0.9954	65				
Cl_4Ti; Titanium chloride *	410	KM	293;303	No	-21816	89.226	-	-				

† See evaluation of CCl_4 by W. Gerrard. b.p. - boiling point at 1 atm pressure. v.p. - vapor pressure.

* Solubilities at 1 atm estimated from values at other pressures.

COMPONENTS:	EVALUATOR:
1. Chlorine; Cl_2; [7782-50-5] 2. Organic liquids	P. G. T. Fogg, Department of Chemistry, Polytechnic of North London, Holloway, London, U.K. September 1980.

TABLE II

Quantitative evaluation of measurements of the solubility of chlorine in liquids.

Solvent	b.p. /K	Authors	Temp. /K	Allowance for v.p. of solvent	x_{Cl_2} (1 atm)	x_{Cl_2} (corrected) (1 atm)
C_3H_7NO; *N*,*N*-dimethyl-formamide	422	DuP	273	-	0.569	
C_6H_6; Benzene	353	PO	293	No	0.2081*	0.2254
C_7H_{16}; Heptane	372	TH	273	Yes		
Cl_2O_2S; Sulfuryl chloride	343	Sch	273	No	0.206	
Cl_4Si; Tetrachloro-silane	331	TH	273	Yes	0.288	
1,2-Benzene-dicarboxylic acid esters:		PMBK	323	-		
$C_{10}H_{10}O_4$ dimethyl ester	557				0.116*	
$C_{12}H_{14}O_4$ diethyl ester	571				0.124*	
$C_{16}H_{22}O_4$ dibutyl ester	613				0.144*	
$C_{24}H_{38}O_4$ dioctyl ester					0.175*	
$C_{26}H_{42}O_4$ dinonyl ester					0.180*	
$C_{30}H_{50}O_4$ diundecyl ester					0.192*	

* estimated from measurements made at other pressures.
b.p. boiling point at 1 atm pressure.

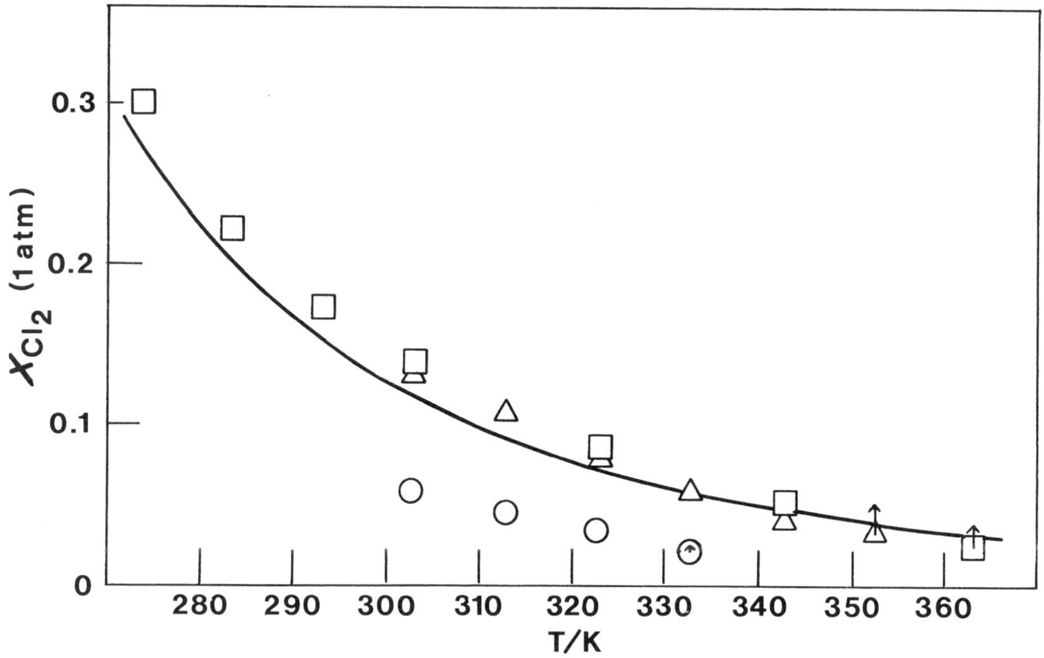

Figure 2. Experimental values of x_{Cl_2} (1 atm) compared with the reference line.

□ tetrachloroethene

○ acetic acid

△ methylbenzene.

The effects of "correction" of one experimental value for each compound are indicated by vertical arrows.

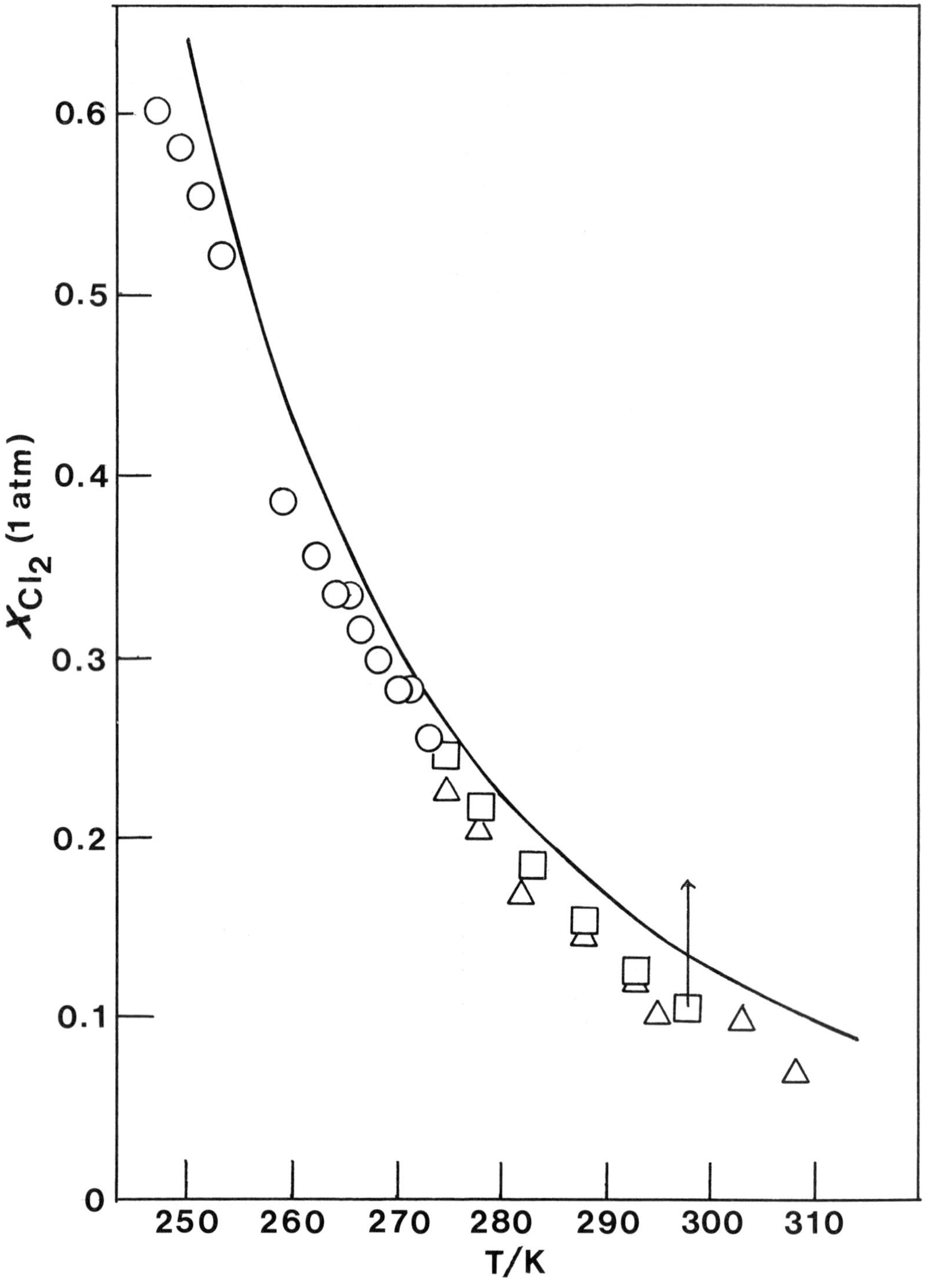

Figure 3. Experimental values of x_{Cl_2} (1 atm) compared with the reference line.

□ trichloromethanesulfenyl chloride

○ carbon disulfide

△ dichlorodioxochromium.

The effect of "correction" of one experimental value for carbon disulfide is indicated by a vertical arrow.

COMPONENTS:

1. Chlorine; Cl_2; [7782-50-5]

2. Heptane; C_7H_{16}; [142-82-5]

ORIGINAL MEASUREMENTS:

Taylor, N. W.; Hildebrand, J. H.
J. Amer. Chem. Soc. 1923, *45*, 682-694.

VARIABLES:

PREPARED BY:

P. G. T. Fogg

EXPERIMENTAL VALUES:

T/K	g Cl_2/g of solution	p_{Cl_2}/mmHg	Mole fractions x_{Cl_2}	x_{Cl_2} (1 atm)
273.16	0.2036	750	0.265	0.270

760 mmHg = 1 atm

1 atm = 1.013×10^5 Pascal

AUXILIARY INFORMATION

METHOD/APPARATUS/PROCEDURE:

Saturator tube (1). Chlorine was bubbled through liquid. Samples of saturated liquid were run into a weighed flask containing aqueous KI, and iodine was titrated with standard solution of thiosulfate. Temperature control was manual to 0.1 K.
Saturation was at barometric pressure, p_{total}. The value of $p_{C_7H_{16}}$ was estimated by Raoult's law from vapor pressure data for C_7H_{16} in ref. (2). Hence:

$$p_{Cl_2} = p_{total} - p_{C_7H_{16}}$$

SOURCE AND PURITY OF MATERIALS:

1. Cl_2: from an electrolytic supply; contained less than 0.1% of free acid.

2. C_7H_{16}: not specified.

ESTIMATED ERROR:

REFERENCES:

1. Bichowsky, F. R.; Storch, H. *J. Amer. Chem. Soc.* 1915, *37*, 2695.

2. Landolt-Börnstein Tabellen.

COMPONENTS:

1. Chlorine; Cl_2; [7782-50-5]

2. Heptane; C_7H_{16}; [142-82-5]

ORIGINAL MEASUREMENTS:

Semb, J.

J. Am. Pharm. Assoc. 1935, *24*, 547-553.

VARIABLES:

Temperature

PREPARED BY:

P. G. T. Fogg

EXPERIMENTAL VALUES:

T/K	g Cl_2/100 cm^3 solution	Mole fraction,* x_{Cl_2}
254.2	32.9	0.394
262.2	25.9	0.340
273.2	17.9	0.265
284.2	12.0	0.197
293.2	10.1	0.173
303.2	7.7	0.139
313.2	5.4	0.103
313.7	5.6	0.106
333.2	3.17	0.0646
334.2	2.95	0.0604
349.2	2.25	0.0478
350.2	1.79	0.0384
358.2	1.65	0.0359

The pressure of chlorine plus heptane was equal to barometric pressure (unspecified).

* Calculated by compiler using published densities of heptane (1) and assuming that dissolution of chlorine causes no change of volume of the liquid phase.

AUXILIARY INFORMATION

METHOD/APPARATUS/PROCEDURE:

Chlorine was passed into about 5 cm^3 of heptane in a 25 cm^3 graduated cylinder until the solution was saturated. The volume of solution was recorded and the dissolved chlorine determined iodometrically. Light was excluded as far as possible but the formation of HCl was detected in all cases.

SOURCE AND PURITY OF MATERIALS:

1. Cl_2: from a cylinder purified and dried.

2. C_7H_{16}: isolated from a commercial oil and purified by chemical treatment and fractional distillation; b.p. 96.8-97.4 °C (743 mmHg).

ESTIMATED ERROR:

REFERENCES:

1. J. Timmermans, *"Physico-Chemical Constants of Pure Organic Compounds"*, Vol. 2, Elsevier, London. 1965.

COMPONENTS:

1. Chlorine; Cl_2; [7782-50-5]

2. Heptane; C_7H_{16}; [142-82-5]

ORIGINAL MEASUREMENTS:

Zetkin, V. I.; Kosorotov, V. I.; Zakharov, E. V.; Martynov, Yu. M.; Dzhagatspanyan, R. V.
Zh. Fiz. Khim. 1970, *44*, 830-831.
(Complete article deposited at VINITI. No. 1480-70, dep. from 24 Feb. 1970.)

VARIABLES:

Temperature, pressure

PREPARED BY:

P. G. T. Fogg

EXPERIMENTAL VALUES:

T/K	Mole fraction, x_{Cl_2} (1 atm)
283	0.195
293	0.157
303	0.125
313	0.104
323	0.085

760 mmHg = 1 atm

1 atm = 1.013×10^5 Pascal

The authors gave the following equation:

$$\log_{10}(x_{Cl_2}/p_{Cl_2}) = (930/(T/K)) - 3.97$$

where p_{Cl_2} is the pressure of Cl_2 in atm.

This equation cannot be tested because only measurements made at a pressure of chlorine equal to one atmosphere were reported.

AUXILIARY INFORMATION

METHOD/APPARATUS/PROCEDURE:

The apparatus is described in ref. (1). Chlorine at a pressure of 1 atm ± 10 mmHg was absorbed by heptane in a thermostatted vessel of capacity 100 cm^3. Chlorination of the heptane was prevented by addition of phenol. Samples were withdrawn periodically for iodometric analysis until consistent results were obtained.

It was stated that the system obeys Henry's law as shown by the variation of the absorption when the partial pressure of chlorine was varied by dilution with air. However only measurements made at a partial pressure of one atmosphere were reported.

SOURCE AND PURITY OF MATERIALS:

1. Cl_2: to USSR standard GOST 6718-53. Dried by passing through H_2SO_4.

2. C_7H_{16}: "Chemically pure" grade dried with molecular sieve.

ESTIMATED ERROR:

REFERENCES:

1. Zetkin, V. I.; Kolesnikov, I. M.; Zakharov, E. V.; Dzhagatspanyan, R.V.
Khim. Promst. (Moscow) 1966, *42* (8), 624.

COMPONENTS:

1. Chlorine; Cl_2; [7782-50-5]

2. Octane; C_8H_{18}; [111-65-9]

ORIGINAL MEASUREMENTS:

Zetkin, V. I.; Kosorotov, V. I.; Zakharov, E. V.; Martynov, Yu. M.; Dzhagatspanyan, R. V.
Zh. Fiz. Khim. 1970, *44*, 830-831.
(Complete article deposted at VINITI. No. 1480-70, dep. from 24 Feb. 1970.)

VARIABLES:

Temperature, pressure

PREPARED BY:

P. G. T. Fogg

EXPERIMENTAL VALUES:

T/K	Mole fraction, x_{Cl_2} (1 atm)
283	0.210
293	0.168
303	0.134
313	0.107
323	0.092

760 mmHg = 1 atm

1 atm = 1.013×10^5 Pascal

The authors gave the following equation:

$$\log_{10}(x_{Cl_2}/p_{Cl_2}) = (950/(T/K)) - 4.02$$

where p_{Cl_2} is the pressure of Cl_2 in atm.

This equation cannot be tested because only measurements made at a pressure of chlorine equal to one atmosphere were reported.

AUXILIARY INFORMATION

METHOD/APPARATUS/PROCEDURE:

The apparatus is described in ref. (1). Chlorine at a pressure of 1 atm ± 10 mmHg was absorbed by octane in a thermostatted vessel of capacity 100 cm^3. Chlorination of the octane was prevented by addition of phenol. Samples were withdrawn periodically for iodometric analysis until consistent results were obtained.

It was stated that the system obeys Henry's law as shown by the variation of the absorption when the partial pressure of chlorine was varied by dilution with air. However only measurements made at a partial pressure of one atmosphere were reported.

It is not clear whether allowances were made for the contribution of the vapor pressure of octane to the total pressure.

SOURCE AND PURITY OF MATERIALS:

1. Cl_2: to USSR standard GOST 6718-53. Dried by passing through H_2SO_4.

2. C_8H_{18}: "Chemically pure" grade dried with molecular sieve.

ESTIMATED ERROR:

REFERENCES:

1. Zetkin, V. I.; Kolesnikov, I. M.; Zakharov, E. V.; Dzhagatspanyan, R.V.
Khim. Promst. (Moscow) 1966, *42* (8), 624.

COMPONENTS:

1. Chlorine; Cl_2; [7782-50-5]

2. 2,2,4-Trimethylpentane; C_8H_{18}; [540-84-1]

ORIGINAL MEASUREMENTS:

Zetkin, V. I.; Kosorotov, V. I.; Zakharov, E. V.; Martynov, Yu. M.; Dzhagatspanyan, R. V.
Zh. Fiz. Khim. 1970, *44*, 830-831.
(Complete article deposited at VINITI No. 1480-70, dep. from 24 Feb. 1970.)

VARIABLES:

Temperature, pressure

PREPARED BY:

P. G. T. Fogg

EXPERIMENTAL VALUES:

T/K	Mole fraction, x_{Cl_2} (1 atm)
283	0.187
293	0.154
303	0.123
313	0.101
323	0.086

760 mmHg = 1 atm

1 atm = 1.013×10^5 Pascal

The authors gave the following equation:

$$\log_{10}(x_{Cl_2}/p_{Cl_2}) = (870/(T/K)) - 3.77$$

where p_{Cl_2} is the pressure of Cl_2 in atm.

This equation cannot be tested because only measurements made at a pressure of chlorine equal to one atmosphere were reported.

AUXILIARY INFORMATION

METHOD/APPARATUS/PROCEDURE:

The apparatus is described in ref. (1). Chlorine at a pressure of 1 atm ± 10 mmHg was absorbed by iso-octane in a thermostatted vessel of capacity 100 cm^3. Chlorination of the iso-octane was prevented by addition of 0.5 wt-% of phenol. Samples were withdrawn periodically for iodometric analysis until consistent results were obtained.

It was stated that the system obeys Henry's law as shown by the variation of the absorption when the partial pressure of chlorine was varied by dilution with air. However only measurements made at a partial pressure of 1 atm were reported.

It is not clear whether allowances were made for the contribution of the vapor pressure of iso-octane to the total pressure

SOURCE AND PURITY OF MATERIALS:

1. Cl_2: to USSR standard GOST 6718-53. Dried by passing through H_2SO_4.

2. C_8H_{18}: "Chemically pure" grade dried with molecular sieve.

ESTIMATED ERROR:

REFERENCES:

1. Zetkin, V. I.; Kolesnikov, I. M.; Zakharov, E. V.; Dzhagatspanyan, P. B.
Khim. Promst. (Moscow) 1966, *42* (8), 624.

COMPONENTS:

1. Chlorine; Cl_2; [7782-50-5]

2. Decane; $C_{10}H_{22}$; [124-18-5]

ORIGINAL MEASUREMENTS:

Zetkin, V. I.; Kosorotov, V. I.; Zakharov, E. V.; Martynov, Yu. M.; Dzhagatspanyan, R. V.
Zh. Fiz. Khim. 1970, *44*, 830-831.

(Complete article deposited at VINITI. No. 1480-70, dep. from 24 Feb. 1970.)

VARIABLES:

Temperature, pressure

PREPARED BY:

P. G. T. Fogg

EXPERIMENTAL VALUES:

T/K	p_{Cl_2}/mmHg	Mole fraction x_{Cl_2}	T/K	p_{Cl_2}/mmHg	Mole fraction x_{Cl_2}
283	150	0.058	303	550	0.115
	275	0.105		760	0.151
	400	0.150	313	150	0.031
	550	0.190		275	0.053
	760	0.237		400	0.067
293	150	0.050		550	0.092
	275	0.083		760	0.119
	400	0.101	323	150	0.021
	550	0.147		275	0.035
	760	0.192		400	0.052
303	150	0.034		550	0.074
	275	0.058		760	0.100
	400	0.088			

The authors have given the following equation:

$$\log_{10}(x_{Cl_2}/p_{Cl_2}) = (1010/(T/K)) - 4.16$$

where p_{Cl_2} is the pressure of Cl_2 in atm.

The compiler has calculated that experimental values of x_{Cl_2} for this equation with a standard deviation of 0.008.

760 mmHg = 1 atm   1 atm = 1.013×10^5 Pascal

AUXILIARY INFORMATION

METHOD/APPARATUS/PROCEDURE:

The apparatus is described in ref. (1). Chlorine was absorbed by decane in a thermostatted vessel of capacity 100 cm^3. Chlorination of the decane was prevented by addition of 0.5 wt-% of phenol. Samples were withdrawn periodically for iodometric analysis until consistent results were obtained. The partial pressure of chlorine was varied by dilution with air. Pressures were controlled to ± 10 mmHg.

It is not clear whether allowances were made for the contribution of the vapor pressure of decane to the total pressure.

SOURCE AND PURITY OF MATERIALS:

1. Cl_2: to USSR standard GOST 6718-53. Dried by passing through H_2SO_4.

2. $C_{10}H_{22}$: "Chemically pure" grade dried with molecular sieve.

ESTIMATED ERROR:

REFERENCES:

1. Zetkin, V. I.; Kolesnikov, I. M.; Zakharov, E. V.; Dzhagatspanyan, P. B.
Khim. Promst. (Moscow) 1966, *42* (8), 624.

COMPONENTS:	ORIGINAL MEASUREMENTS:
1. Chlorine; Cl_2; [7782-50-5] 2. Tetradecane; $C_{14}H_{30}$; [629-59-4]	Zetkin, V. I.; Kosorotov, V. I.; Zakharov, E. V.; Martynov, Yu. M.; Dzhagatspanyan, R. V. *Zh. Fiz. Khim.* 1970, *44*, 830-831. (Complete article deposited at VINITI. No. 1480-70, dep. from 24 Feb. 1970.)
VARIABLES:	PREPARED BY:
Temperature, pressure	P. G. T. Fogg

EXPERIMENTAL VALUES:

T/K	Mole fraction, x_{Cl_2} (1 atm)
283	0.316
293	0.254
303	0.202
313	0.158
323	0.132

760 mmHg = 1 atm

1 atm = 1.013×10^5 Pascal

The authors gave the following equation:

$$\log_{10}(x_{Cl_2}/p_{Cl_2}) = (1090/(T/K)) - 4.32$$

where p_{Cl_2} is the pressure of Cl_2 in atm.

This equation cannot be tested because only measurements made at a pressure of chlorine equal to one atmosphere were reported.

AUXILIARY INFORMATION

METHOD/APPARATUS/PROCEDURE:

The apparatus is described in ref. (1). Chlorine at a pressure of 1 atm ± 10 mmHg was absorbed by tetradecane in a thermostatted vessel of capacity 100 cm^3. Chlorination of the tetradecane was prevented by addition of 0.5 wt-% of phenol. Samples were withdrawn periodically for iodometric analysis until consistent results were obtained.

It was stated that the system obeys Henry's law as shown by the variation of the absorption when the partial pressure of chlorine was varied by dilution with air. However only measurements made at a partial pressure of one atmosphere were reported.

It is not clear whether allowances were made for the contribution of the vapor pressure of tetradecane to the total pressure.

SOURCE AND PURITY OF MATERIALS:

1. Cl_2: to USSR standard GOST 6718-53.

2. $C_{14}H_{30}$: "Chemically pure" grade dried with molecular sieve.

ESTIMATED ERROR:

REFERENCES:

1. Zetkin, V. I.; Kolesnikov, I. M.; Zakharov, E. V.; Dzhagatspanyan, P. B.
Khim. Promst. (Moscow) 1966, *42* (8), 624.

COMPONENTS:

1. Chlorine; Cl_2; [7782-50-5]

2. Cyclohexane; C_6H_{12}; [110-82-7]

ORIGINAL MEASUREMENTS:

Tsiklis, D. S.; Svetlova, G. M.;
Zh. Fiz. Khim. 1958, *32*, 1476.

VARIABLES:

Temperature, pressure

PREPARED BY:

P. G. T. Fogg

EXPERIMENTAL VALUES:

T/K	283.2	293.2	313.2
p_{Cl_2}/mmHg	Mole fraction x_{Cl_2}	Mole fraction x_{Cl_2}	Mole fraction x_{Cl_2}
100	0.0129	0.0092	0.0061
200	0.0287	0.0200	0.0134
300	0.0465	0.0321	0.0213
400	0.0692	0.0472	0.0300
500	0.110	0.0699	0.0400
600		0.111	0.0515
700			0.0665
800			0.0867

760 mmHg = 1 atm
1 atm = 1.013×10^5 Pascal

The authors gave the following equation:

$$\log_{10}\left\{\frac{p_{Cl_2}/\text{mmHg}}{x_{Cl_2}}\right\} = \log_{10}K - \beta(1 - x^2_{C_6H_{12}})$$

where K has the values 8240, 12180 and 17400 at 283.2 K, 293.2 K and 313.2 K, respectively and β has the values 1.250, 1.706 and 1.700 at 283.2 K, 293.2 K and 313.2 K, respectively.
The compiler has calculated that values of p_{Cl_2}/mmHg fit the equation with a standard deviation of 8.
The authors have given the head of solution of chlorine calculated from the variation of solubility with temperature as -4500 cal mol^{-1} (± 10-15%).

AUXILIARY INFORMATION

METHOD/APPARATUS/PROCEDURE:

A known weight of chlorine was dissolved in a known weight of cyclohexane and the resulting total pressure was measured by a manometer. The pressure of chlorine was equal to the total pressure minus the vapor pressure of the cyclohexane. Temperatures were controlled to ± 0.1 °C.

SOURCE AND PURITY OF MATERIALS:

Not stated.

ESTIMATED ERROR:

REFERENCES:

COMPONENTS:	ORIGINAL MEASUREMENTS:
1. Chlorine; Cl_2; [7782-50-5] 2. Benzene; C_6H_6; [71-43-2]	Pozin, M. E.; Opykhtina, M. A. *Zh. Prik. Khim. (Leningrad)* 1947, *20*, 523-531.
VARIABLES:	PREPARED BY:
Pressure	P. G. T. Fogg

EXPERIMENTAL VALUES:

p_{Cl_2}/mmHg	Mole fraction, x_{Cl_2}
48.8	0.0152
53.6	0.0115
123.4	0.0367
263.5	0.0687
282.5	0.0684
311.5	0.0775
416	0.1091
460	0.1287
494	0.1378
663	0.1708
693	0.1809
730	0.1978
763	0.2089

Temperature = 293.2 K

760 mmHg = 1 atm

1 atm = 1.013×10^5 Pascal

The authors gave the Henry's law constant as:

$$H = 3.8 \times 10^3 \text{ mmHg}/x_{Cl_2}$$

The compiler has calculated that the mean value of $(p_{Cl_2}/\text{mmHg})/x_{Cl_2}$ is 3788 with a standard deviation of 362.

AUXILIARY INFORMATION

METHOD/APPARATUS/PROCEDURE:

The volume of chlorine absorbed by a measured quantity of benzene was determined by use of apparatus described in ref. (1). The absorbing liquid was stirred and maintained at a temperature constant to within 2 °C.

SOURCE AND PURITY OF MATERIALS:

2. C_6H_6: b.p. 80.5 °C (barometric pressure unspecified); density (20 °C) 0.8796 g cm^{-3}.

ESTIMATED ERROR:

REFERENCES:

1. Pozin, M. E.
Zh. Prik. Khim. (Leningrad) 1947, *20*, 345.

COMPONENTS:

1. Chlorine; Cl_2; [7782-50-5]

2. Benzene; C_6H_6; [71-43-2]

ORIGINAL MEASUREMENTS:

Krivonos, F. F.

J. Appl. Chem. USSR 1958, *31*, 487-491.

VARIABLES:

Temperature

PREPARED BY:

P. G. T. Fogg

EXPERIMENTAL VALUES:

T/K	Bunsen coefficient	Mole fraction, x_{Cl_2}* (1 atm) (calculated by compiler)
283.2	95	0.274
293.2	69	0.217
303.2	57	0.189
313.2	49	0.168
323.2	42	0.149

* Values of x_{Cl_2} (1 atm) have been calculated by the compiler on the assumption that the Bunsen coefficient is independent of pressure. This is equivalent to assuming the validity of Henry's law (mole ratio form).

1 atm = 1.013×10^5 Pascal

AUXILIARY INFORMATION

METHOD/APPARATUS/PROCEDURE:

A mixture of chlorine and air was admitted into a round bottomed flask of known volume. The quantities of the two were calculated from the total weight, total volume and total pressure of the mixture (measured by a manometer). A measured volume of benzene was introduced and the total pressure measured after equilibrium had been reached. The partial pressure of chlorine, p_{Cl_2}, was calculated from:

$$p_{total} = p_{C_6H_6} + p_{air} + p_{Cl_2}$$

It was assumed that $p_{C_6H_6}$ was equal to the vapor pressure of pure benzene. Temperatures were controlled by a thermostat bath and light was excluded as far as possible. The authors claimed that reproducible results indicated that no irreversible reaction between benzene and chlorine took place.

SOURCE AND PURITY OF MATERIALS:

Not specified

ESTIMATED ERROR:

REFERENCES:

COMPONENTS:	ORIGINAL MEASUREMENTS:
1. Chlorine; Cl_2; [7782-50-5] 2. Benzene, C_6H_6; [71-43-2]	Cervinka, M. *Chem. Prum.* 1960, *10*, 249-251.
VARIABLES: Pressure, temperature	PREPARED BY: P. G. T. Fogg

EXPERIMENTAL VALUES:

T/K	H/mmHg	Mole fraction, x_{Cl_2} (1 atm) (calculated by compiler)
298.3	5.74×10^3	0.132
308.2	7.33×10^3	0.104
314.2	7.60×10^3	0.100
329.2	9.45×10^3	0.0804
341.2	11.8×10^3	0.0644

The Henry's law constant, H, was taken to be the slope of a plot of pressure of Cl_2 in the gas phase against mole fraction of Cl_2 in the liquid phase. The author has given the following equation:

$$\log_{10}H = 6.1227 - 701.9/(T/K)$$

The compiler has calculated that the experimental values of H quoted above fit this equation with a standard deviation of 0.27×10^3.

The heat of solution of chlorine, calculated by the author from the variation of H with temperature, is -3.20 kcal mol^{-1}.

760 mmHg = 1 atm
1 atm = 1.013×10^5 Pascal

AUXILIARY INFORMATION

METHOD/APPARATUS/PROCEDURE:

The absorption of chlorine at partial pressures up to 1 atm was determined by passing mixtures of chlorine and air into benzene in an absorption vessel. The apparatus is described in ref. (1). Temperatures were controlled to ± 0.5 K. The chlorine content of the gas phase and the liquid phase were determined by iodometry.

SOURCE AND PURITY OF MATERIALS:

1. Cl_2 & air: dried with concentrated H_2SO_4.

2. C_6H_6: distilled before use.

ESTIMATED ERROR:

REFERENCES:

1. Balej, J.; Regner, A. *Chem. Listy* 1956, *50*, 1374.

COMPONENTS:	ORIGINAL MEASUREMENTS:
1. Chlorine; Cl_2; [7782-50-5] 2. Benzene; C_6H_6; [71-43-2]	Zetkin, V.I.; Kosorotov, V.I.; Stul, B.Ya.; Dzhagatspanyan, R.V.; *Khim. Prom.* 1971, *47*, 102-103.
VARIABLES:	PREPARED BY:
Temperature	P.G.T. Fogg

EXPERIMENTAL VALUES:

T/K	x_{Cl_2}
283.2	0.209
293.2	0.154
303.2	0.107
313.2	0.083
323.2	0.063
333.2	0.045

The pressure of Cl_2 plus C_6H_6 was apparently equal to barometric pressure.

The following equation was given :

$$\log_{10} x_{Cl_2} = 1250/(T/K) - 5.08.$$

The heat of solution of Cl_2 calculated from the variation of solubility with temperature was given as - 5700 cal mol^{-1}.

AUXILIARY INFORMATION

METHOD/APPARATUS/PROCEDURE:

The method and apparatus were d scribed in ref. (1). Chlorine was bubbled through a measured quantity of benzene in a thermostatted vessel of capacity 100 cm^3. Chlorination of the benzene was prevented by addition of 0.5 wt % of phenol. Samples were withdrawn periodically for iodometric analysis until consistent results were obtained.

SOURCE AND PURITY OF MATERIALS:

Cl_2 : to USSR standard GOST 6718-53. Dried by passing through H_2SO_4.

C_6H_6 : 99% pure. Dried with $CaCl_2$.

ESTIMATED ERROR:

REFERENCES:

1. Zetkin, V.I. et al.
Zh. Fiz. Khim. 1970, *44*, 830.

COMPONENTS:

1. Chlorine; Cl_2; [7782-50-5]

2. Benzene; C_6H_6; [71-43-2]

ORIGINAL MEASUREMENTS:

Lohse, M.; Deckwer, W.-D.

J. Chem. Eng. Data

1981, *26*, 159-161.

VARIABLES:

Temperature

PREPARED BY:

C. L. Young

EXPERIMENTAL VALUES:

T/K	Solubility,§ S/mol dm^{-3} (soln.)
288.15	3.590
298.15	2.521
308.15	1.780
318.15	1.110
328.15	0.841
338.15	0.503

§ at a partial pressure of Cl_2 of 101.3 kPa.

AUXILIARY INFORMATION

METHOD/APPARATUS/PROCEDURE:

Solvent saturated with chlorine in dark (∿1% phenol added to prevent photochemical chlorination). Samples withdrawn added to known amount of additional solvent and then 250 cm^3 of potassium iodide solution. Released iodine titrated with thiosulfate.

SOURCE AND PURITY OF MATERIALS:

No details given.

ESTIMATED ERROR:

$\delta T/K = \pm 0.1$; $\delta S = \pm 3\%$

(estimated by compiler).

REFERENCES:

COMPONENTS:

1. Chlorine; Cl_2; [7782-50-5]

2. Methylbenzene; C_7H_8; [108-88-3]

ORIGINAL MEASUREMENTS:

Egunov, A. V.; Konobeev, B. I.; Ryabov, E. A.; Gubanova, T. I.

Zh. Prik. Khim. (Leningrad) 1973, *46*, 1855-1856.

VARIABLES:

Temperature

PREPARED BY:

P. G. T. Fogg

EXPERIMENTAL VALUES:

T/K	Mole fraction, x_{Cl_2}
303.2	0.132
313.2	0.106
323.2	0.080
333.2	0.054
343.2	0.039
353.2	0.032

The partial pressure of chlorine plus partial pressure of toluene was apparently equal to barometric pressure (unspecified).

AUXILIARY INFORMATION

METHOD/APPARATUS/PROCEDURE:

Chlorine was bubbled through a weighed portion of the solvent (15-20 g) until a differential manometer registered no pressure drop when the chlorine supply was discontinued. The chlorine content of the solution was determined from the increase in weight and also argentometrically. Loss of solvent in the gas stream was shown to be negligible in comparison with other errors which the authors estimate to be ± 5%. Light was excluded and 1-2 wt-% of phenol was added to inhibit chlorination of toluene. The chlorination products which were formed were estimated by gas chromatography and appropriate corrections made during calculation of solubility.

SOURCE AND PURITY OF MATERIALS:

1. Cl_2: from a cylinder; dried over H_2SO_4.

2. C_7H_8: "chemically pure" grade.

ESTIMATED ERROR:

REFERENCES:

COMPONENTS:

1. Chlorine; Cl_2; [7782-50-5]

2. Methylbenzene (toluene); C_7H_8; [108-88-3]

ORIGINAL MEASUREMENTS:

Lohse, M.; Deckwer, W.-D.
J. Chem. Eng. Data
1981, *26*, 159-161.

VARIABLES:

Temperature

PREPARED BY:

C. L. Young

EXPERIMENTAL VALUES:

T/K	Solubility,[§] S/mol dm^{-3} (soln.)
288.15	2.570
298.15	1.905
308.15	1.479
318.15	1.122
328.15	0.771
338.15	0.611
348.15	0.464

§ at a partial pressure of Cl_2 of 101.3 kPa.

AUXILIARY INFORMATION

METHOD/APPARATUS/PROCEDURE:

Solvent saturated with chlorine in dark (∿1% phenol added to prevent photochemical chlorination). Samples withdrawn added to known amount of additional solvent and then 250 cm^3 of potassium iodide solution. Released iodine titrated with thiosulfate.

SOURCE AND PURITY OF MATERIALS:

No details given.

ESTIMATED ERROR:

$\delta T/K = \pm 0.1$; $\delta S = \pm 3\%$
(estimated by compiler).

REFERENCES:

COMPONENTS:

1. Chlorine; Cl_2; [7782-50-5]

2. 1,2-Dimethylbenzene; C_8H_{10}; [95-47-6]

ORIGINAL MEASUREMENTS:

Lohse, M.; Deckwer, W.-D.
J. Chem. Eng. Data
1981, *26*, 159-161.

VARIABLES:

Temperature

PREPARED BY:

C. L. Young

EXPERIMENTAL VALUES:

T/K	Solubility[§], S/mol dm^{-3} (soln.)
308.15	1.124
328.15	0.765

[§]at a partial pressure of 101.3 kPa.

AUXILIARY INFORMATION

METHOD/APPARATUS/PROCEDURE:

Solvent saturated with chlorine in dark (∿1% phenol added to prevent photochemical chlorination). Samples withdrawn added to known amount of additional solvent and then 250 cm^3 of potassium iodide solution. Released iodine titrated with thiosulfate.

SOURCE AND PURITY OF MATERIALS:

No details given.

ESTIMATED ERROR:

$\delta T/K = \pm 0.1$; $\delta S = \pm 3\%$
(estimated by compiler).

REFERENCES:

COMPONENTS:

1. Chlorine; Cl_2; [7782-50-5]

2. 1,3-Dimethylbenzene; C_8H_{10}; [108-38-3]

ORIGINAL MEASUREMENTS:

Lohse, M.; Deckwer, W.-D.
J. Chem. Eng. Data
1981, *26*, 159-161.

VARIABLES:

Temperature

PREPARED BY:

C. L. Young

EXPERIMENTAL VALUES:

T/K	Solubility,[§] S/mol dm^{-3} (soln.)
298.15	1.373
318.15	0.915
338.15	0.627

[§] at a partial pressure of 101.3 kPa.

AUXILIARY INFORMATION

METHOD/APPARATUS/PROCEDURE:

Solvent saturated with chlorine in dark (∿1% phenol added to prevent photochemical chlorination). Samples withdrawn added to known amount of additional solvent and then 250 cm^3 of potassium iodide solution. Released iodine titrated with thiosulfate.

SOURCE AND PURITY OF MATERIALS:

No details given.

ESTIMATED ERROR:

$\delta T/K = \pm 0.1$; $\delta S = \pm 3\%$ (estimated by compiler).

REFERENCES:

COMPONENTS:	ORIGINAL MEASUREMENTS:
1. Chlorine; Cl_2; [7782-50-5] 2. 1,4-Dimethylbenzene (*p*-xylene); C_8H_{10}; [106-42-3]	Lohse, M.; Deckwer, W.-D. *J. Chem. Eng. Data* 1981, *26*, 159-161.
VARIABLES:	PREPARED BY:
Temperature	C. L. Young

EXPERIMENTAL VALUES:

T/K	Solubility[§], S/mol dm^{-3} (soln.)
288.15	1.693
298.15	1.397
308.15	1.127
318.15	0.921
328.15	0.767
338.15	0.623
348.15	0.484

[§] at a partial pressure of Cl of 101.3 kPa.

AUXILIARY INFORMATION

METHOD/APPARATUS/PROCEDURE:

Solvent saturated with chlorine in dark (∿ 1% phenol added to prevent photochemical chlorination). Samples withdrawn added to known amount of additional solvent and then 250 cm^3 of potassium iodide solution. Released iodine titrated with thiosulfate.

SOURCE AND PURITY OF MATERIALS:

No details given.

ESTIMATED ERROR:

$\delta T/K = \pm 0.1$; $\delta S = \pm 3\%$ (estimated by compiler).

REFERENCES:

COMPONENTS:

1. Chlorine; Cl_2; [7782-50-5]

2. Ethylbenzene; C_8H_{10}; [100-41-4]

ORIGINAL MEASUREMENTS:

Lohse, M.; Deckwer, W.-D.

J. Chem. Eng. Data

1981, *26*, 159-161.

VARIABLES:

Temperature

PREPARED BY:

C. L. Young

EXPERIMENTAL VALUES:

T/K	Solubility[§] S/mol dm^{-3} (soln.)
288.15	2.193
298.15	1.657
308.15	1.228
318.15	0.980
328.15	0.779
338.15	0.660
348.15	0.528

§ at a partial pressure of Cl_2 of 101.3 kPa.

AUXILIARY INFORMATION

METHOD/APPARATUS/PROCEDURE:

Solvent saturated with chlorine in dark (∿1% phenol added to prevent photochemical chlorination). Samples withdrawn added to known amounts of additional solvent and then 250 cm^3 of potassium iodide solution. Released iodine titrated with thiosulfate.

SOURCE AND PURITY OF MATERIALS:

No details given.

ESTIMATED ERROR:

$\delta T/K = \pm 0.1$; $\delta S = \pm 3\%$ (estimated by compiler).

REFERENCES:

COMPONENTS:	ORIGINAL MEASUREMENTS:
1. Chlorine; Cl_2; [7782-50-5] 2. Acetic acid; $C_2H_4O_2$; [64-19-7]	Jones, W. J. *J. Chem. Soc.* <u>1911</u>, *99*, 392-404.
VARIABLES:	**PREPARED BY:**
Concentration	P. G. T. Fogg

EXPERIMENTAL VALUES:

T/K	S	p_{Cl_2}/Pa*	Mole fraction,* x_{Cl_2}
289.2	36.6	948	0.00086
289.2	36.9	1038	0.00095
289.2	36.9	1739	0.00160
289.2	36.4	2395	0.00218

$$S = \frac{\text{concentration of } Cl_2 \text{ in mol dm}^{-3} \text{ of the solution}}{\text{concentration of } Cl_2 \text{ in mol dm}^{-3} \text{ of the gas phase}}$$

* Calculated by the compiler from the volumes of titrant given in the original paper. It has been assumed that the volume of the solution is equal to the volume of the original liquid at these low concentrations.

Extrapolation to p_{Cl_2} = 1 atm is not valid.

AUXILIARY INFORMATION

METHOD APPARATUS/PROCEDURE:

The apparatus consisted of a bubbler and wash bottles. The concentration of the original solution was determined by iodometry. Chlorine, carried away from the solution in a stream of air, was also determined by iodometry. The results were given as a partition coefficient, S. S was deemed to be constant and independent of pressure p_{Cl_2}.

SOURCE AND PURITY OF MATERIALS:

1. Cl_2: not specified.

2. $C_2H_4O_2$: some prepared in the laboratory; some obtained from Kahlbaum. Both samples said to be "specially purified".

ESTIMATED ERROR:

REFERENCES:

COMPONENTS:

1. Chlorine; Cl_2; [7782-50-5]
2. Acetic acid; $C_2H_4O_2$; [64-19-7]

ORIGINAL MEASUREMENTS:

Kumar, S.; Gehlawat, J.K.;
J. Chem. Technol. Biotechnol. 1979 *29*, 353-360.

VARIABLES:

Temperature

PREPARED BY:

P.G.T. Fogg.

EXPERIMENTAL VALUES:

T/K	$mol_{Cl_2} dm^{-3}$ *	Mole fraction** x_{Cl_2}
303	1.06	0.058
313	0.81	0.045
323	0.59	0.034
333	0.38	0.022

Total pressure = 101 kPa.

* It is not clear whether values refer to $mol_{Cl_2} dm^{-3}$ (solution) or to $mol_{Cl_2} dm^{-3}$ (solvent).

** Calculated by the compiler using published values of densities of acetic acid (ref. 1) and assuming that dissolution of chlorine causes no change in the volume of the liquid phase.

The authors stated that the solubility data were unlikely to be in error by more than 15%. No basis for this statement was given.

AUXILIARY INFORMATION

METHOD/APPARATUS/PROCEDURE:

The authors stated that the solutions were saturated with chlorine at a specified temperature and the dissolved chlorine was determined by standard iodometric titration. No further details were given.

SOURCE AND PURITY OF MATERIALS:

Cl_2 : analytical grade.

$C_2H_4O_2$: analytical grade.

ESTIMATED ERROR:

REFERENCES:

1. J. Timmermans, "Physico-Chemical Constants of Pure Organic Compounds", Vol. 2, Elsevier, London. 1965.

COMPONENTS:

1. Chlorine; Cl_2; [7782-50-5]

2. Acetic acid, anhydride; $C_4H_6O_3$; [108-24-7]

ORIGINAL MEASUREMENTS:

Jones, W. J.

J. Chem. Soc. 1911, *99*, 392-404.

VARIABLES:

Concentration

PREPARED BY:

P. G. T. Fogg

EXPERIMENTAL VALUES:

T/K	conc. Cl_2 in soln. /mol dm^{-3}	S	p_{Cl_2}/Pa*	Mole fraction, x_{Cl_2} *
288.2	0.01574	39.6	953	0.00148
288.2	0.01015	40.1	606	0.00095
288.2	0.00403	38.8	249	0.00038

$$S = \frac{\text{concn. } Cl_2 \text{ in mol dm}^{-3} \text{ solution}}{\text{concn. } Cl_2 \text{ in mol dm}^{-3} \text{ gas phase}}$$

* Calculated by the compiler on the assumption that the volume of the solution is equal to that of the original liquid at these low concentrations.

Extrapolation to p_{Cl_2} = 1 atm is not valid.

AUXILIARY INFORMATION

METHOD/APPARATUS/PROCEDURE:

The apparatus consisted of a bubbler and wash bottles. The concentration of the original solution was determined by iodometry. Chlorine, carried away from the solution in a stream of air, was also determined by iodometry. The results were given as a partition coefficient, S. S was deemed to be constant and independent of pressure p_{Cl_2}.

SOURCE AND PURITY OF MATERIALS:

1. Cl_2: not specified.

2. $C_4H_6O_3$: supplied by Kahlbaum.

ESTIMATED ERROR:

REFERENCES:

COMPONENTS:

1. Chlorine; Cl_2; [7782-50-5]

2. 1,2-Benzenedicarboxylic acid esters.

ORIGINAL MEASUREMENTS:

Prokop'eva, M. F.; Mar'yanovskaya, K. Sh.; Bukina, V. K.; Korol', A. N. *Zh. Fiz. Khim.* 1971, *45*, 722.

(Complete article deposited at VINITI. No.2400-70, dep. from 17 Dec. 1970).

VARIABLES:

None

PREPARED BY:

P. G. T. Fogg

EXPERIMENTAL VALUES:

Temperature = 323.2 K

Stationary phase	Specific retention volume	Mole fraction,* x_{Cl_2} (1 atm)
Dimethyl ester, $C_{10}H_{10}O_4$ [131-11-3]	14.92	0.116
Diethyl ester, $C_{12}H_{14}O_4$ [84-66-2]	14.10	0.124
Dibutyl ester, $C_{16}H_{22}O_4$ [84-74-2]	13.34	0.144
Dioctyl ester, $C_{24}H_{38}O_4$ [117-84-0]	12.00	0.175
Dinonyl ester, $C_{26}H_{42}O_4$ [84-76-4]	11.59	0.180
Diundecyl ester, $C_{30}H_{50}O_4$ [3648-20-2]	11.01	0.192

* Values of x_{Cl_2} (1 atm) have been calculated by the compiler on the assumption that specific retention volumes given above are in units of $cm^3\ g^{-1}$ and that these values can be equated with the volume of gas, corrected to 273.2 K, absorbed by 1 g of solvent (see ref. 1). It has also been assumed that this volume is independent of pressure. This last assumption may have introduced appreciable error and the values of x_{Cl_2} (1 atm) given above can only be thought of as approximate values.

AUXILIARY INFORMATION

METHOD/APPARATUS/PROCEDURE:

Retention volumes for chlorine were determined for the various stationary phases using conventional gas chromatographic techniques.

SOURCE AND PURITY OF MATERIALS:

No information.

ESTIMATED ERROR:

REFERENCES:

1. Ambrose, D.; Keulemans, A. I. M.; Purnell, J. H. *Anal. Chem.* 1958, *30*, 1582.

COMPONENTS:	ORIGINAL MEASUREMENTS:
1. Chlorine; Cl_2; [7782-50-5] 2. *N,N*-Dimethyl formamide; C_3H_7NO; [68-12-2]	DuPont de Nemours and Co. (Inc.); *Chem. Eng. News* 1955, *33*, 2366.
VARIABLES:	**PREPARED BY:** P. G. T. Fogg

EXPERIMENTAL VALUES:

T/K	p_{Cl_2}	Vol of Cl_2 absorbed per vol. of DMF	Mole fraction*, x_{Cl_2} (1 atm)
273.2	1 atm	385	0.569

$$1\ \text{atm} = 1.013 \times 10^5\ \text{Pascal}$$

* Calculated by compiler on the assumption that the volume of gas absorbed is independent of pressure and that the density of DMF at 273.2 K is given by a linear extrapolation of published values at 293.2 K (1) and 298.2 K (2).

AUXILIARY INFORMATION

METHOD/APPARATUS/PROCEDURE:

Not stated.

SOURCE AND PURITY OF MATERIALS:

Not stated.

ESTIMATED ERROR:

REFERENCES:

1. *"Handbook of Chemistry & Physics"* (57th edition), C.R.C. Press, Cleveland, Ohio. 1976.
2. *"Lange's Handbook of Chemistry"* (12th edition), McGraw-Hill, New York. 1979.

COMPONENTS:

1. Chlorine; Cl_2; [7782-50-5]
2. Nitrobenzene; $C_6H_5NO_2$; [98-95-3]

ORIGINAL MEASUREMENTS:

Zetkin, V.I.; Kolesnikov, I.M.; Zakharov, E.V.; Dzhagatspanyan, R.V.

Khim. Prom. 1966, *42*, 624-626

VARIABLES:

Temperature

PREPARED BY:

P.G.T. Fogg

EXPERIMENTAL VALUES:

T/K	$mol_{Cl_2}/mol_{C_6H_5NO_2}$	$g_{Cl_2}/100g_{C_6H_5NO_2}$	Mole fraction* x_{Cl_2}
313	0.0891	5.138	0.0818
333	0.0610	3.518	0.0575
353	0.0408	2.353	0.0392
373	0.0264	1.522	0.0257
393	0.0190	1.096	0.0187
413	0.0143	0.825	0.0141
433	0.0111	0.646	0.0110

The total pressure was apparently equal to barometric pressure (unspecified).

The heat of solution of Cl_2 calculated from the variation of solubility with temperature was given by the authors as - 4680 cal mol^{-1}.

* Calculated by compiler.

AUXILIARY INFORMATION

METHOD/APPARATUS/PROCEDURE:

Chlorine was bubbled through nitrobenzene in a thermostatted vessel of capacity 100 cm^3. Samples were withdrawn periodically for iodometric analysis until consistent results were obtained.

SOURCE AND PURITY OF MATERIALS:

Cl_2 : to USSR standards GOST 6718-53. Dried with H_2SO and with $CaCl_2$.

$C_6H_5NO_2$: Dried with $CaCl_2$. Distilled through a 20-plate column before use.

ESTIMATED ERROR:

REFERENCES:

COMPONENTS:	ORIGINAL MEASUREMENTS:
1. Chlorine; Cl_2; [7782-50-5] 2. Carbon disulfide; CS_2; [75-15-0]	Raskina, A.D.; Zetkin, V.I.; Zakharov, E.V.; Kolesnikov, I.M. Kosorotov, V.I.; *Zh. Prik. Khim. (Leningrad)*, 1972, *45*, 1333-1334.
VARIABLES: Temperature	PREPARED BY: P.G.T. Fogg

EXPERIMENTAL VALUES:

T/K	g_{Cl_2}/100 g of solution	Mole fraction x_{Cl_2}
275	23.0	0.246
278	20.4	0.218
283	17.2	0.184
288	14.4	0.154
293	11.7	0.125
298	9.9	0.106

The total pressure was apparently equal to barometric pressure (unspecified).

The authors have given the following equation:

$$\log_{10}(100x_{Cl_2}) = (1250/(T/K)) - 3.18$$

The compiler has calculated that values of x_{Cl_2} quoted above fit this equation with a standard deviation of 0.011. The compiler has also calculated that the equation:

$\log_{10}(100x_{Cl_2}) = (1304/(T/K)) - 3.55$ fits the data better (standard deviation 0.002).

The authors have given the heat of solution of chlorine, calculated from the variation of solubility with temperature, as - 5.7 kcal mol^{-1}.

AUXILIARY INFORMATION

METHOD/APPARATUS/PROCEDURE:

Chlorine or chlorine/nitrogen mixtures were bubbled through carbon disulfide until the concentration of chlorine in the liquid phase, as determined by iodometry, was constant. The apparatus was protected from light. The total pressure of gas was apparently equal to barometric pressure although this is not explicitly stated. Variations of solubility when chlorine was diluted with nitrogen is said to indicate that Henry's law is obeyed although these experimental measurements are not reported. No allowances seem to have been made for the partial pressure of the carbon disulfide.

SOURCE AND PURITY OF MATERIALS:

Cl_2 : to USSR standard GOST 6718-53; additionally dried over H_2SO_4.

N_2 : to GOST 9293-59; dried over H_2SO_4.

CS_2 : to GOST 1541-42; distilled and dried over ignited $CaCl_2$.

ESTIMATED ERROR:

REFERENCES:

COMPONENTS:

1. Chlorine; Cl_2; [7782-50-5]
2. Dichloromethane; CH_2Cl_2; [75-09-2]

ORIGINAL MEASUREMENTS:

Vdovichenko, V. T.; Kondratenko, V. I.

Khim. Promst. 1967, *43*, 290-291.

VARIABLES:

Temperature

PREPARED BY:

P. G. T. Fogg

EXPERIMENTAL VALUES:

T/K	Mole fraction, x_{Cl_2} (1 atm)
254.2	0.338
273.2	0.198
283.2	0.152
298.2	0.102

1 atm = 1.013×10^5 Pascal

AUXILIARY INFORMATION

METHOD/APPARATUS/PROCEDURE:

The method described by Taylor and Hildebrand was used (1).

SOURCE AND PURITY OF MATERIALS:

1. Cl_2: to USSR standard GOST 6718-53.
2. CH_2Cl_2: b.p. 40.0-40.3 °C (barometric pressure unspecified); density (20 °C) 1.3260 g cm^{-3}; n_D (20 °C) 1.4245.

ESTIMATED ERROR:

REFERENCES:

1. Taylor, N. W.; Hildebrand, J. H. *J. Amer. Chem. Soc.* 1923, *45*, 682.

COMPONENTS:

1. Chlorine; Cl_2; [7782-50-5]

2. Trichloromethane; $CHCl_3$; [67-66-3]

ORIGINAL MEASUREMENTS:

Vdovichenko, V. T.; Kondratenko, V. I.

Khim. Promst. 1967, *43*, 290-291.

VARIABLES:

Temperature

PREPARED BY:

P. G. T. Fogg

EXPERIMENTAL VALUES:

T/K	Mole fraction, x_{Cl_2} (1 atm)
257.2	0.399
273.2	0.259
283.2	0.178
298.2	0.114

1 atm = 1.013×10^5 Pascal

AUXILIARY INFORMATION

METHOD/APPARATUS/PROCEDURE:

The method described by Taylor and Hildebrand was used (1).

SOURCE AND PURITY OF MATERIALS:

1. Cl_2: to USSR standard GOST 6718-53.

2. $CHCl_3$: b.p. 61.1-61.2 °C; density (293.2 K) 1.4892 g cm^{-3}; n_D (293.2 K) 1.4464.

ESTIMATED ERROR:

REFERENCES:

1. Taylor, N. W.; Hildebrand, J. H. *J. Amer. Chem. Soc.* 1923, *45*, 682.

COMPONENTS:

1. Chlorine; Cl_2; [7782-50-5]

2. Tetrachloromethane; CCl_4; [56-23-5]

EVALUATOR:

W. Gerrard, Department of Chemistry,
The Polytechnic of North London,
Holloway,
London, N7 8DB
UNITED KINGDOM

CRITICAL EVALUATION:

Only five sets of values need be considered in arriving at an average plot of x_{Cl_2} vs. T/K from which the recommended values are taken. The original data are given as referring to a partial pressure of 1 atm. i.e. p_{Cl_2} = 1 atm. The uncertainty in these values for p_{Cl_2} = 1 atm, lies in two assumptions. Firstly, the total pressure $p_T = p_{Cl_2} + p_{CCl_4}$ was measured, and p_{CCl_4} was estimated by assuming Raoult's law, although this is not explicitly stated in every paper. Secondly, the derived p_{Cl_2} value is taken to adjust the corresponding x_{Cl_2} value to that at 1 atm, on the assumption that Henry's law, mole fraction form, holds; but this assumption is frequently not valid, although it may be used with caution.

Data used in the construction of the average curve are shown in Fig. 1, and are taken from the following contributors: Taylor and Hildebrand (1) (T-H), Smith (2) (S), Kogan, Kol'tsov, and Litvinov (3) (K-K-L), Vdovichenko and Kondratenko (4) (V-K), Raskina, Zetkin, Zakharov, Kolesnikov, and Kosorotov (5) (not shown on diagram because the numbers are identical with the K-K-L values, although three T/K values are slightly different). The derived plot is shown in Fig. 2.

Data by Curda and Holas (6) are for a total pressure, p_T, about 1 atm, see line C-H in Fig. 1. and are excluded from the mean line. Data by Blair and Yost (7) should not be used for extrapolation beyond the highest p_{Cl_2} = 73.4 mm Hg. Data by Jones (8) should not be used for extrapolation beyond p_{Cl_2} = 13 mm Hg. Jakowkin's isolated value (9) is for similar low p_{Cl_2}. Perkin's (10) and Egunov et al. (11) data are for total pressure about 1 atm. and serve to support the average line. Value by Schwab and Hantke (12) is out-of-line.

REFERENCES:

1. Taylor, N.W.; Hildebrand, J.H. *J.Amer.Chem.Soc.* 1923, *45*, 682.
2. Smith, T.L. *J.Phys. Chem.* 1955, *59*, 188.
3. Kogan, L.M.; Kol'tsov, N.S.; Litvinov, N.D. *Zh.Fiz.Khim.* 1963, *37*, 1914.
4. Vdovichenko, V.T.; Kondratenko, V.I. *Khim.Prom.* 1967, *43*, 290.
5. Raskina, A.D.; Zetkin, V.I.; Zakharov, E.V.; Kolesnikov, I.M.; Kosorotov, V.I. *Zh.Prikl.Khim.* 1972, *45*, 1333.
6. Curda, M; Holas, J. *Chem. Prumysl.* 1964, *14*, 547.
7. Blair, M.C. Jr.; Yost, D.M. *J. Amer. Chem. Soc.* 1933, *55*, 4489.
8. Jones, W.J. *J. Chem. Soc.* 1911, *99*, 392.
9. Jakowkin, A.A. *Z. Phys. Chem.* 1899, *29*, 613.
10. Perkin, W.H. *J. Chem. Soc.* 1894, *65*, 20.
11. Egunov, A.V.; Konobeev, B.I.; Ryabov, E.A.; Gubanova, T.I. *Zh. Prikl. Khim.* 1973, *46*, 1855.
12. Schwab, G.M.; Hantke, G. *Z. Phys. Chem.* 1924, *114*, 251.

COMPONENTS:	EVALUATOR:
1. Chlorine; Cl_2; [7782-50-5] 2. Tetrachloromethane; CCl_4; [56-23-5]	W. Gerrard, Department of Chemistry, The Polytechnic of North London, Holloway, London, N7 8DB. UNITED KINGDOM:

CRITICAL EVALUATION:

RECOMMENDED NUMERICAL VALUES

T/K	x_{Cl_2} (1 atm)
253.16	0.600
263.16	0.434
273.16	0.305
283.16	0.262
293.16	0.181
303.16	0.142
313.16	0.112
323.16	0.088
333.16	0.072
343.16	0.058
353.16	0.050

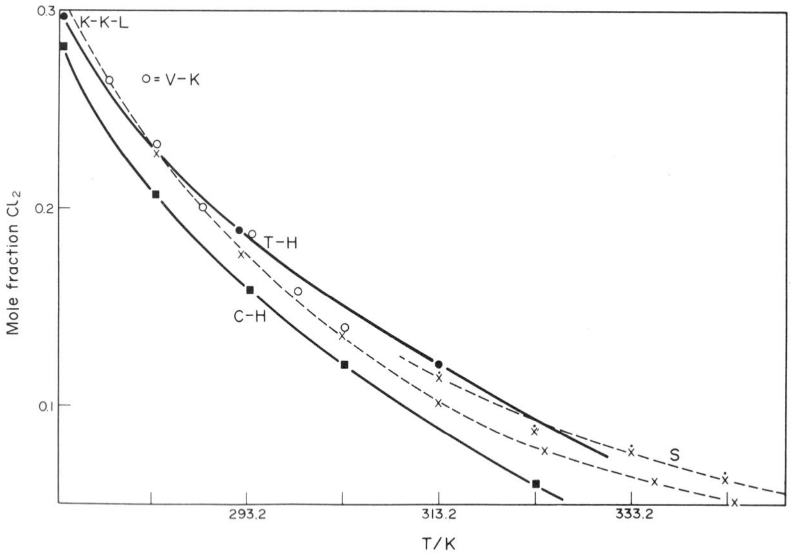

COMPONENTS:	EVALUATOR:
1. Chlorine; Cl_2; [7782-50-5] 2. Tetrachloromethane; CCl_4; [56-23-5]	W. Gerrard, Department of Chemisty, The Polytechnic of North London, Holloway, London, N7 8DB UNITED KINGDOM

CRITICAL EVALUATION:

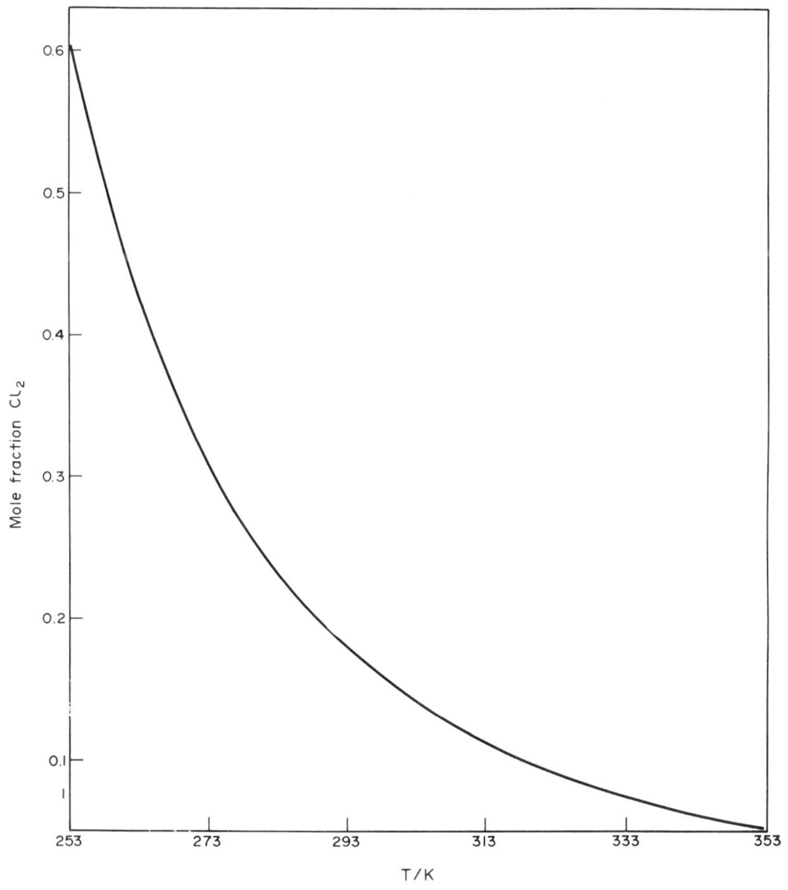

COMPONENTS:

1. Chlorine; Cl_2; [7782-50-5]
2. Tetrachloromethane; CCl_4; [56-23-5]

ORIGINAL MEASUREMENTS:

Perkin, W. H.

J. Chem. Soc.

1894, *65*, 20-28.

VARIABLES:

PREPARED BY:

W. Gerrard

EXPERIMENTAL VALUES:

T/K	wt-%	B*	x_{Cl_2}
286.16	10.29	4.02	0.199
	9.88	4.204	0.192

1 atm = 1.013×10^5 Pascal

* see note below.

AUXILIARY INFORMATION

METHOD/APPARATUS/PROCEDURE:

No details except as below.
The amount of chlorine taken up by the solvent was determined by introducing a weighed quantity of the solution into a bottle containing potassium iodide, and titrating with sodium thiosulfate. The pressure was not mentioned; it was presumably barometric.

NOTE:
The pressure was presumably total (approximately 1 atm). The solubility was expressed as a mole ratio: $1\text{-}Cl_2 + B\text{-}CCl_4$, B being the number of moles of CCl_4 for each mole of Cl_2. The wt-% was also given.

SOURCE AND PURITY OF MATERIALS:

Not specified.

ESTIMATED ERROR:

REFERENCES:

COMPONENTS:	ORIGINAL MEASUREMENTS:
1. Chlorine; Cl_2; [7782-50-5] 2. Tetrachloromethane; CCl_4; [56-23-5]	Jakowkin, A.A. *Z. Phys. Chem.* 1899, *29*, 613.
VARIABLES:	PREPARED BY: W. Gerrard

EXPERIMENTAL VALUES:

T/K	Absorption coefficient,	x_{Cl_2} (approximate, assuming p_{Cl_2} is about 1 atm and the coefficient is independent of pressure).
273.16	83.3 (83.3 = 1/0.012)	0.267

760 mmHg = 1 atm = 1.013×10^5 Pa

Solubility expressed as absorption coefficient:

$$\frac{\text{concn. in gas}}{\text{concn. in liquid}} = 0.012 \text{ (average)}$$

AUXILIARY INFORMATION

METHOD/APPARATUS/PROCEDURE:

Two bulbs, one connected to an ordinary pipette carrying a tap. Absorption was at a pressure of about 15 mm Hg. Amount absorbed was determined by a titration (potassium iodide-thiosulfate).

SOURCE AND PURITY OF MATERIALS:

Cl_2 : Self-prepared. Purity not specified.

CCl_4 : Not specified.

ESTIMATED ERROR:

REFERENCES:

COMPONENTS:

1. Chlorine; Cl_2; [7782-50-5]
2. Tetrachloromethane; CCl_4; [56-23-5]

ORIGINAL MEASUREMENTS:

Jones, W. J.
J. Chem. Soc.
1911, *99*, 392-404.

VARIABLES:

Pressure

PREPARED BY:

W. Gerrard

EXPERIMENTAL VALUES:

T/K	S	p_{Cl_2}/mmHg	x_{Cl_2}
288.16	50.7	2.34	0.000634
	51.4	7.03	0.00193
	53.7	8.29	0.00237
	51.0	12.14	0.00331

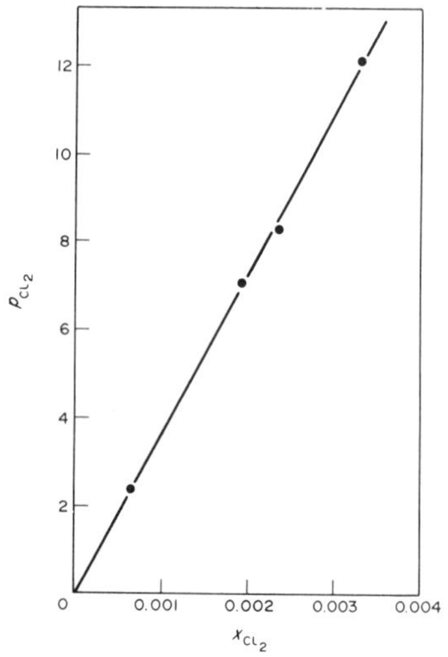

(Extrapolation to p_{Cl_2} = 1 atm is not valid.)

760 mmHg = 1 atm = 1.013×10^5 Pascal

AUXILIARY INFORMATION

METHOD/APPARATUS/PROCEDURE:

Bubbler and wash bottles. Concentration of the original solution determined iodometrically. Chlorine, carried away from the solution in a stream of air, was determined by iodometric titration. The results were given as a partition coefficient,

$$S = \frac{\text{concn. } Cl_2 \text{ in moles/dm}^3 \text{ solution}}{\text{concn. } Cl_2 \text{ in moles/dm}^3 \text{ gas phase}}.$$

S was deemed to be constant and independent of pressure, p_{Cl_2}.

NOTES: Compiler has given the approximate p_{Cl_2} and x_{Cl_2} values on the assumption that the volume of solution is equal to that of the original liquid for these low concentrations.

SOURCE AND PURITY OF MATERIALS:

1. Cl_2 : not specified.
2. CCl_4 : redistilled.

ESTIMATED ERROR:

REFERENCES:

COMPONENTS:

1. Chlorine; Cl_2; [7782-50-5]
2. Tetrachloromethane; CCl_4; [56-23-5]

ORIGINAL MEASUREMENTS:

Taylor, N. W.; Hildebrand, J. H.
J. Amer. Chem. Soc.
1923, *45*, 682-694.

VARIABLES:

Temperature

PREPARED BY:

W. Gerrard

EXPERIMENTAL VALUES:

T/K	g Cl_2/g of solution	p_{gas} /mmHg	x_{gas}	x_{gas} (1 atm)
273.16	0.156	729	0.286	0.298
292.16	0.0848	680	0.163	0.187
313.16	0.0433	557	0.0892	0.1215

760 mmHg = 1 atm = 1.013×10^5 Pascal

AUXILIARY INFORMATION

METHOD/APPARATUS/PROCEDURE:

Saturator tube (1). Chlorine was bubbled through liquid. Samples of saturated liquid were run into a weighed flask containing aqueous KI, and iodine was titrated with standard solution of thiosulfate. Temperature control was manual to 0.1 K. Saturation at barometric pressure, p_{total}, partial pressure, p_{gas}, estimated by Raoult's law to give p_{CCl_4}. Hence $p_{gas} = p_{total} - p_{CCl_4}$. x_{gas} (1 atm) estimated by Henry's law, mole fraction form.

SOURCE AND PURITY OF MATERIALS:

1. Chlorine was from an electrolytic supply; it contained less than 0.1% of free acid.
2. Purity of CCl_4 not specified. Vapor pressure data from ref. 2.

ESTIMATED ERROR:

REFERENCES:

1. Bichowsky, F. R.; Storch, H. *J. Amer. Chem. Soc.* 1915, *37*, 2695.
2. Young, S. *J. Chem. Soc.* 1891, *59*, 911.

COMPONENTS:

1. Chlorine; Cl_2; [7782-50-5]
2. Tetrachloromethane; CCl_4; [56-23-5]

ORIGINAL MEASUREMENTS:

Schwab, G.M.; Hantke, G.
Z. Phys. Chem. 1924, *114*, 251.

VARIABLES:

PREPARED BY:

W. Gerrard

EXPERIMENTAL VALUES:

T/K	s	$1/s$	Approx. x_{Cl_2} (1 atm)
292.16	0.031	32.26	0.116

760 mmHg = 1 atm = 1.013×10^5 Pa

Solubility was expressed as an absorption coefficient

$$s = \frac{\text{concn. of } Cl_2 \text{ in gas}}{\text{concn. of } Cl_2 \text{ in liquid}}$$

Actual pressure was about 700 mm Hg.

AUXILIARY INFORMATION

METHOD/APPARATUS/PROCEDURE:

Liquid held in small bulb of 1 cm^3 capacity connected to manometer assembly. Diagram given in original paper.

Absorption coefficient determined by measurement of the decrease of pressure with time in an apparatus of known volume connected with a small bulb containing a small amount of CCl_4.

SOURCE AND PURITY OF MATERIALS:

Cl_2 : Gas from liquid passed through calcium chloride.

CCl_4 : Not specified.

ESTIMATED ERROR:

±10% (Compiler)

REFERENCES:

COMPONENTS:

1. Chlorine; Cl_2; [7782-50-5]
2. Tetrachloromethane; CCl_4; [56-23-5]

ORIGINAL MEASUREMENTS:

Blair, M.C. Jr.; Yost, D.M.

J. Amer. Chem. Soc. 1933, *55*, 4489-96.

VARIABLES:

Pressure

PREPARED BY:

W. Gerrard.

EXPERIMENTAL VALUES:

p_{Cl_2} /mmHg	x_{Cl_2} at p_{Cl_2}	Henry's Constant* H_{Cl_2}
18.6	0.00394	4730
22.8	0.00448	5080
28.0 (a)	0.00609	4600
30.4	0.00615	4940
32.2	0.00690	4670
39.8	0.00781	5100
64.6	0.01245	5180
69.0	0.01363	5070
73.4	0.01391	5280

(a) Given as 2.80 in the original. The lowest value of H_{Cl_2} gives x_{Cl_2} = 0.165; the highest value gives x_{Cl_2} = 0.144, for p_{Cl_2} = 1 atm.

760 mmHg = 1 atm.
1 atm = 1.013×10^5 Pascal.

* H_{Cl_2} defined as $H_{Cl_2} = p_{Cl_2}/x_{Cl_2}$

AUXILIARY INFORMATION

METHOD /APPARATUS/PROCEDURE:

The vapor pressures, p_{Cl_2}, were determined by the iodometric determination of chlorine carried away in a stream of nitrogen (see ref. 1), at 298.16 K. The pressures were not greater than 74 mmHg.

SOURCE AND PURITY OF MATERIALS:

All reagents were stated to be of the highest purity obtainable.

ESTIMATED ERROR:

REFERENCES:

1. Bichowsky, F.R.; Storch, H.

J. Amer. Chem. Soc. 1915, *37*, 2696.

COMPONENTS:

1. Chlorine; Cl_2; [7782-50-5]
2. Tetrachloromethane; CCl_4; [56-23-5]

ORIGINAL MEASUREMENTS:

Smith, T. L.
J. Phys. Chem.
1955, *59*, 188-189.

VARIABLES:

Temperature

PREPARED BY:

W. Gerrard

EXPERIMENTAL VALUES:

T/K	Total pressure p_T/atm	Moles Cl_2/(kg solution)/atm Cl_2	x_{Cl_2}/atm Cl_2 [a]
313.16	2.02	0.797	
	2.36	0.789	0.115
323.16	2.02	0.590	
	2.36	0.602	0.0875
333.16	2.02	0.522	
	2.36	0.522	0.0771
343.16	2.36	0.417	
	2.36	0.433	0.0632
353.16	2.36	0.340	
	2.70	0.329	0.0502
363.16	2.70	0.269	0.0444

1 atm = 1.013×10^5 Pascal

[a] at 1 atm partial pressure of chlorine

AUXILIARY INFORMATION

METHOD/APPARATUS/PROCEDURE:

Two bulbs separated by stopcock. Bulbs are detachable for weighing. Bourdon gauge. Chlorine passed into liquid in upper bulb until constant gauge reading attained. Sample then run into lower bulb for weighing and iodometric titration. Saturation at a total pressure p_T of about 2 atm. Partial pressure p_{Cl_2} derived from $p_{Cl_2} = p_T - p_{CCl_4}$, assuming Raoult's law. Basis of adjustment to p_{Cl_2} = 1 atm probably Henry's law, mole fraction form.

SOURCE AND PURITY OF MATERIALS:

1. Cl_2 : from cylinder (The Matheson Co.).
2. CCl_4 : distilled.

ESTIMATED ERROR:

REFERENCES:

COMPONENTS:	ORIGINAL MEASUREMENTS:
1. Chlorine; Cl_2; [7782-50-5] 2. Tetrachloromethane; CCl_4; [56-23-5]	Kogan, L.M.; Kol'tsov, N.S.; Litvinov, N.D. *Zh. Fiz. Khim.* 1963, *37*, 1914-7.
VARIABLES:	PREPARED BY:
Temperature	W. Gerrard

EXPERIMENTAL VALUES:

T/K	Mole fraction of chlorine, x_{Cl_2}
253.16	0.603
262.16	0.456
273.16	0.309
283.46	0.228
292.36	0.177
303.16	0.136
313.16	0.102
324.16	0.0755
335.66	0.0625
344.16	0.0518

1 atm = 1.013 x 10^5 Pascal

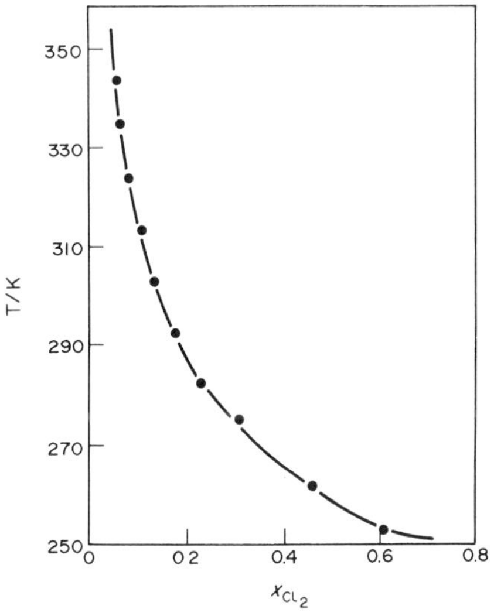

AUXILIARY INFORMATION

METHOD /APPARATUS/PROCEDURE:

Thermostatted double gas buret, null manometer, magnet-striker and stirrer. Diagram of elaborate assembly is given in the original paper.

A weighed amount of liquid in a sealed tube was placed in the absorption vessel. The system was evacuated, filled with gas, and the sealed tube was broken. The volume of gas absorbed was measured at a fixed pressure (total). Light was excluded.

Volume of chlorine measured at constant pressure, which was not stated. Henry's law was used to calculate the mole fraction values for p_{Cl_2} = 1 atm; but details were not disclosed. Allowance for p_{CCl_4} was not mentioned.

SOURCE AND PURITY OF MATERIALS:

Cl_2 conformed with USSR standard GOST 6718-53; purity stated to be 99.97.

CCl_4 was dried, distilled and checked by physical data.

ESTIMATED ERROR:

Temperature control was ±0.02 to ±0.05 °C. Volume could be read to 0.05 cm^3. The reproducibility of the volume measurements was stated to be within ±0.25%.

COMPONENTS:

1. Chlorine; Cl_2; [7782-50-5]
2. Tetrachloromethane; CCl_4; [56-23-5]

ORIGINAL MEASUREMENTS:

Curda, M., Holas, J.

Chem. Prumysl. 1964, *14*, 547-8.

VARIABLES:

Temperature

PREPARED BY:

W. Gerrard

EXPERIMENTAL VALUES:

T/K	Cl_2 wt % (p_T about 1 atm)	x_{Cl_2} (Calc. by compiler)
273.16	15.4	0.283
283.16	10.8	0.208
293.16	8.0	0.159
303.16	6.0	0.122
323.16	2.9	0.0609
343.16	0.6	0.0129

1 atm = 1.013×10^5 Pascal.

AUXILIARY INFORMATION

METHOD/APPARATUS/PROCEDURE:

Jacketted bubbler tube. Diagram in original paper.

Sample removed by a tap at the bottom of the tube. Chlorine determined iodometrically. Temperature measured within 0.5°C.

Saturation at barometric pressure, total.

SOURCE AND PURITY OF MATERIALS:

Cl_2, of technical grade, of 99.2% purity.

CCl_4, distilled, b.p. 77°C.

ESTIMATED ERROR:

REFERENCES:

COMPONENTS:

1. Chlorine; Cl_2; [7782-50-5]
2. Tetrachloromethane; CCl_4; [56-23-5]

ORIGINAL MEASUREMENTS:

Vdovichenko, V. T.; Kondratenko, V. I.
Khim. Prom.
1967, *43*, 290-291.

VARIABLES:

Temperature

PREPARED BY:

W. Gerrard

EXPERIMENTAL VALUES:

T/K	x_{Cl_2}
253.16	0.592
261.16	0.464
273.16	0.303
278.16	0.264
283.16	0.232
288.16	0.201
293.16	0.186
298.16	0.158
303.16	0.139

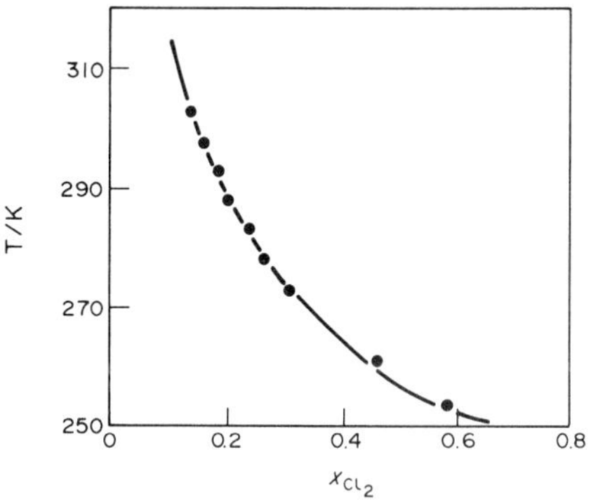

Note: Mole fraction values, x_{Cl_2} are for p_{Cl_2} = 1 atm.

1 atm = 1.013×10^5 Pascal

AUXILIARY INFORMATION

METHOD/APPARATUS/PROCEDURE:

The method described by Taylor and Hildebrand was used (1).

SOURCE AND PURITY OF MATERIALS:

1. Cl_2 : GOST 6718-53 (All Union Standard).
2. CCl_4 : b.p., d_4^{20}, and n_D^{20} given as 76.7-76.8 °C, 1.5970 and 1.4602, respectively.

ESTIMATED ERROR:

REFERENCES:

1. Taylor, N. W.; Hildebrand, J. H. *J. Amer. Chem. Soc.* 1923, *45*, 682.

COMPONENTS:

1. Chlorine; Cl_2; [7782-50-5]

2. Tetrachloromethane; CCl_4; [56-23-5]

ORIGINAL MEASUREMENTS:

Raskina, A.D; Zetkin, V.I.; Zakharov, E.V.; Kolesnikov, I.M.; Kosorotov, V.I.

Zh. Prikl. Khim. 1972, *45*, 1333-4.

VARIABLES:

Temperature

PREPARED BY:

W. Gerrard.

EXPERIMENTAL VALUES:

T/K		x_{Cl_2}
273.16		0.309
283.16	(283.46)	0.228
292.16	(292.36)	0.177
303.16	(303.06)	0.136
313.16		0.102
324.16		0.0755

1 atm = 1.013×10^5 Pascal

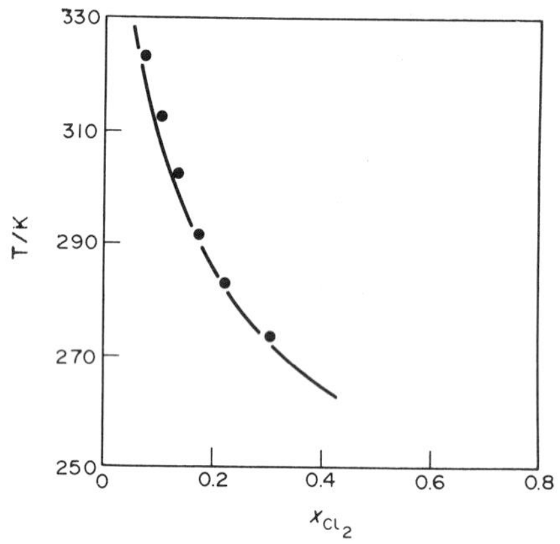

AUXILIARY INFORMATION

METHOD /APPARATUS/PROCEDURE:

Chlorine bubbled through liquid until saturated at total pressure, p_T, apparently atmospheric.

Thermostatted cylinder, 100 cm^3 capacity protected from light. Reflux condenser, bubbler fitted Schott porous plate. Cl_2 or (Cl_2+ N_2) passed through liquid (70 cm^3). Samples withdrawn periodically for iodometric analysis, until constant result obtained. Three successive analyses stated to show satisfactory reproducibility.

From the variation of the partial pressure, p_{Cl_2}, by dilution with nitrogen it was concluded that Henry's law was obeyed. Allowance for the partial pressure of CCl_4 was not mentioned. Results were given as mole %, presumably for p_{Cl_2} = 1 atm, but this was not explicitly stated. The numbers are identical with those of ref. 1, cited by authors, but three T/K ( ) are slightly different.

SOURCE AND PURITY OF MATERIALS:

Cl : GOST 6318-55 standard. Dried by sulfuric acid.

CCl : technical grade, GOST 4-65, dried and distilled.

ESTIMATED ERROR:

REFERENCES:

1. Kogan, L.M.; Kol'tsov, N.S.; Litvinov, N.D.; *Zh. Fiz. Khim.* 1963, *37*, 1914.

COMPONENTS:

1. Chlorine; Cl_2; [7782-50-5]
2. Tetrachloromethane; CCl_4; [56-23-5]

ORIGINAL MEASUREMENTS:

Egunov, A. V.; Konobeev, B. I.; Ryabov, E. A.; Gubanova, T. I.
Zh. Prikl. Khim.
1973, *46*, 1855-6.

VARIABLES:

Temperature

PREPARED BY:

W. Gerrard

EXPERIMENTAL VALUES:

T/K	wt-%	$x_{Cl_2}^{*}$
313.16	5.27	0.108
323.16	3.9	0.081
333.16	3.5	0.073

$^{*}p_{total} \approx 1$ atm $= 1.013 \times 10^5$ Pascal (calculated by compiler)

AUXILIARY INFORMATION

METHOD/APPARATUS/PROCEDURE:

Bubbler tube, differential manometer, and rheometer. Diagram given in paper. Chlorine from cylinder was passed through concentrated sulfuric acid and a rheometer into a known weight of liquid (15-20 g) until equilibrium was reached as indicated by a differential manometer.
Amount absorbed was determined by increase in weight. Light was excluded.
Chlorine bubbled through liquid until equilibrium attained at a total pressure (barometric, unspecified).

SOURCE AND PURITY OF MATERIALS:

1. Cl_2: Unspecified.
2. CCl_4: Chemically pure grade.

ESTIMATED ERROR:

REFERENCES:

COMPONENTS:	ORIGINAL MEASUREMENTS:
1. Chlorine; Cl_2; [7782-50-5] 2. Trichloromethanesulfenyl chloride; CCl_4S; [594-42-3]	Raskina, A.D.; Zetkin, V.I.; Zakharov, E.V.; Kolesnikov, I.M.; Kosorotov, V.I.; *Zh. Prik. Khim. (Leningrad)*, 1972, *42*, 1333-1334.
VARIABLES:	PREPARED BY:
Temperature	P.G.T. Fogg

EXPERIMENTAL VALUES:

T/K	g_{Cl_2}/100g of solution	Mole fraction x_{Cl_2}
275	16.0 *	0.225
278	8.9	0.204
283	7.4	0.168
288	6.0	0.145
293	5.0	0.121
295	4.1	0.102
303	3.4 *	0.098
308	2.8	0.070

The total pressure was apparently equal to barometric pressure (unspecified).

* these values are not consistent with the corresponding values of the mole fraction given by the authors.

The authors have given the following equation :

$$\log_{10} (100x_{Cl_2}) = (1250/(T/K)) - 3.20$$

The compiler has calculated that values of x_{Cl_2} quoted above fit this equation with a standard deviation of 0.007.

The authors have given the heat of solution of chlorine, calculated from the variation of solubility with temperature, as - 5.7 kcal mol^{-1}

AUXILIARY INFORMATION

METHOD/APPARATUS/PROCEDURE:

Chlorine or chlorine/nitrogen mixtures were bubbled through CCl_4S until the concentration of chlorine in the liquid phase, as determined by iodometry, was constant. The apparatus was protected from light. The total pressure of gas was apparently equal to barometric pressure although this is not explicitly stated. Variations of solubility when chlorine was diluted with nitrogen is said to indicate that Henry's law is obeyed although these experimental measurements are not reported. No allowances seem to have been made for the partial pressure of the CCl_4S.

SOURCE AND PURITY OF MATERIALS:

Cl_2 : to USSR standard GOST 6718-53 dried over H_2SO_4.

N_2 : to GOST 9293-59; dried over H_2SO_4.

CCl_4S : made by chlorination of CS_2; purified by distillation to 99.5%

ESTIMATED ERROR:

REFERENCES:

COMPONENTS:

1. Chlorine; Cl_2; [7782-50-5]
2. Carbon disulfide; CS_2; [75-15-0]
3. Trichloromethanesulfenyl chloride; CCl_4S; [594-42-3]
4. Tetrachloromethane; CCl_4; [56-23-5]

ORIGINAL MEASUREMENTS:

Raskina, A.D.; Zetkin, V.I. Zakharov, E.V.; Kolesnikov, I.M.; Kosorotov, V.I.

Zh. Prik. Khim. (Leningrad), 1972, *45,* 1333-1334.

VARIABLES:

Composition

PREPARED BY:

P.G.T. Fogg

EXPERIMENTAL VALUES:

Composition before passage of chlorine (mole %) CS_2	CCl_4S	CCl_4	g Cl_2/100g solution	Mole fraction x_{Cl_2}
15.6	18.4	66.0	14.3	0.263
15.6	66.0	18.4	12.2	0.240
62.5	7.5	30.0	21.8	0.263(0.297)*
62.5	18.7	18.8	20.1	0.250(0.283)*
88.0	5.4	6.6	21.4	0.248
42.7	23.0	34.3	16.4	0.263

Temperature : 275 K

The total pressure was apparently equal to barometric pressure (unspecified).

* values recalculated by the compiler from wt % of Cl_2 given by the authors.

AUXILIARY INFORMATION

METHOD/APPARATUS/PROCEDURE:

Chlorine was bubbled through the liquid until the concentration of chlorine in the liquid phase, as determined by iodometry, was constant. The apparatus was protected from light.

The total pressure of gas was apparently equal to barometric pressure although this is not explicitly stated.

SOURCE AND PURITY OF MATERIALS:

Cl_2 : to USSR standard GOST 6718-53; additionally dried over H_2SO_4.

CS_2 : to GOST 1541-42; distilled and dried over ignited $CaCl_2$.

CCl_4S: made by chlorination of CS_2; purified by distillation to 99.5%

CCl_4 : technical grade (GOST 4-65); dried over ignited $CaCl_2$.

ESTIMATED ERROR:

REFERENCES:

COMPONENTS:	ORIGINAL MEASUREMENTS:
1. Chlorine; Cl_2; [7782-50-5] 2. 1,2-Dibromoethane; $C_2H_4Br_2$; [106-93-4]	Taylor, N. W.; Hildebrand, J. H. *J. Amer. Chem. Soc.* 1923, *45*, 682-694.
VARIABLES:	**PREPARED BY:**
Temperature	P. G. T. Fogg

EXPERIMENTAL VALUES:

T/K	g Cl_2/g of solution	p_{Cl_2}/mmHg	Mole fractions x_{Cl_2}	x_{Cl_2}(1 atm)
293.16	0.0813	752	0.190	0.192
313.16	0.0469	736	0.1151	0.119

760 mmHg = 1 atm

1 atm = 1.013×10^5 Pascal

AUXILIARY INFORMATION

METHOD/APPARATUS/PROCEDURE:

Saturator tube (1). Chlorine was bubbled through liquid. Samples of saturated liquid were run into a weighed flask containing aqueous KI, and iodine was titrated with standard solution of thiosulfate. Temperature control was manual to 0.1 K.
Saturation was at barometric pressure, p_{total}. The value of $p_{C_2H_4Br_2}$ was estimated by Raoult's law from vapor pressure data given in ref. (2). Hence:

$$p_{Cl_2} = p_{total} - p_{C_2H_4Br_2}$$

The authors then used Henry's law (mole fraction form) to calculate x_{Cl_2} (1 atm) from x_{Cl_2}.

SOURCE AND PURITY OF MATERIALS:

1. Cl_2: from an electrolytic supply; contained less than 0.1% of free acid.

2. $C_2H_4Br_2$: not specified.

ESTIMATED ERROR:

REFERENCES:

1. Bichowsky, F. R.; Storch, H. *J. Amer. Chem. Soc.* 1915, *37*, 2695.

2. Landolt-Börnstein Tabellen.

COMPONENTS:	ORIGINAL MEASUREMENTS:
1. Chlorine; Cl_2; [7782-50-5] 2. Tetrachloroethene; C_2Cl_4; [127-18-4]	Curda, M.; Holas, J. *Chem. Prum.* 1964, *14*, 547-548.
VARIABLES:	PREPARED BY:
Temperature	P. G. T. Fogg

EXPERIMENTAL VALUES:

T/K	g $_{Cl_2}$/100 g soln.	Mole fraction,* x_{Cl_2}
273.2	15.4	0.299
283.2	10.8	0.221
293.2	8.1	0.171
303.2	6.4	0.138
323.2	3.8	0.085
343.2	2.2	0.050
363.2	1.0	0.023

The pressure of Cl_2 plus C_2Cl_4 was equal to barometric pressure.

* Calculated by compiler.

AUXILIARY INFORMATION

METHOD/APPARATUS/PROCEDURE:

Chlorine was passed into the C_2Cl_4 in a jacketted bubbler tube (diagram in original paper). Samples were withdrawn through a tap at the bottom of the tube. Temperature was controlled to ± 0.5 °C. The total pressure was equal to barometric pressure (unspecified).

SOURCE AND PURITY OF MATERIALS:

1. Cl_2: technical grade, 99.2% pure.

2. C_2Cl_4: technical grade; distilled; b.p. 120 °C.

ESTIMATED ERROR:

REFERENCES:

COMPONENTS:

1. Chlorine; Cl_2; [7782-50-5]

2. 1,1,2,2-Tetrachloroethane; $C_2H_2Cl_4$; [79-34-5]

ORIGINAL MEASUREMENTS:

Kalz, G.; Naumann, A.
Plaste Kautsch. 1971, *18*, 500-501.

VARIABLES:

Temperature

PREPARED BY:

P. G. T. Fogg

EXPERIMENTAL VALUES:

T/K	g $Cl_2 dm^{-3}$	Mole fraction*, x_{Cl_2}
298.2	51.77	0.072
323.2	50.25	0.071
348.2	33.05	0.049
363.2	18.72	0.029
383.2	17.87	0.028

The pressure of Cl_2 plus $C_2H_2Cl_4$ was apparently equal to barometric pressure (unspecified).

* Approximate values calculated by compiler on the assumption that dissolution of Cl_2 causes no change in volume of the liquid phase. Densities of $C_2H_2Cl_4$ at 293.2 K and 298.2 K are given in ref. (1). Densities at other temperatures were obtained by extrapolation.

AUXILIARY INFORMATION

METHOD/APPARATUS/PROCEDURE:

Chlorine from a cylinder was passed into 400 cm^3 of the solvent in a flask fitted with stirrer and reflux condenser. Temperatures were controlled by a thermostat. The chlorine which dissolved was determined by iodometry. The extent of irreversible reaction between chlorine and tetrachloroethane was measured by estimation of chloride ions produced. At 90 °C about 3% of the tetrachloroethane reacted but at other temperatures there was less than 0.2% reaction.

SOURCE AND PURITY OF MATERIALS:

1. Cl_2: from a cylinder.

2. $C_2H_2Cl_4$: b.p. 146.2 °C (barometric pressure unspecified); density (20 °C) 1.5950 g cm^{-3}; n_D (20 °C) 1.4938.

ESTIMATED ERROR:

REFERENCES:

1. J. Timmermans, "*Physico-Chemical Constants of Pure Organic Compounds*", Vol. 2, Elsevier, London. 1965.

COMPONENTS:

1. Chlorine; Cl_2; [7782-50-5]

2. 1,1,2,3,4,4-Hexachloro-1,3-butadiene; C_4Cl_6; [87-68-3]

ORIGINAL MEASUREMENTS:

Kogan, L. M.; Kol'tsov, N. S.; Litvinov, N. D. *Zh. Fiz. Khim.* 1963, *37*, 1014-1016.

VARIABLES:

Temperature, pressure

PREPARED BY:

P. G. T. Fogg

EXPERIMENTAL VALUES:

T = 303.2 K

p_{Cl_2}/mmHg	Mole fraction x_{Cl_2}
408.4	0.0844
600.8	0.125
745.7	0.155
773.7	0.160
926.2	0.191
974.2	0.202

p_{Cl_2} = 760 mmHg

T/K	Mole fraction x_{Cl_2}
253.5	0.607
261.7	0.467
266.0	0.421
271.8	0.358
272.8	0.353
283.0	0.250
293.4	0.199
303.2	0.158
334.4	0.083
352.9	0.059
365.3	0.052

760 mmHg = 1 atm

1 atm = 1.013×10^5 Pascal

AUXILIARY INFORMATION

METHOD/APPARATUS/PROCEDURE:

The volumes of gas absorbed by weighed quantities of hexachlorobutadiene were determined by a static method using a thermostatted gas burette connected to absorption vessels and a vacuum line as described in ref. (1). Pressures were measured by a mercury manometer. Chlorine did not come into contact with any metal surfaces. Temperatures were controlled to ± 0.02 K in the range 283 K - 353 K and to ± 0.05 outside this range. The apparatus was shielded from light.

SOURCE AND PURITY OF MATERIALS:

1. Cl_2: to USSR standard GOST 6718-53.

2. C_4Cl_6: m.p. 18.7 °C; density (20 °C) 1.6793 g cm^{-3}; n_D (20 °C) 1.5558.

ESTIMATED ERROR:

REFERENCES:

1. Kogan, L. M.; Kol'tsov, N. S.; Litvinov, N. D. *Zh. Fiz. Khim.* 1963, *37*, 1040.

COMPONENTS:

1. Chlorine; Cl_2; [7782-50-5]

2. Hexadecafluoroheptane; (Perfluoroheptane); C_7F_{16}; [335-57-9]

ORIGINAL MEASUREMENTS:

Gjaldbaek, J. C.; Hildebrand, J. H.

J. Amer. Chem. Soc. 1950, *72*, 609-611.

VARIABLES:

Temperature

PREPARED BY:

P. G. T. Fogg

EXPERIMENTAL VALUES:

T/K	Bunsen coefficient	Mole fraction, x_{Cl_2} (1 atm)
273.16	20.1	0.164[a]
288.76		0.1165
288.96		0.1174
292.96		0.1101
293.16	12.3	0.110[b]
295.16		0.1051
298.16	10.7	0.0977
298.46		0.0975

[a] Extrapolated

[b] Interpolated

1 atm = 1.013×10^5 Pascal

AUXILIARY INFORMATION

METHOD/APPARATUS/PROCEDURE:

Chlorine was bubbled through solvent for about 1 hour and the chlorine in a weighed sample of the solution estimated by iodometry. Temperature control was manual to 0.1 K. Saturation was at barometric pressure, p_{total}. The value of $p_{C_7F_{16}}$ was estimated by Raoult's law from vapor pressure data for C_7F_{16} in ref. (1). Hence:

$$p_{Cl_2} = p_{total} - p_{C_7F_{16}}$$

The authors then used Henry's law (mole fraction form) to calculate x_{Cl_2} (1 atm).

SOURCE AND PURITY OF MATERIALS:

1. Cl_2: from a cylinder (Ohio Chem. Co.)

2. C_7F_{16}: distilled, b.p. 355.49 - 355.58 K (753.8 mmHg).

ESTIMATED ERROR:

REFERENCES:

1. Fowler, R. D.; Hamilton, J. M.; Kasper, J. S.; Weber, C. E.; Burford, W. B.; Anderson, H. C. *Ind. Eng. Chem.* 1947, *39*, 375.

COMPONENTS:

1. Chlorine; Cl_2; [7782-50-5]
2. 1,2-Dichlorobenzene; $C_6H_4Cl_2$; [95-50-1]

ORIGINAL MEASUREMENTS:

Zetkin, V.I.; Kosorotov, V.I.; Stul, B.Ya.; Dzhagatspanyan, R.V.;

Khim. Prom. 1971, *47*, 102-103.

VARIABLES:

Temperature

PREPARED BY:

P.G.T. Fogg.

EXPERIMENTAL VALUES:

T/K	x_{Cl_2}
288.2	0.249
293.2	0.230
303.2	0.158
313.2	0.120
323.2	0.100
333.2	0.083

The pressure of Cl_2 plus $C_6H_4Cl_2$ was apparently equal to barometric pressure.

The following equation was given :

$$\log_{10} x_{Cl_2} = (960/(T/K)) - 4.02.$$

The heat of solution of Cl_2 calculated from the variation of solubility with temperature was given as - 4400 cal mol^{-1}.

AUXILIARY INFORMATION

METHOD/APPARATUS/PROCEDURE:

The method and apparatus were described in ref. (1). Chlorine was bubbled through a measured quantity of dichlorobenzene in a thermostatted vessel of capacity 100 cm^3. Samples were withdrawn periodically for iodometric analysis until consistent results were obtained.

SOURCE AND PURITY OF MATERIALS:

Cl_2 : to USSR standard GOST 6718-53.

$C_6H_4Cl_2$: 99% pure. Dried with $CaCl_2$

ESTIMATED ERROR:

REFERENCES:

1. Zetkin, V.I. et. al.

Zh. Fiz. Khim. 1970, *44*, 830.

COMPONENTS:

1. Chlorine; Cl_2; [7782-50-5]
2. 1,2,4-Trichlorobenzene; $C_6H_3Cl_3$; [120-82-1]

ORIGINAL MEASUREMENTS:

Zetkin, V.I. Kosorotov, V.I.; Stul, B.Ya.; Dzhagatspanyan, R.V.;

Khim. Prom. 1971, *47*, 102-103.

VARIABLES:

Temperature

PREPARED BY:

P.G.T. Fogg.

EXPERIMENTAL VALUES:

T/K	x_{Cl_2}
288.2	0.254
293.2	0.210
303.2	0.151
313.2	0.117
323.2	0.091
333.2	0.070

The pressure of Cl_2 plus $C_6H_3Cl_3$ was apparently equal to barometric pressure.

The following equation was given :

$$\log_{10} x_{Cl_2} = (1050/(T/K)) - 4.29$$

The heat of solution of Cl_2 calculated from the variation of solubility with temperature was given as - 4800 cal mol^{-1}.

AUXILIARY INFORMATION

METHOD/APPARATUS/PROCEDURE:

The method and apparatus were described in ref. (1). Chlorine was bubbled through a measured quantity of trichlorobenzene in a thermostatted vessel of capacity 100 cm^3. Samples were withdrawn periodically for iodometric analysis until consistent results were obtained.

SOURCE AND PURITY OF MATERIALS:

Cl_2 : to USSR standard GOST 6718-53.

$C_6H_3Cl_3$: 99% pure. Dried with $CaCl_2$

ESTIMATED ERROR:

REFERENCES:

1. Zetkin, V.I. et. al.; *Zh. Fiz. Khim.* 1970, *44*, 830.

COMPONENTS:

1. Chlorine; Cl_2; [7782-50-5]
2. Chlorobenzene; C_6H_5Cl; [108-90-7]

ORIGINAL MEASUREMENTS:

Silberstein, B.; Date, R. V.; Butt, J. B.
J. Chem. Engng. Data 1969, *14*, 253-4.

VARIABLES:

PREPARED BY:

C. L. Young

EXPERIMENTAL VALUES:

T/K	Solubility†, S/mol dm^{-3}	Mole fraction†* of chlorine in liquid, x_{Cl_2}
303.15	1.56	0.138
318.15	1.10	0.103
333.15	0.78	0.076

† at a partial pressure of 101.3 kPa.

* calculated by compiler assuming solubility is given as mol dm^{-3} (solvent).

AUXILIARY INFORMATION

METHOD/APPARATUS/PROCEDURE:

Chlorobenzene saturated with gas by bubbling chlorine through at 283 K. Temperature then raised to experimental temperature and chlorine allowed to equilibriate. Then stannic chloride in chlorobenzene solution added to cause chlorine to react with chlorobenzene to give dichlorobenzene.

SOURCE AND PURITY OF MATERIALS:

1. Matheson high purity grade; purity better than 99.5 mole per cent.
2. Fisher Scientific certified reagent, dried over $CaCl_2$, distilled middle fraction stored over calcium hydride in dark.

ESTIMATED ERROR:

$\delta T/K = \pm 0.05$; $\delta x_{Cl_2} = \pm 3\%$.

REFERENCES:

COMPONENTS:

1. Chlorine; Cl_2; [7782-50-5]

2. Chlorobenzene; C_6H_5Cl; [108-90-7]

ORIGINAL MEASUREMENTS:

Cervinka, M.
Chem. Prum. 1960, *10*, 249-251.

VARIABLES:

Pressure, temperature

PREPARED BY:

P. G. T. Fogg

EXPERIMENTAL VALUES:

T/K	H/mmHg	Mole fraction, x_{Cl_2} (1 atm) (calculated by compiler)
298.2	6.61×10^3	0.115
317.2	8.32×10^3	0.0913
338.2	12.4×10^3	0.0613

The Henry's law constant, H, was taken to be the slope of a plot of pressure of Cl_2 in the gas phase against mole fraction of Cl_2 in the liquid phase. The author has given the following equation:

$$\log_{10}H = 6.0234 - 659.8/(T/K)$$

The compiler has calculated that the experimental values of H quoted above fit this equation with a standard deviation of 0.54×10^3.

The heat of solution of chlorine, calculated by the author from the variation of H with temperature, is -3.01 kcal mol^{-1}.

760 mmHg = 1 atm
1 atm = 1.013×10^5 Pascal

AUXILIARY INFORMATION

METHOD/APPARATUS/PROCEDURE:

The absorption of chlorine at partial pressures up to 1 atm was determined by passing mixtures of chlorine and air into chlorobenzene in an absorption vessel. The apparatus is described in ref. (1). Temperatures were controlled to ± 0.5 K. The chlorine content of the gas phase and the liquid phase were determined by iodometry.

SOURCE AND PURITY OF MATERIALS:

1. Cl_2 & air: dried with concentrated H_2SO_4.

2. C_6H_5Cl: distilled before use.

ESTIMATED ERROR:

REFERENCES:

1. Balej, J.; Regner, A.
Chem. Listy 1956, *50*, 1374.

COMPONENTS:

1. Chlorine; Cl_2; [7782-50-5]
2. Chlorobenzene; C_6H_5Cl; [108-90-7]

ORIGINAL MEASUREMENTS:

Zetkin, V.I.; Kosorotov, V.I.; Stul, B.Ya.; Dzhagatspanyan, R.V.;

Khim. Prom. 1971, *47*, 102-103.

VARIABLES:

Temperature

PREPARED BY:

P.G.T. Fogg.

EXPERIMENTAL VALUES:

T/K	x_{Cl_2}
283.2	0.250
293.2	0.178
303.2	0.141
313.2	0.112
323.2	0.079
333.2	0.071

The pressure of Cl_2 plus C_6H_5Cl was apparently equal to barometric pressure.

The following equation was given :

$$\log_{10} x_{Cl_2} = 980/(T/K) - 4.08$$

The heat of solution of Cl_2 calculated from the variation of solubility with temperature was given as - 4500 cal mol^{-1}.

AUXILIARY INFORMATION

METHOD/APPARATUS/PROCEDURE:

The method and apparatus were described in ref. (1). Chlorine was bubbled through a measured quantity of chlorobenzene in a thermostatted vessel of capacity 100 cm^3. Samples were withdrawn periodically for iodometric analysis until consistent results were obtained.

SOURCE AND PURITY OF MATERIALS:

Cl_2 : to USSR standard GOST 6718-53 Dried by passing through H_2SO_4.

C_6H_5Cl:99% pure. Dried with $CaCl_2$.

ESTIMATED ERROR:

REFERENCES:

1. Zetkin, V.I.; et al.; *Zh. Fiz. Khim.* 1970, *44*, 830

COMPONENTS:

1. Chlorine; Cl_2; [7782-50-5]

2. Chlorobenzene; C_6H_5Cl; [108-90-7]

ORIGINAL MEASUREMENTS:

Krentsel, L. B.; Litmanovich, A. D.; Shtern, V. Ya.

Zh. Prik. Khim. (Leningrad) 1972, *45*, 1875-1876.

VARIABLES:

Pressure, temperature

PREPARED BY:

P. G. T. Fogg

EXPERIMENTAL VALUES:

T/K	H/mmHg	Mole fraction, x_{Cl_2} (1 atm) (calculated by compiler)
313.2	7.6×10^3	0.100
323.2	9.3×10^3	0.082
333.2	11.4×10^3	0.067
343.2	13.5×10^3	0.056

760 mmHg = 1 atm
1 atm = 1.013×10^5 Pascal

The Henry's law constant, H, was taken to be the slope of a plot of pressure of Cl_2 in the gas phase against mole fraction of Cl_2 in the liquid phase. The authors stated that values fit the equation:

$$\log_{10}H = 6.83 - 925/(T/K)).$$

The compiler has calculated that the values of H which have been given fit this equation with a standard deviation of 0.09×10^3.

The heat of solution of chlorine in chlorobenzene, calculated by the authors from the variation of H with T, was given as -4.3 kcal mol^{-1}.

AUXILIARY INFORMATION

METHOD/APPARATUS/PROCEDURE:

A high vacuum assembly was used. The amount of chlorine dissolved by a measured volume of chlorobenzene was calculated from the decrease in pressure of a known volume of chlorine. Pressures were measured by a mercury manometer but direct contact between mercury and chlorine was prevented by the use of a glass Bourdon type gauge as a null instrument. Benzoquinone (0.1 wt-%) was added to the chlorobenzene and experiments were conducted in a darkened room.

The authors stated that reproducible results were obtained and that this was evidence that there was no irreversible reaction between chlorine and chlorobenzene under the conditions of the experiment.

SOURCE AND PURITY OF MATERIALS:

1. Cl_2: condensed and degassed under vacuum.

2. C_6H_5Cl: degassed under vacuum; vapor pressure agreed with literature values in ref. (1).

ESTIMATED ERROR:

REFERENCES:

1. Stull, D. R. *Vapor Pressures of Pure Substances* (Russian translation), *Izd. Inostr. Lit.* 1949.

COMPONENTS:

1. Chlorine; Cl_2; [7782-50-5]
2. (Trifluoromethyl)benzene; (Benzotrifluoride); $C_7H_5F_3$; [98-08-8]

ORIGINAL MEASUREMENTS:

Ushakov, A.A.; Kosorotov, V.I.; Stul, B. Ya.; Motsarev, G.V.; Dzhagatspanyan, R.V.

Zhur. Prikl. Khim. 1977, *50*, 425-7.

VARIABLES:

Temperature

PREPARED BY:

C.L. Young

EXPERIMENTAL VALUES:

T/K	Mole ratio solubility	Wt. % solubility g Cl_2/g $C_7H_5F_3$	Mole fraction of chlorine in liquid, x_{Cl_2} +
279	0.2550	12.40	0.2035
298	0.1406	6.84	0.1235
313	0.0896	4.36	0.0824
323	0.0646	3.14	0.0608
333	0.0489	2.38	0.0468
353	0.0276	1.30	0.0261

\+ calculated by compiler

AUXILIARY INFORMATION

METHOD/APPARATUS/PROCEDURE:

Solvent saturated with gas by bubbling chlorine through at a rate of 0.3 mol h^{-1}. Samples withdrawn and concentration of chlorine determined by iodometric titration.

SOURCE AND PURITY OF MATERIALS:

1. Dried with sulfuric acid and calcium chloride.

2. Distilled, purity not less than 99.5 mole per cent.

ESTIMATED ERROR:

$\delta T/K = \pm 0.3$; $\delta x_{Cl_2} = \pm 2\%$.

(estimated by compiler).

REFERENCES:

COMPONENTS:

1. Chlorine; Cl_2; [7782-50-5]

2. (Trichloromethyl)benzene; $C_7H_5Cl_3$; [98-07-7]

ORIGINAL MEASUREMENTS:

Egunov, A. V.; Konobeev, B. I.; Ryabov, E. A.; Gubanova, T. I.

Zh. Prik. Khim. (Leningrad) 1973, *46*, 1855-1856.

VARIABLES:

Temperature

PREPARED BY:

P. G. T. Fogg

EXPERIMENTAL VALUES:

T/K	Mole fraction, x_{Cl_2}
303.2	0.124
313.2	0.105
323.2	0.084
333.2	0.071
343.2	0.060
353.2	0.050
363.2	0.047
373.2	0.035
393.2	0.022
423.2	0.014

The partial pressure of chlorine plus the partial pressure of benzotrichloride was apparently equal to barometric pressure (unspecified).

AUXILIARY INFORMATION

METHOD/APPARATUS/PROCEDURE:

Chlorine was bubbled through a weighed portion of the solvent (15-20 g) until a differential manometer registered no pressure drop when the chlorine supply was discontinued. The chlorine absorbed was found from the increase in weight of the solution. Loss of solvent in the gas stream was shown to be negligible in comparison with other experimental errors which the authors estimate to be ± 5%. Light was excluded. Products of further chlorination were estimated by gas chromatography and appropriate corrections made.

SOURCE AND PURITY OF MATERIALS:

1. Cl_2: from a cylinder; dried over H_2SO_4.

2. $C_7H_5Cl_3$: "pure" grade.

ESTIMATED ERROR:

REFERENCES:

COMPONENTS:

1. Chlorine; Cl_2; [7782-50-5]

2. (Dichloromethyl)benzene; $C_7H_6Cl_2$; [98-87-3]

ORIGINAL MEASUREMENTS:

Egunov, A. V.; Konobeev, B. I.; Ryabov, E. A.; Gubanova, T. I.

Zh. Prik. Khim. (Leningrad) 1973, *46*, 1855-1856.

VARIABLES:

Temperature

PREPARED BY:

P. G. T. Fogg

EXPERIMENTAL VALUES:

T/K	Mole fraction, x_{Cl_2}
303.2	0.124
313.2	0.107
323.2	0.089
333.2	0.072
343.2	0.067
353.2	0.053
363.2	0.045
373.2	0.034
383.2	0.029
403.2	0.020

The partial pressure of chlorine plus partial pressure of benzal chloride was apparently equal to barometric pressure (unspecified).

AUXILIARY INFORMATION

METHOD/APPARATUS/PROCEDURE:

Chlorine was bubbled through a weighed portion of the solvent (15-20 g) until a differential manometer registered no pressure drop when the chlorine supply was discontinued. The chlorine absorbed was found from the increase in weight of the solution. Loss of solvent in the gas stream was shown to be negligible in comparison with other experimental errors which the authors estimate to be ± 5%. Light was excluded. Products of further chlorination were estimated by gas chromatography and appropriate corrections made.

SOURCE AND PURITY OF MATERIALS:

1. Cl_2: from a cylinder; dried over H_2SO_4.

2. $C_7H_6Cl_2$: by fractionation of products of chlorination of toluene; contained 1.2 wt-% of benzotrichloride.

ESTIMATED ERROR:

REFERENCES:

COMPONENTS:	ORIGINAL MEASUREMENTS:
1. Chlorine; Cl_2; [7782-50-5] 2. (Chloromethyl)benzene; C_7H_7Cl ; [100-44-7]	Egunov, A. V.; Konobeev, B. I.; Ryabov, E. A.; Gubanova, T. I. *Zh. Prik. Khim. (Leningrad)* 1973, *46*, 1855-1856.
VARIABLES:	PREPARED BY:
Temperature	P. G. T. Fogg

EXPERIMENTAL VALUES:

T/K	Mole fraction, x_{Cl_2}
303.2	0.125
313.2	0.108
323.2	0.087
333.2	0.074
343.2	0.062
353.2	0.052
363.2	0.044
373.2	0.037
383.2	0.027

The partial pressure of chlorine plus the partial pressure of benzyl chloride was apparently equal to barometric pressure (unspecified).

AUXILIARY INFORMATION

METHOD/APPARATUS/PROCEDURE:

Chlorine was bubbled through a weighed portion of the solvent (15-20 g) until a differential manometer registered no pressure drop when the chlorine supply was discontinued. The chlorine absorbed was found from the increase in weight of the solution. Loss of solvent in the gas stream was shown to be negligible in comparison with other experimental errors which the authors estimate to be ± 5%. Light was excluded. Products of further chlorination were estimated by gas chromatography and appropriate corrections made during calculation of solubility.

SOURCE AND PURITY OF MATERIALS:

1. Cl_2: from a cylinder; dried over H_2SO_4.

2. C_7H_7Cl : "pure" grade.

ESTIMATED ERROR:

REFERENCES:

COMPONENTS:

1. Chlorine; Cl_2; [7782-50-5]
2. 1-Chloro-2-methylbenzene (2-Chlorotoluene); C_7H_7Cl; [95-49-8]

ORIGINAL MEASUREMENTS:

Lohse, M.; Deckwer, W.-D.
J. Chem. Eng. Data 1981, *26*, 159-161.

VARIABLES:

Temperature

PREPARED BY:

C. L. Young

EXPERIMENTAL VALUES:

T/K	Solubility[§] S/mol dm^{-3} (soln.)
288.15	2.306
298.15	1.745
308.15	1.311
318.15	1.036
328.15	0.842
338.15	0.674
348.15	0.547

§ at a partial pressure of 101.3 kPa.

AUXILIARY INFORMATION

METHOD/APPARATUS/PROCEDURE:

Solvent saturated with chlorine in dark (~1% phenol added to prevent photochemical chlorination). Samples withdrawn added to known amount of additional solvent and then 250 cm^3 of potassium iodide solution. Released iodine titrated with thiosulfate.

SOURCE AND PURITY OF MATERIALS:

No details given.

ESTIMATED ERROR:

$\delta T/K = \pm 0.1$; $\delta S = \pm 3\%$ (estimated by compiler).

REFERENCES:

COMPONENTS:

1. Chlorine; Cl_2; [7782-50-5]
2. 1-Chloro-3-methylbenzene (3-Chlorotoluene); C_7H_7Cl; [108-41-8]

ORIGINAL MEASUREMENTS:

Lohse, M.; Deckwer, W.-D.
J. Chem. Eng. Data
1981, *26*, 159-161.

VARIABLES:

Temperature

PREPARED BY:

C. L. Young

EXPERIMENTAL VALUES:

T/K	Solubility§, S/mol dm^{-3} (soln.)
298.15	1.398
308.15	1.180
318.15	0.940
338.15	0.624
348.15	0.499

§ at a partial pressure of 101.3 kPa.

AUXILIARY INFORMATION

METHOD/APPARATUS/PROCEDURE:

Solvent saturated with chlorine in dark (∿1% phenol added to prevent photochemical chlorination). Samples withdrawn added to known amount of additional solvent and then 250 cm^3 of potassium iodide solution. Released iodine titrated with thiosulfate.

SOURCE AND PURITY OF MATERIALS:

No details given.

ESTIMATED ERROR:

$\delta T/K = \pm 0.1$; $\delta S = \pm 3\%$ (estimated by compiler).

REFERENCES:

COMPONENTS:

1. Chlorine; Cl_2; [7782-50-5]
2. 1-Chloro-4-methylbenzene (4-Chlorotoluene); C_7H_7Cl; [106-43-4]

ORIGINAL MEASUREMENTS:

Lohse, M.; Deckwer, W.-D.
J. Chem. Eng. Data
1981, *26*, 159-161.

VARIABLES:

Temperature

PREPARED BY:

C. L. Young

EXPERIMENTAL VALUES:

T/K	Solubility§, S/mol dm^{-3} (soln.)
298.15	1.386
318.15	0.884
328.15	0.741
338.15	0.552
348.15	0.479

§ at a partial pressure of 101.3 KPa.

AUXILIARY INFORMATION

METHOD/APPARATUS/PROCEDURE:

Solvent saturated with chlorine in dark (~1% phenol added to prevent photochemical chlorination). Samples withdrawn added to known amount of additional solvent and then 250 cm^3 of potassium iodide solution. Released iodine titrated with thiosulfate.

SOURCE AND PURITY OF MATERIALS:

No details given.

ESTIMATED ERROR:

$\delta T/K = \pm 0.1$; $\delta S = \pm 3\%$ (estimated by compiler).

REFERENCES:

COMPONENTS:

1. Chlorine; Cl_2; [7782-50-5]
2. (Trifluoromethyl) benzene, (Benzotrifluoride); $C_7H_5F_3$; [98-08-8]
3. Chloro(trifluoromethyl)benzene; $C_7H_4Cl_3F$; [52181-51-8]
4. Dichloro(trifluoromethyl)benzene; $C_7H_3Cl_2F_3$;

ORIGINAL MEASUREMENTS:

Ushakov, A.A.; Kosorotov, V.I. Stul, B. Ya., Motsarev, G.V.; Dzhagatspanyan, R.V.

Zhur. Prikl. Khim. 1977, *50*, 425-7

VARIABLES:

Temperature, composition

PREPARED BY:

C.L. Young

EXPERIMENTAL VALUES:

T/K	Mole ratio solubility	Wt. % solubility g Cl_2/100g(2+3+4)	Mole fraction[+] of chlorine in liquid, x_{Cl_2}
Mixture containing 58.29 Wt % $C_7H_5F_3$, 39.47 Wt % $C_7H_4ClF_3$ and 2.24 Wt % $C_7H_3Cl_2F_3$.			
298	0.1543	6.80	0.1325
313	0.1040	4.58	0.0933
323	0.0799	3.52	0.0733
333	0.0604	2.66	0.0563
353	0.0381	1.68	0.0363
Mixture containing 22.82 Wt % $C_7H_5F_3$, 60.86 Wt % $C_7H_4ClF_3$, and 16.32 Wt % $C_7H_3Cl_2F_3$.			
298	0.1727	6.88	0.1457
313	0.1255	5.00	0.1102
323	0.1024	4.08	0.0918
333	0.0833	3.32	0.0760
353	0.0587	2.20	0.0517

+ calculated by compiler.

AUXILIARY INFORMATION

METHOD/APPARATUS/PROCEDURE:

Solvent saturated with gas by bubbling chlorine through at a rate of 0.3 mol h^{-1}. Samples withdrawn and concentration of chlorine determined by iodometric titration.

SOURCE AND PURITY OF MATERIALS:

1. Dried with sulfuric acid and calcium chloride.
2. Distilled, purity not less than 99.5 mole per cent.
3. and 4. Prepared by chlorination of benzotrifluoride in presence of ferric chloride at 60°C.

ESTIMATED ERROR:

$\delta T/K = \pm 0.3$; $\delta x_{Cl_2} = \pm 2\%$.

(estimated by compiler).

REFERENCES:

COMPONENTS:

1. Chlorine; Cl_2; [7782-50-5]
2. (Trifluoromethyl)benzene; (Benzotrifluoride); $C_7H_5F_3$; [98-08-8]
3. Chloro(trifluoromethyl)benzene; $C_7H_4ClF_3$

ORIGINAL MEASUREMENTS:

Ushakov, A.A.; Kosorotov, V.I.; Stul, B.Ya.; Motsarev, G.V. Dzhagatspanyan, R.V.

Zhur. Prikl. Khim. 1977, *50*, 425-7.

VARIABLES:

Temperature

PREPARED BY:

C.L. Young

EXPERIMENTAL VALUES:

Mixture containing 84.84 weight percent of benzotrifluoride before passage of chlorine

T/K	Mole ratio solubility	Wt. % solubility g Cl_2/g (2+3)	Mole fraction of chlorine in liquid, x_{Cl_2} +
298	0.1340	6.60	0.1229
313	0.0870	4.58	0.0886
323	0.0660	3.62	0.0713
333	0.0510	2.80	0.0561
353	0.0324	1.70	0.0348

\+ calculated by compiler

AUXILIARY INFORMATION

METHOD/APPARATUS/PROCEDURE:

Solvent saturated with gas by bubbling chlorine through at a rate of 0.3 mol h^{-1}. Samples withdrawn and concentration of chlorine determined by iodometric titration.

SOURCE AND PURITY OF MATERIALS:

1. Dried with sulfuric acid and calcium chloride.
2. Distilled, purity not less than 99.5 mole per cent.
3. Prepared by chlorination of benzotrifluoride in presence of ferric chloride at 60°C.

ESTIMATED ERROR:

$\delta T/K = \pm 0.3$; $\delta x_{Cl_2} = \pm 2\%$.

(estimated by compiler)

REFERENCES:

COMPONENTS:

1. Chlorine; Cl_2; [7782-50-5]
2. 3,4,5-Trichloro-2-(trichloro-methyl)-pyridine; C_6HCl_6N; [1201-30-5]

ORIGINAL MEASUREMENTS:

Kosorotov, V. I.; Stul, B. Ya.; Dzhagatspanyan, R. V.

Zh. Prikl. Khim. 1978, *51*, 887-889 (*J. Appl. Chem. USSR* 1978, *51*, 858-860).

VARIABLES:

Temperature

PREPARED BY:

C. L. Young

EXPERIMENTAL VALUES:

Pressure, p = 101.3 kPa

T/K	Solubility, S/mol dm^{-3}
378.2	0.21
393.2	0.17
408.2	0.13
423.2	0.11
473.2	0.056

AUXILIARY INFORMATION

METHOD/APPARATUS/PROCEDURE:

Chlorine was bubbled through the liquid until equilibrium was attained at a total pressure equal to the barometric pressure. The results were corrected to 101.3 kPa pressure. The chlorine in solution was determined by iodometry. Reaction of chlorine and solvent was minimised by 0.2 mole per cent phenol.

SOURCE AND PURITY OF MATERIALS:

1. No details given.

2. Purity 95-99 mole per cent as determined by chromatography. 0.2 mole per cent phenol added as a chlorination inhibitor.

ESTIMATED ERROR:

$\delta T/K = \pm 0.1$; $\delta S/S = \pm 0.07$.

REFERENCES:

COMPONENTS:	ORIGINAL MEASUREMENTS:
1. Chlorine; Cl_2; [7782-50-5] 2. 3,4,5-Trichloro-2-(dichloromethyl)pyridine; $C_6H_2Cl_5N$; [7041-22-7]	Kosorotov, V. I.; Stul, B. Ya.; Dzhagatspanyan, R. V. *Zh. Prikl. Khim.* 1978, *51*, 887-889 (*J. Appl. Chem. USSR* 1978, *51*, 858-860).
VARIABLES:	PREPARED BY:
Temperature	C. L. Young

EXPERIMENTAL VALUES:

Pressure, p = 101.3 kPa

T/K	Solubility, S/mol dm^{-3}
333.2	0.51
363.2	0.26
378.2	0.19
408.2	0.12

AUXILIARY INFORMATION

METHOD/APPARATUS/PROCEDURE:

Chlorine was bubbled through the liquid until equilibrium was attained at a total pressure equal to the barometric pressure. The results were corrected to 101.3 kPa pressure. The chlorine in solution was determined by iodometry. Reaction of chlorine and solvent was minimised by 0.2 mole per cent phenol.

SOURCE AND PURITY OF MATERIALS:

1. No details given.
2. Purity 95-99 mole per cent as determined by chromatography. 0.2 mole per cent phenol added as a chlorination inhibitor.

ESTIMATED ERROR:

$\delta T/K = \pm 0.1$; $\delta S/S = \pm 0.05$.

REFERENCES:

COMPONENTS:

1. Chlorine; Cl_2; [7782-50-5]
2. 3,5-Dichloro-2-(trichloromethyl)-pyridine; $C_6H_2Cl_5N$; [1128-16-1]

ORIGINAL MEASUREMENTS:

Kosorotov, V. I.; Stul, B. Ya.; Dzhagatspanyan, R. V.

Zh. Prikl. Khim. 1978, *51*, 887-889 (*J. Appl. Chem. USSR* 1978, *51*, 858-860).

VARIABLES:

Temperature

PREPARED BY:

C. L. Young

EXPERIMENTAL VALUES:

Pressure, p = 101.3 kPa

T/K	Solubility, S/mol dm^{-3}
333.2	0.50
363.2	0.26
393.2	0.15
423.2	0.10

AUXILIARY INFORMATION

METHOD/APPARATUS/PROCEDURE:

Chlorine was bubbled through the liquid until equilibrium was attained at a total pressure equal to the barometric pressure. The results were corrected to 101.3 kPa pressure. The chlorine in solution was determined by iodometry. Reaction of chlorine and solvent was minimised by 0.2 mole per cent phenol.

SOURCE AND PURITY OF MATERIALS:

1. No details given.
2. Purity 95-99 mole per cent as determined by chromatography. 0.2 mole per cent phenol added as a chlorination inhibitor.

ESTIMATED ERROR:

$\delta T/K = \pm 0.1$; $\delta S/S = \pm 0.05$.

REFERENCES:

COMPONENTS:

1. Chlorine; Cl_2; [7782-50-5]
2. 1,2-Dichloro-4-nitrobenzene; $C_6H_3Cl_2NO_2$; [99-54-7]

ORIGINAL MEASUREMENTS:

Zetkin, V.I.; Kolesnikov, I.M. Zakharov, E.V.; Dzhagatspanyan, R.V.

Khim. Prom. 1966, *42*, 624-626.

VARIABLES:

Temperature

PREPARED BY:

P.G.T. Fogg

EXPERIMENTAL VALUES:

T/K	$mol_{Cl_2}/mol_{C_6H_3Cl_2NO_2}$	$g_{Cl_2}/100g_{C_6H_3Cl_2NO_2}$	Mole fraction* x_{Cl_2}
333	0.0517	1.911	0.0492
353	0.0363	1.342	0.0351
373	0.0258	0.954	0.0252
393	0.0188	0.695	0.0185
413	0.0146	0.540	0.0144
433	0.0114	0.421	0.0113

The total pressure was apparently equal to barometric pressure (unspecified).

The heat of solution of Cl_2, calculated from the variation of solubility with temperature, was given by the authors as - 4340 cal mol^{-1}.

* Calculated by compiler.

AUXILIARY INFORMATION

METHOD/APPARATUS/PROCEDURE:

Chlorine was bubbled through the solvent in a thermostatted vessel of capacity 100 cm^3. Samples were withdrawn periodically for iodometric analysis until consistent results were obtained.

SOURCE AND PURITY OF MATERIALS:

Cl_2 : to USSR standard GOST 6718-53

$C_6H_3Cl_2NO_2$: prepared by chlorination of 1-chloro-4-nitrobenzene; recrystallised before use.

ESTIMATED ERROR:

REFERENCES:

COMPONENTS:

1. Chlorine; Cl_2; [7782-50-5]
2. 1,4-Dichloro-2-nitrobenzene; $C_6H_3Cl_2NO_2$; [89-61-2]

ORIGINAL MEASUREMENTS:

Zetkin, V.I.; Kolesnikov, I.M.; Zakharov, E.V.; Dzhagatspanyan, R.V.

Khim. Prom. 1966, *42*, 624-626.

VARIABLES:

Temperature

PREPARED BY:

P.G.T. Fogg

EXPERIMENTAL VALUES:

T/K	$mol_{Cl_2}/mol_{C_6H_3Cl_2NO_2}$	$g_{Cl_2}/100g_{C_6H_3Cl_2NO_2}$	Mole fraction* x_{Cl_2}
333	0.0675	2.495	0.0632
353	0.0463	1.712	0.0443
373	0.0316	1.168	0.0306
393	0.0233	0.861	0.0228
413	0.0177	0.654	0.0174
433	0.0137	0.506	0.0135

The total pressure was apparently equal to barometric pressure (unspecified).

The heat of solution of Cl_2 , calculated from the variation of solubility with temperature, was given by the authors as - 4580 cal mol^{-1}.

* Calculated by compiler.

AUXILIARY INFORMATION

METHOD/APPARATUS/PROCEDURE:

Chlorine was bubbled through the solvent in a thermostatted vessel of capacity 100 cm^3. Samples were withdrawn periodically for iodometric analysis until consistent results were obtained.

SOURCE AND PURITY OF MATERIALS:

Cl_2 : to USSR standard GOST 6718-53. Dried with H_2SO_4 and with $CaCl_2$.

$C_6H_3Cl_2NO_2$: to standard TU GAP-U 539-54. Recrystallised before use.

ESTIMATED ERROR:

REFERENCES:

COMPONENTS:

1. Chlorine; Cl_2; [7782-50-5]
2. 2-Chloro-6-(trichloromethyl)-pyridine; $C_6H_3Cl_4N$; [1929-82-4]

ORIGINAL MEASUREMENTS:

Kosorotov, V. I.; Stul, B. Ya.; Dzhagatspanyan, R. V.
Zh. Prikl. Khim. 1978, *51*, 887-889 (*J. Appl. Chem. USSR* 1978, *51*, 858-860).

VARIABLES:

Temperature

PREPARED BY:

C. L. Young

EXPERIMENTAL VALUES:

Pressure, p = 101.3 kPa

T/K	Solubility, S/mol dm^{-3}
363.2	0.13
393.2	0.10
423.2	0.08
443.2	0.066
473.2	0.05

AUXILIARY INFORMATION

METHOD/APPARATUS/PROCEDURE:

Chlorine was bubbled through the liquid until equilibrium was attained at a total pressure equal to the barometric pressure. The results were corrected to 101.3 kPa pressure. The chlorine in solution was determined by iodometry. Reaction of chlorine and solvent was minimised by 0.2 mole per cent phenol.

SOURCE AND PURITY OF MATERIALS:

1. No details given.
2. Purity 95-99 mole per cent as determined by chromatography. 0.2 mole per cent phenol added as a chlorination inhibitor.

ESTIMATED ERROR:

$\delta T/K = \pm 0.1$; $\delta S/S = \pm 0.07$.

REFERENCES:

COMPONENTS:

1. Chlorine; Cl_2; [7782-50-5]
2. 5-Chloro-2-(trichloromethyl)-pyridine; $C_6H_3Cl_4N$; [1197-03-1]
3. 3,5-Dichloro-2-(dichloromethyl)-pyridine; $C_6H_3Cl_4N$; [7041-25-0]

ORIGINAL MEASUREMENTS:

Kosorotov, V. I.; Stul, B. Ya.; Dzhagatspanyan, R. V.

Zh. Prikl. Khim. 1978, *51*, 887-889 (*J. Appl. Chem. USSR* 1978, *51*, 858-860).

VARIABLES:

Temperature

PREPARED BY:

C. L. Young

EXPERIMENTAL VALUES:

Pressure, p = 101.3 kPa

T/K	Solubility, S/mol dm^{-3}
333.2	0.57
363.2	0.36
393.2	0.26
423.2	0.17
443.2	0.12

AUXILIARY INFORMATION

METHOD/APPARATUS/PROCEDURE:

Chlorine was bubbled through the liquid until equilibrium was attained at a total pressure equal to the barometric pressure. The results were corrected to 101.3 kPa pressure. The chlorine in solution was determined by iodometry. Reaction of chlorine and solvent was minimised by 0.2 mole per cent phenol.

SOURCE AND PURITY OF MATERIALS:

1. No details given.
2. Purity 95-99 mole per cent as determined by chromatography. 0.2 mole per cent phenol added as a chlorination inhibitor.
3. As in 2 above.

ESTIMATED ERROR:

$\delta T/K = \pm 0.1$; $\delta S/S = \pm 0.05$.

REFERENCES:

COMPONENTS:

1. Chlorine; Cl_2; [7782-50-5]
2. 1-Chloro-2-nitrobenzene; $C_6H_4ClNO_2$; [88-73-3]

ORIGINAL MEASUREMENTS:

Zetkin, V.I.; Kolesnikov, I.M.; Zakharov, E.V.; Dzhagatspanyan, R.V.

Khim. Prom. 1966, *42*, 624-626.

VARIABLES:

Temperature

PREPARED BY:

P.G.T. Fogg

EXPERIMENTAL VALUES:

T/K	$mol_{Cl_2}/mol_{C_6H_4ClNO_2}$	$g_{Cl_2}/100g_{C_6H_4ClNO_2}$	Mole fraction* x_{Cl_2}
313	0.0732	3.298	0.0682
333	0.0482	2.172	0.0460
353	0.0337	1.518	0.0326
373	0.0231	1.041	0.0226
393	0.0176	0.793	0.0173
413	0.0134	0.604	0.0132
433	0.0106	0.478	0.0105

The total pressure was apparently equal to barometric pressure (unspecified).

The heat of solution of Cl_2 calculated from the variation of solubility with temperature was given by the authors as - 4340 cal mol^{-1}.

* Calculated by compiler.

AUXILIARY INFORMATION

METHOD/APPARATUS/PROCEDURE:

Chlorine was bubbled through the solvent in a thermostatted vessel of capacity 100 cm^3. Samples were withdrawn periodically for iodometric analysis until consistent results were obtained.

SOURCE AND PURITY OF MATERIALS:

Cl_2 : to USSR standard GOST 6718-53. Dried with H_2SO_4 and with $CaCl_2$.

$C_6H_4ClNO_2$: to standard TU GAP-U 451-53. Recrystallised before use.

ESTIMATED ERROR:

REFERENCES:

COMPONENTS:

1. Chlorine; Cl_2; [7782-50-5]
2. 1-Chloro-3-nitrobenzene; $C_6H_4ClNO_2$; [121-73-3]

ORIGINAL MEASUREMENTS:

Zetkin, V.I.; Kolesnikov, I.M.; Zakharov, E.V.; Dzhagatspanyan, R.V.

Khim. Prom. 1966, *42*, 624-626.

VARIABLES:

Temperature

PREPARED BY:

P.G.T. Fogg

EXPERIMENTAL VALUES:

T/K	$mol_{Cl_2}/mol_{C_6H_4ClNO_2}$	$g_{Cl_2}/100g_{C_6H_4ClNO_2}$	Mole fraction* x_{Cl_2}
333	0.0468	2.109	0.0447
353	0.0321	1.446	0.0311
373	0.0219	0.987	0.0214
393	0.0157	0.707	0.0154
413	0.0117	0.527	0.0116
433	0.0091	0.410	0.0090

The total pressure was apparently equal to barometric pressure (unspecified).

The heat of solution of Cl_2 calculated from the variation of solubility with temperature was given by the authors as - 4700 cal mol^{-1}.

* Calculated by compiler.

AUXILIARY INFORMATION

METHOD/APPARATUS/PROCEDURE:

Chlorine was bubbled through the solvent in a thermostatted vessel of capacity 100 cm^3. Samples were withdrawn periodically for iodometric analysis until consistent results were obtained.

SOURCE AND PURITY OF MATERIALS:

Cl_2 : to USSR standard GOST 6718-53. Dried with H_2SO_4 and with $CaCl_2$.

$C_6H_4ClNO_2$: prepared by chlorination of nitrobenzene. Purified by crystallisation.

ESTIMATED ERROR:

REFERENCES:

COMPONENTS:

1. Chlorine; Cl_2; [7782-50-5]
2. 1-Chloro-4-nitrobenzene; $C_6H_4ClNO_2$; [100-00-5]

ORIGINAL MEASUREMENTS:

Zetkin, V.I. Kolesnikov, I.M.; Zakharov, E.V.; Dzhagatspanyan, R.V.;

Khim. Prom. 1966, *42*, 624-626.

VARIABLES:

Temperature

PREPARED BY:

P.G.T. Fogg

EXPERIMENTAL VALUES:

T/K	$mol_{Cl_2}/mol_{C_6H_4ClNO_2}$	$g_{Cl_2}/100g_{C_6H_4ClNO_2}$	Mole fraction* x_{Cl_2}
373	0.0274	1.235	0.0267
393	0.0185	0.834	0.0182
413	0.0132	0.595	0.0130
433	0.0097	0.437	0.0096

The total pressure was apparently equal to barometric pressure (unspecified).

The heat of solution of Cl_2 calculated from the variation of solubility with temperature was given by the authors as - 5560 cal mol^{-1}.

* Calculated by compiler.

AUXILIARY INFORMATION

METHOD APPARATUS/PROCEDURE:

Chlorine was bubbled through the solvent in a thermostatted vessel of capacity 100 cm^3. Samples were withdrawn periodically for iodometric analysis until consistent results were obtained.

SOURCE AND PURITY OF MATERIALS:

Cl_2 : to USSR standard GOST 6718-53. Dried with H_2SO_4 and with $CaCl_2$.

$C_6H_4ClNO_2$: to standard TU 9728-61. Recrystallised before use.

ESTIMATED ERROR:

REFERENCES:

COMPONENTS:

1. Chlorine; Cl_2; [7782-50-5]
2. 2-(Trichloromethyl)-pyridine; $C_6H_4Cl_3N$; [4377-37-1]

ORIGINAL MEASUREMENTS:

Kosorotov, V.I.; Stul, B. Ya.; Dzhagatspanyan, R. V.
Zh. Prikl. Khim. 1978, *51*, 887-889 (*J. Appl. Chem. USSR* 1978, *51*, 858-860).

VARIABLES:

Temperature

PREPARED BY:

C. L. Young

EXPERIMENTAL VALUES:

Pressure, p = 101.3 kPa

T/K	Solubility, S/mol dm^{-3}
333.2	0.63
363.2	0.38
393.2	0.26
423.2	0.19
443.2	0.15

AUXILIARY INFORMATION

METHOD APPARATUS/PROCEDURE:

Chlorine was bubbled through the liquid until equilibrium was attained at a total pressure equal to the barometric pressure. The results were corrected to 101.3 kPa pressure. The chlorine in solution was determined by iodometry. Reaction of chlorine and solvent was minimised by 0.2 mole per cent phenol.

SOURCE AND PURITY OF MATERIALS:

1. No details given.

2. Purity 95-99 mole per cent as determined by chromatography. 0.2 mole per cent phenol added as a chlorination inhibitor.

ESTIMATED ERROR:

$\delta T/K = \pm 0.1$; $\delta S/S = \pm 0.05$.

REFERENCES:

COMPONENTS:

1. Chlorine; Cl_2; [7782-50-5]
2. Dichlorodioxochromium; Cl_2CrO_2; [14977-61-8]

ORIGINAL MEASUREMENTS:

Roozeboom, H.W.B.;

Rec. Trav. Chim. 1885, *4*, 379-381.

VARIABLES:

Temperature, concentration

PREPARED BY:

P.G.T. Fogg.

EXPERIMENTAL VALUES:

T/K	Quantity of chlorine *en atomes.*	p_{Cl_2}/mmHg	Mole fraction* Cl_2
273.2	1.70	1302	0.459
273.2	1.43	1208	0.417
273.2	1.40	1198	0.412
273.2	1.35	1189	0.403
273.2	1.30	1150	0.394
273.2	1.25	1125	0.385
273.2	1.20	1101	0.375
273.2	1.16	1069	0.367
273.2	1.12	1051	0.359
273.2	1.04	1016	0.342
273.2	1.00	984	0.333
273.2	0.91	923	0.313
273.2	0.78	824	0.281
273.2	0.68	753	0.254
247.2	3.08	765	0.606
249.2	3.08	791	0.606
251.2	3.08	844	0.606
251.2	2.77	820	0.581
251.2	2.31	734	0.536
253.2	3.08	896	0.606
253.2	2.77	862	0.581
253.2	2.31	780	0.536
255.2	2.77	905	0.581
255.2	2.31	831	0.536
257.2	2.00	851	0.500
258.2	2.77	1002	0.581

AUXILIARY INFORMATION

METHOD APPARATUS/PROCEDURE:

The pressure over a known weight of Cl_2 dissolved in a known weight of Cl_2CrO_2 in a small bulb at 247.2K to 273.2K was measured by a mercury manometer in which the mercury was protected by sulphuric acid. The concentration of Cl_2 was changed by removing gas through a side tap and collecting it over brine saturated with chlorine. The weight of Cl_2 removed, and the weight remaining in the flask, were calculated from the volume of gas collected. No allowance was made for the vapor pressure of Cl_2CrO_2.

SOURCE AND PURITY OF MATERIALS:

Cl_2 : not specified.

Cl_2CrCO_2: fractionally distilled under dry CO_2; b.p.117°C (barometric pressure unspecified.)

ESTIMATED ERROR:

REFERENCES:

COMPONENTS:

1. Chlorine; Cl_2; [7782-50-5]

2. Dichlorodioxochromium; Cl_2CrO_2; [14977-61-8]

ORIGINAL MEASUREMENTS:

Roozeboom, H.W.B.; *Rec. Trav. Chim.* 1885, *4*, 379-381.

EXPERIMENTAL VALUES:

T/K	Quantity of chlorine *en atomes*	p_{Cl_2}/mmHg	Mole fraction* x_{Cl_2}
259.2	2.00	902	0.500
259.2	1.24	754	0.383
261.2	2.77	1105	0.581
262.2	1.70	952	0.459
262.2	1.24	818	0.383
263.2	2.00	1015	0.500
264.2	1.70	995	0.459
264.2	1.24	870	0.383
265.2	1.39	930	0.410
266.2	1.39	948	0.410
266.2	1.24	921	0.383
268.2	1.24	975	0.383
270.2	1.24	1034	0.383
271.2	1.70	1232	0.459
273.2	1.70	1295	0.459
273.2	1.39	1185	0.410
273.2	1.24	1118	0.383
273.2	1.07	1031	0.349
273.2	0.76	825	0.275

760 mmHg = 1 atm; 1 atm = 1.013×10^5 Pascal.

* calculated by the compiler on the assumption that concentration *en atomes* is the ratio of atoms of dissolved chlorine to molecules of Cl_2CrO_2.

COMPONENTS:	ORIGINAL MEASUREMENTS:
1. Chlorine; Cl_2; [7782-50-5] 2. Sulfuryl chloride; Cl_2O_2S; [7791-25-5]	Schulze, H. *J. Prakt. Chem.* 1881, *24*, 168.
VARIABLES:	PREPARED BY: P. G. T. Fogg

EXPERIMENTAL VALUES:

T/K	Wt. of Cl_2 absorbed per unit wt. of Cl_2O_2S	Vol. of Cl_2 absorbed per vol. of Cl_2O_2S	Mole fraction,* x_{Cl_2}
273.2	0.136	71	0.206

The pressure of chlorine was apparently equal to barometric pressure (unspecified).

* Calculated by compiler.

AUXILIARY INFORMATION

METHOD APPARATUS/PROCEDURE:	SOURCE AND PURITY OF MATERIALS:
Chlorine was bubbled through a weighed quantity of sulfuryl chloride in a stoppered cylinder fitted with inlet and outlet tube. Passage of gas was continued until bubbles of gas no longer appeared to be absorbed. The amount of chlorine absorbed was found by weighing.	No details given. ESTIMATED ERROR: REFERENCES:

COMPONENTS:

1. Chlorine; Cl_2; [7782-50-5]

2. Tetrachlorosilane, (Silicon tetrachloride); $SiCl_4$; [10026-04-7]

ORIGINAL MEASUREMENTS:

Taylor, N. W.; Hildebrand, J. H.
J. Amer. Chem. Soc. 1923, *45*, 682-694.

VARIABLES:

PREPARED BY:

P. G. T. Fogg

EXPERIMENTAL VALUES:

T/K	g Cl_2/g of solution	p_{Cl_2}/mmHg	Mole fractions x_{Cl_2}	x_{Cl_2}(1 atm)
273.16	0.131	681	0.266	0.288

760 mmHg = 1 atm

1 atm = 1.013×10^5 Pascal

AUXILIARY INFORMATION

METHOD/APPARATUS/PROCEDURE:

Saturator tube (1). Chlorine was bubbled through liquid. Samples of saturated liquid were run into a weighed flask containing aqueous KI, and iodine was titrated with standard solution of thiosulfate. Temperature control was manual to 0.1 K.
Saturation was at barometric pressure, p_{total}. The value of p_{SiCl_4} was estimated by Raoult's law from vapor pressure data for $SiCl_4$ given in ref. (2). Hence:

$$p_{Cl_2} = p_{total} - p_{SiCl_4}$$

The authors also estimated x_{Cl_2} (1 atm) from x_{Cl_2} by Henry's law (mole fraction form).

SOURCE AND PURITY OF MATERIALS:

1. Cl_2: from an electrolytic supply; contained less than 0.1% of free acid.

2. $SiCl_4$: not specified.

ESTIMATED ERROR:

REFERENCES:

1. Bichowsky, F. R.; Storch, H. *J. Amer. Chem. Soc.* 1915, *37*, 2695.

2. Landolt Börnstein Tabellen.

COMPONENTS:	ORIGINAL MEASUREMENTS:
1. Chlorine; Cl_2; [7782-50-5] 2. Titanium chloride; Cl_4Ti; [7550-45-0]	Krieve, W. F.; Mason, D. M. *J. Phys. Chem.* 1956, *60*, 374.
VARIABLES:	PREPARED BY:
Temperature, pressure	P. G. T. Fogg

EXPERIMENTAL VALUES:

T/K	p_{Cl_2}/atm	Mole fraction x_{Cl_2}	T/K	p_{Cl_2}/atm	Mole fraction x_{Cl_2}
293.2	1.50	0.2513	303.2	3.39	0.4227
	2.20	0.3585		3.93	0.4844
	2.71	0.4358		4.63	0.5571
	3.13	0.4964		5.41	0.6412
	3.67	0.5674		6.14	0.7107
	4.24	0.6499			
	4.76	0.7229			

The recovery of Cl_2 by dissolution in Cl_4Ti on an industrial scale has been reported (1).

AUXILIARY INFORMATION

METHOD APPARATUS/PROCEDURE:

A known weight of chlorine, determined by the change in pressure in the bomb containing the gas, was added to a known weight of titanium tetrachloride in a second bomb. When equilibrium had been attained the pressure in the second bomb was recorded. All pressures were measured by calibrated stainless steel Bourdon gauge to ± 0.5%. Temperatures were controlled to ± 0.1 °C. Solubilities were calculated from the pressure data using published gas density data (2).

SOURCE AND PURITY OF MATERIALS:

1. Cl_2: A commercial sample was purified by fractional distillation to remove air and CO_2.

2. $TiCl_4$: from National Bureau of Standards; 99.999% pure.

ESTIMATED ERROR:

REFERENCES:

1. Anon.
Res. Discl. 1977, *157*, 15.
2. Hulme, R.E.; Tillman, A.B.
Chem. Eng. 1949, *56*, 99.

COMPONENTS:

1. Fluorine; F_2; [7782-41-4]
2. Hydrogen fluoride; HF; [7664-39-3]

ORIGINAL MEASUREMENTS:

Rewick, R. T.; Tolberg, W. E.; Hill, M. E.

J. Chem. Engng. Data 1970, *15*, 527-530.

VARIABLES:

Temperature, pressure

PREPARED BY:

C. L. Young

EXPERIMENTAL VALUES:

T/K	Total pressure P/kPa	Partial pressure P/kPa	Conc. of fluorine /mol kg^{-1}	Mole fraction of fluorine, x_{F_2} *
292.95	280.0	177.3	0.0356	0.000713
	538.6	434.6	0.0925	0.00185
	799.9	695.9	0.150	0.00300
	1064	959.9	0.204	0.00410
	1088	983.9	0.206	0.00414
272.94	364.0	316.0	0.0690	0.00138
	370.6	322.6	0.0697	0.00139
	729.3	681.3	0.148	0.00296
	950.6	903.0	0.196	0.00393
	1192	1144	0.247	0.00496
249.95	815.9	798.6	0.194	0.00389
207.95	637.3	634.6	0.187	0.00375
195.95	204.0	202.6	0.0675	0.00135
	209.3	208.0	0.0678	0.00136
	440.0	438.6	0.144	0.00288
	735.9	734.6	0.243	0.00486

* calculated by compiler

AUXILIARY INFORMATION

METHOD/APPARATUS/PROCEDURE:

Static equilibrium cell made of Monel metal. Known amounts of fluorine added to known amount of hydrogen fluoride in cell. Quantity of gas absorbed calculated from equilibrium pressure and cell volume.

SOURCE AND PURITY OF MATERIALS:

1. Purity 98 mole per cent obtained from Allied Chemical Corp. Fractionally distilled final purity better than 99 mole per cent. Major impurities nitrogen and oxygen.
2. Anhydrous sample obtained from Olin Corp. Conductivity 3×10^5 ohm^{-1} cm^{-1}.

ESTIMATED ERROR:

$\delta T/K = \pm 0.02$; $\delta x_{F_2} = \pm 3\%$ (estimated by compiler).

REFERENCES:

COMPONENTS:

1. Chlorine monoxide; Cl_2O; [7791-21-1]
2. Water; H_2O; [7732-18-5]

ORIGINAL MEASUREMENTS:

Ourisson, M.M.J.; Kastner, M.

Bull. Soc. Chim. France. 1939, *6*, 1307-11.

VARIABLES:

Temperature

PREPARED BY:

R. Battino.

EXPERIMENTAL VALUES:

T/K [a]	m^b_{HOCl}/mol 1000g H_2O^{-1}	P_{Cl_2}/mm [c]	P_{Cl_2O}/mm [d]	T/K	K_1	K_2
283.15	0.26	0.097	0.014	283.15	0.21	0.24
	0.57	0.29	0.068	293.15	0.65	0.70
	0.70	0.36	0.103			
	0.90	0.56	0.170			
	1.05	0.67	0.232			
	1.23	1.05	0.318			
	1.55	1.59	0.505			
	1.65	1.65	0.572			
	1.92	2.08	0.774			
	2.00	2.16	0.840			
	2.03	2.10	0.865			
293.15	0.148	0.16	0.014			
	0.149	0.15	0.014			
	0.178	0.18	0.021			
	0.175	0.13	0.020			
	0.189	0.15	0.023			
	0.222	0.23	0.032			
	0.287	0.21	0.054			
	0.287	0.19	0.023			
	0.327	0.36	0.070			
	0.545	0.68	0.19			
	0.553	0.74	0.20			
	0.641	1.11	0.27			
	0.651	1.2	0.28			

AUXILIARY INFORMATION

METHOD/APPARATUS/PROCEDURE:

The amount of HOCl in solution is determined by chemical titration.

SOURCE AND PURITY OF MATERIALS:

No details given.

ESTIMATED ERROR:

$\delta m_{HOCl}/m_{HOCl} = 0.03.$

compiler's estimate.

REFERENCES:

COMPONENTS:	ORIGINAL MEASUREMENTS
1. Chlorine monoxide; Cl_2O;	Ourisson, M.M.J.; Kastner, M.
2. Water; H_2O; [7732-18-5]	*Bull. Soc. Chim. France.* 1939, *6*, 1307-11.

EXPERIMENTAL VALUES

T/K [a]	m_{HOCl}/mol 1000g H_2O^{-1}	P_{Cl_2}/mm	P_{Cl_2O}/mm
293.15	0.672	1.2	0.29
	0.747	1.2	0.36
	0.771	1.0	0.39
	0.812	1.1	0.43
	0.842	1.2	0.46
	0.943	1.4	0.58
	0.941	1.5	0.58
	1.22	2.7	0.97
	1.52	4.1	1.50
	2.55	10.3	4.23
	2.68	10.9	4.67
283.15	60.16 [e]		760. [e]
293.15	34.19 [e]		760. [e]

a. Temperature reported as 10 and 20° C.

b. Molality (moles HOCl/1000g H_2O) in solution.

c. P_{Cl_2} is defined as $2P_{Cl_2O} + P_{HOCl}$. P_{Cl_2} is also defined as $P_{Cl_2} = 2K_1[HOCl]^2 + K_2[HOCl]$.
$K_1 = P_{Cl_2O}/[HOCl]^2$. $K_2 = P_{HOCl}/[HOCl]$.

d. P_{Cl_2O} calculated from $P_{Cl_2O} = K\,[HOCl]^2$

e. Extrapolated values to $P_{Cl_2O} = 1$ atm. Since $[Cl_2O]$ in solution were not given it was not possible to calculate a mole fraction solubility for Cl_2O.

COMPONENTS:

(1) Chlorine monoxide; Cl_2O; [7791-21-1]

(2) Water; H_2O; [7732-18-5]

ORIGINAL MEASUREMENTS:

Secoy, C.H.; Cady, G.H.
J. Amer. Chem. Soc. 1941, *63*, 2504-8.

VARIABLES:

T/K: 276-293
p/mm Hg: 1-75

PREPARED BY:

R. Battino

EXPERIMENTAL VALUES:

Solubility[a] of Chlorine Monoxide as g Cl_2O/100 g H_2O

p_{Cl_2O}/mm	0°C[b]	10°C	20°C
1	7.6	5.2	3.5
5	16.7	12.3	8.0
10	23.7	17.4	11.1
20	33.6	24.2	15.4
30	40.7	29.2	18.7
40	46.9	33.0	21.3
50	52.4	36.2	23.7
60	56.7	39.0	25.8
70	60.2	41.4	27.7
75	61.6	42.5	28.6

a. Actual measurements were made at 3.46°C (1.27 to 247.6 mm), 6.00°C (15.17 to 15.70 mm), 9.92°C (0.918 to 75.62 mm), 16.00°C (32.38 to 33.66 mm), and 19.98°C (1.11 to 87.49 mm). The numbers in parentheses give the range of partial pressure of Cl_2O at each temperature. The authors condensed these values to the above solubility table.

b. Temperatures reported to 0.01°C.

contd.

AUXILIARY INFORMATION

METHOD/APPARATUS/PROCEDURE:

Two bulbs are connected by stop-cocks. A solution of hypochlorous acid is introduced in one and the apparatus rotated so the solution moves through both bulbs. After equilibration samples of the gas and liquid phases are chemically analysed. The analysis of the gas samples used the method of Spinks (1). The pressure was altered during the run.

SOURCE AND PURITY OF MATERIALS:

(1) Cl_2O - no comment by authors.

(2) H_2O - no comment by authors.

ESTIMATED ERROR:

$\delta S/S$ = 0.02 to 0.05 (estimated by compiler).

REFERENCES:

(1) Spinks *J. Amer. Chem. Soc.* 1931, *53*, 3015.

COMPONENTS:	ORIGINAL MEASUREMENTS:
(1) Chlorine monoxide; Cl_2O; [7791-21-1] (2) Water; H_2O; [7732-18-5]	Secoy, C.H.; Cady, G.H. *J. Amer. Chem. Soc.* 1941, *63*, 2504-8.

Solubility[a] of Chlorine Monoxide

T/K	x_1[b]	L[c]	S[d]
273.15	0.5259	1375	533.3
283.15	0.4273	962	359.9
293.15	0.3353	672	243.4

a. All values calculated by compiler.
b. Mole fraction solubility (extrapolated) at 1 atm partial pressure of gas.
c. Ostwald coefficient (extrapolated).
d. Solubility as g Cl_2O/100 g H_2O at 1 atm partial pressure of gas (extrapolated).

Equilibrium Constant K[a]

t/°C	K
3.46	0.41 ± 0.01
6.00	0.48 ± 0.01
9.92	0.62 ± 0.01
16.00	1.01 ± 0.01
19.98	1.47 ± 0.01

a. $K = P_{Cl_2O} \times a_{H_2O}/(a_{HOCl})^2$ for $2\ HOCl(aq) = Cl_2O(g) + H_2O(l)$.

Assume $a_{H_2O} = 1$ and $K^1 = P_{Cl_2O}/m^2$ where m is the molal concentration of HOCl. The K^1 values are plotted against m and extrapolated to infinite dilution to give the values of K in this table.

COMPONENTS:	ORIGINAL MEASUREMENTS:
(1) Chlorine monoxide; Cl_2O; [7791-21-1] (2) Water; H_2O; [7732-18-5]	Roth, W.A. *Z. Physik. Chem.* 1942, *A191*, 248-50.
VARIABLES: Concentration: molality HOCl, 1.8 to 5.1	PREPARED BY: R. Battino

EXPERIMENTAL VALUES:

Solubility data at 3.46°C

m^a	m_{HOCl}	m_{Cl_2O}	p_{Cl_2O}/mmHg	K(Henry)[b]
1.76	1.754	0.012	1.27	0.00945
2.49	2.478	0.024	2.54	0.00945
2.84	2.824	0.031	3.40	0.0091
3.60	3.575	0.050	5.31	0.0094
3.89	3.861	0.058	6.31	0.0092
4.48	4.441	0.078	8.27	0.0094
4.57	4.530	0.080	8.46	0.00945
4.88	4.834	0.092	10.10	0.0091
5.07	5.0205	0.099	10.62	0.00935
			Average	0.00937

a. $m = m_{HOCl} + 0.5\ m_{Cl_2O}$

b. Henry's Law constant $(= m_{Cl_2O}/p_{Cl_2O})$ refers to undissociated Cl_2O only.

c. The extrapolated solubility at p_{Cl_2O} = 1 atm is 7.12 moles Cl_2O/litre H_2O or a mole fraction of 0.1137 for 1 atm partial pressure of gas. This is a mole fraction of 0.454 for <u>total</u> amount of Cl_2O dissolved at p_{Cl_2O} = 1 atm

$$K = m_{HOCl}^{2}/m_{Cl_2O} = 255$$

At 1 atm m_{Cl_2O} = 7.21 and m_{HOCl} = 42.61, therefore total Cl_2O = 28.42 moles/litre H_2O, therefore x = 0.454.

AUXILIARY INFORMATION

METHOD/APPARATUS/PROCEDURE:

Henry's Law constants determined from measuring hydrolysis constant of hydrolysis constant of hypochlorous acid solutions at different molalities. Author points out systematic errors in work of Secoy and Cady (1).

SOURCE AND PURITY OF MATERIALS:

No details given.

ESTIMATED ERROR:

$\delta K/K$ = 0.005, estimate by author.

REFERENCES:

(1) Secoy, C.H.; Cady, G.H. *J. Amer. Chem. Soc.* 1941, *63*, 2504-8.

COMPONENTS:	EVALUATOR:
1. Chlorine dioxide; ClO_2; 2. Water; H_2O; [7732-18-5]	Rubin Battino, Department of Chemistry, Wright State University, Dayton, Ohio, 45431, U.S.A.

CRITICAL EVALUATION:

Both the data of Kepinski and Trzeszczynski (1) and that of Haller and Northgraves (2) were used for the smoothing equation. This comprised six data points from the former and three from the latter. The standard deviation of the fit was 4.6% indicating rather poor data (but the only data available for this system). The fitting equation is:

$$\ln x_1 = 7.91625 + 0.479133/(T/100K) - 11.0593 \ln (T/100K) \quad ...(1)$$

where x_1 is the mole fraction solubility at 101.325 kPa (1 atm) partial pressure of gas. Table 1 gives smoothed values of the mole fraction (at 101.325 kPa) and the Ostwald coefficient at 5K intervals. Table 1 also gives the values of the thermodynamic functions $\Delta\bar{G}^\circ_1$, $\Delta\bar{H}^\circ_1$, $\Delta\bar{S}^\circ_1$ and $\Delta\bar{C}^\circ_{p_1}$ for the transfer of gas from the vapor phase at 101.325 kPa partial gas pressure to the (hypothetical) solution phase of unit mole fraction.

References

1. Kepinski, J.; Trzeszczynski, J. *Rocz. Chem.* 1964, *38*, 201.10.

2. Haller, J. F.; Northgraves, W. W. *Tappi* 1955, *38*, 199-202.

Table 1. Smoothed values of the solubility of chlorine dioxide in water and thermodynamic functions [a] at 101.325 kPa (1 atm) partial pressure of gas.

T/K	$x_1 \times 10^2$ [b]	L [c]	$\Delta\bar{G}^\circ_1$/J mol^{-1}	$\Delta\bar{H}^\circ_1$/J mol^{-1}	$\Delta\bar{S}^\circ_1$/J mol^{-1} K^{-1}
283.15	3.255	43.37	8.06	-26.43	-118.5
288.15	2.674	36.03	8.68	-26.89	-120.2
293.15	2.204	30.04	9.30	-27.35	-121.8
298.15	1.823	25.15	10.56	-27.81	-123.4
303.15	1.513	21.12	11.21	-28.27	-125.0
308.15	1.259	17.79	11.86	-28.73	-126.6
313.15	1.051	15.04	12.52	-29.19	-128.1
318.15	0.880	12.74	13.19	-29.65	-129.6
323.15	0.739	10.83	13.86	-30.11	-131.1
328.15	0.525	9.23	14.54	-30.57	-132.6
333.15	0.445	7.88	15.23	-31.03	-134.0

[a] $\Delta\bar{C}_{p_1}$ was independent of temperature and had the value -92 J K^{-1} mol^{-1}.

[b] Mole fraction solubility at 101.325 kPa partial pressure of gas.

[c] Ostwald coefficient.

COMPONENTS:

1. Chlorine dioxide; ClO_2; [10049-04-4]
2. Water; H_2O; [7732-18-5]

ORIGINAL MEASUREMENTS:

Haller, J.F.; Northgraves, W.W.

Tappi. 1955, *38*, 199-202.

VARIABLES:

Temperature

PREPARED BY:

R. Battino,

EXPERIMENTAL VALUES:

t/°C	P_{ClO_2}/mmHg	Solubility / g l^{-1}
25	34.5	3.01
25	22.1	1.82
25	13.4	1.13
25	8.4	0.69
40	56.2	2.63
40	34.3	1.60
40	18.9	0.83
40	9.9	0.47
60	105.9	2.65
60	53.7	1.18
60	21.3	0.58
60	12.0	0.26

T/K[a]	X,x10^2 [b]	L [c]	Solubility [d]
298.15	1.771	24.41	67.30
313.15	0.956	13.66	35.86
333.15	0.515	7.72	19.06

a. Temperature reported as 24,40,60°C.

b. Mole fraction solubility at 101.325kPa (1 atm) partial pressure of gas. Calculated by compiler. Extrapolated.

c. Ostwald coefficient. Calculated by compiler. Extrapolated.

d. Solubility in g l^{-1}. Extrapolated to 1 atm.

AUXILIARY INFORMATION

METHOD APPARATUS/PROCEDURE:

The data are taken from reference (1). No other details supplied.

SOURCE AND PURITY OF MATERIALS:

ESTIMATED ERROR:

REFERENCES:

1. Kirk, R.E.; Othmer, D.F. "Encyclopedia of Chemical Technology", Vol. 3., p.702. *Interscience Publishers*, (1952)

COMPONENTS:

1. Chlorine dioxide; ClO_2; [10049-04-4]
2. Water; H_2O; [7732-18-5]

ORIGINAL MEASUREMENTS:

Kepinski, J.; Trzeszcynski, J.

Rocz. Chem. 1964, *38*, 201-10.

VARIABLES:

Temperature

PREPARED BY:

R. Battino.

EXPERIMENTAL VALUES:

T/K = 283-318

T/K [a]	$x_1 \times 10^2$ [b]	L [c]	H [d]	Pressure range /mm [e]
283.15	3.271	43.60	0.1666	42-156
288.15	2.664	35.89	0.1347	14-163
293.15	2.208	30.09	0.1111	18-151
298.15	1.827	25.20	0.0914	21-154
308.15	1.296	18.32	0.0634	31-155
318.15	0.962	13.94	0.0474	38-154

a. Temperature reported as 10, 15,...,45°C.

b. Mole fraction solubility at 101.325 kPa (1 atm) partial pressure of gas. Calculated by compiler. Extrapolated.

c. Ostwald coefficient. Calculated by compiler. Extrapolated.

d. Henry's law constant. Units of (g ClO_4/ liter) /mmHg ClO_4.

e. Pressure range of actual measurements.

AUXILIARY INFORMATION

METHOD APPARATUS/PROCEDURE:

Used a dynamic saturation method along with a chemical method of analysis. A drawing and details given in the paper.

SOURCE AND PURITY OF MATERIALS:

1. Chlorine dioxide prepared according to reference (1).
2. Water - no comment by authors.

ESTIMATED ERROR:

$\delta H/H$ = 0.02, compiler's estimate.

REFERENCES:

1. Kepinski, J. *Przemysl. Chem.* 1955, *34*,124.

SYSTEM INDEX

Underlined page numbers refer to evaluation text and those not underlined to compiled tables. All compounds are listed as in Chemical Abstracts. For example, toluene is listed as benzene, methyl- and dimethylsulfoxide is listed as methane, sulfinylbis-.

A

B

(cont.)

(cont.)

(cont.)

(cont.)

M

N

(cont.)

(cont.)

(cont.)

REGISTRY NUMBER INDEX

Underlined page numbers refer to evaluation text and those not underlined to compiled tables.

(cont.)

(cont.)

(cont.)

(cont.)

AUTHOR INDEX

Underlined page numbers refer to evaluation text and those not underlined to compiled tables.

(cont.)

(cont.)

(cont.)

(cont.)